Biosynthesis and its Control
in Plants

Annual Proceedings of the Phytochemical Society

Biosynthesis and its Control in Plants

PROCEEDINGS OF THE
PHYTOCHEMICAL SOCIETY SYMPOSIUM
UNIVERSITY OF KENT AT CANTERBURY AND
SITTINGBOURNE LABORATORIES, SHELL RESEARCH LIMITED
SITTINGBOURNE, KENT
MARCH 1972

Edited by

B. V. MILBORROW

Shell Research Limited
Milstead Laboratory of Chemical Enzymology
Sittingbourne Laboratories
Sittingbourne, Kent, England

1973

ACADEMIC PRESS
LONDON AND NEW YORK

ACADEMIC PRESS INC. (LONDON) LTD.
24/28 Oval Road,
London NW1

United States Edition published by
ACADEMIC PRESS INC.
111 Fifth Avenue
New York, New York 10003

Library of Congress Catalog Card Number: 72-7708
ISBN: 0-12-496150-9

PRINTED IN GREAT BRITAIN BY
WILLIAM CLOWES & SONS LIMITED
LONDON, COLCHESTER AND BECCLES

Contributors

J. W. BRADBEER, *Botany Department, King's College, London, England*

D. E. BRIGGS, *Department of Biochemistry, University of Birmingham, Edgbaston, Birmingham, Warwickshire, England*

J. W. CORNFORTH, *Shell Research Limited, Milstead Laboratory of Chemical Enzymology, Sittingbourne, Kent, England*

D. D. DAVIES, *School of Biological Sciences, University of East Anglia, Norwich, Norfolk, England*

L. FOWDEN, *Department of Botany and Microbiology, University College, London, England*

J. INGLE, *Department of Botany, University of Edinburgh, Scotland*

B. J. MIFLIN, *Department of Plant Science, University of Newcastle upon Tyne, Northumberland, England*

DAPHNE J. OSBORNE, *A.R.C. Unit of Developmental Botany, Cambridge, England*

B. SEDGWICK, *Shell Research Limited, Milstead Laboratory of Chemical Enzymology, Sittingbourne, Kent, England*

H. SMITH, *Department of Physiology and Environmental Studies, University of Nottingham, Nottingham, England*

H. E. STREET, *Botanical Laboratories, School of Biological Sciences, University of Leicester, Leicester, England*

H. TRISTRAM, *Department of Botany and Microbiology, University College, London, England*

C. A. WEST, *Division of Biochemistry, Department of Chemistry, University of California, Los Angeles, California, U.S.A.*

Preface

Living organisms resist the changes imposed on them by the environment—in the words of geneticists "they exhibit homeostasis". They so maintain themselves against outside influences that this property can be used as a criterion to separate living from non-living matter, and yet, paradoxically, the triumph of biochemistry has been, and continues to be, the analysis of organisms and their metabolism in terms of the same chemical and physical processes that operate on inanimate matter. One may justifiably ask how then can life be separated from non-living things, at what level of complexity or organization does life begin? Alternatively, what are the characteristics of the reactions carried out by living organisms which inanimate reactions lack? The reader is advised that clear answers to these questions are not to be found in this book. What he will find is a survey of the current state of our knowledge of the diverse mechanisms used by plants to regulate their biosynthetic metabolism and growth.

During the last fifty years the major metabolic pathways have been largely worked out in sufficient detail to show that the overall reactions do not proceed exactly as predicted from measurements of the kinetics of the individual reactions. Even quite small sequences of reactions that produce cellular products show an unexpected degree of regulation and the more complex growth processes are obviously subject to intricate control.

Although some of the individual steps of many reaction sequences and the enzymes that catalyse them are still undiscovered, a new insight has been obtained into the behaviour of the overall processes during the last fifteen to twenty, and particularly during the last five, years. This new knowledge has begun to give an understanding of how the reactions are controlled within the cell. An earlier volume in this series ("Biosynthetic Pathways in Higher Plants", J. B. Pridham and T. Swain, eds) dealt with the biosynthetic pathways used by plants; in 1972, at the meeting in Canterbury, we returned to the same topic but our discussions had a different emphasis. Obviously, knowledge of this subject is still far from complete but enough is now known of the different kinds of mechanism that regulate growth and metabolism for one to be able to review the subject of control mechanisms which operate in plants.

In the first lecture of the symposium Professor Davies brought out an exciting new concept of metabolic control, arguing that the pH of cells is regulated by the mutually antagonistic activities of carboxylating and decarboxylating enzymes being adjusted as the pH of the cellular environment moved towards or away from their respective optima. He also discusses the other broad concepts of metabolic regulation "energy charge", the NAD/NADH ratio, the pyradoxamine phosphate/pyridoxal phosphate ratio and the

vii

balance and co-ordination of the enzymes of nitrogen and carbohydrate metabolism.

Bacteria, honorary plants for the meeting, regulate their enzymatic activities principally by making new enzyme molecules and the intricacies of the process are described with clarity by Mr Tristram so that the methods of control, described by Dr Miflin, which occur in higher plants (barley roots) are shown in counterpoint. The biosynthesis of amino acids is perhaps better defined than that of any other comparable cellular constituent; consequently, this basic knowledge provides the best framework for investigating regulatory mechanisms. Mr Tristram's review is illustrated by his work using naturally-occurring amino acid analogues (isolated from various higher plants by Professor Fowden and co-workers) as novel inhibitors to probe the capacities of the bacterial biosynthetic system. Similarly, Dr Miflin describes some of his own work to illustrate the interrelationships between the separate pathways by which the amino acids are formed.

The contrasts between the methods of regulation employed by the two kinds of organism are profound: the control of enzyme production used by bacteria can hardly be demonstrated in barley roots where allosteric effects and feedback inhibition are all important and allow the cells to respond more rapidly than if they depended on synthesis and destruction of enzyme protein.

Dr Ingle describes the intricate factors concerned in the production of large quantities of ribosomal RNA in cells, the devices used by plants to make sufficient copies of it and the variety of biochemical, genetic, cytological and analytical techniques that are used to investigate the problem.

Professor Street describes the behaviour of plant cells in tissue culture and shows the potential of the technique for investigating cellular metabolism without the complications of permeation into a cell mass and variations in cell types that bedevil experiments with isolated organs, or even separated tissues. The manipulation of free cells in a chemostat promises to provide a useful addition to the techniques available for biosynthetic and genetic investigations.

In the paper by Dr Daphne Osborne, the molecular events are followed through to their visible, morphological conclusions. She records the effects of ethylene on the favoured test object—the plumular hook of pea seedlings, and describes some of the results. Ethylene is a regulator of plant growth that has some of the attributes of an insect pheromone but its complex action is intimately concerned with the cell wall and its constituent hydroxyproline-rich proteins.

The other paper on the subject of plant hormones was given by Professor West, who reviews gibberellin biosynthesis and what is known of the regulatory processes that operate on it.

In his review Professor Cornforth discusses the influence the isotope effect has on the choice of hydrogen atom removed from a methyl group by isopentenylpyrophosphate isomerase and relates the selection to the strict stereo-specificity of enzyme reactions. The stereospecificity can be considered as a "labour-saving device"—in this case it is simpler to be highly specific. However, the same picture of integration and control is described in other reviews at higher and higher levels of complexity; at the atomic, molecular, enzymatic, pathway, organelle, cell and organ levels of organization there is precision and regulation.

Dr Sedgwick describes what is known of the biosynthesis of fatty acids in plants and how the processes are controlled. The reactions have been more thoroughly studied in animals and several of the regulatory mechanisms are different although recent results obtained by Dr Sedgwick, using chirally-labelled acetate, have shown that the stereochemistry of the reactions is the same.

The papers by Dr Briggs, who discusses the changes which occur during the germination of barley, and Professor Bradbeer, who describes the synthesis of enzymes and other materials which occurs during chloroplast formation, both show that these complex reactions are under precise control. As yet the way the control is operated remains largely unknown.

Professor Smith deals with current research on the regulation of flavonoid synthesis and concentrates in particular on the role of phenylalanine ammonia lyase (PAL). It is a branch-point enzyme for several different families of second-ary products and its activity in extracts is affected by the previous illumination of the plant material from which it is extracted. The present evidence is insufficient to allow any definitive conclusions to be drawn concerning either the enzyme involved with the light-operated switch or even whether the change in overall activity depends on the synthesis of more enzyme protein or on adjustment of the activity of pre-existing molecules.

The final contribution of the Symposium, by Professor Fowden, provides a paradoxical contrast to the foregoing papers which all show evidence for the rigorous control of all aspects of biosynthesis. The non-protein amino acid, azetidine-2-carboxylic acid, occurs widely in the Liliaceae and can be used as a taxonomic marker as it occurs in a very few other species. Recently, as the result of a fortuitous set of circumstances, azetidine-2-carboxylic acid has been identified in sugarbeet where it occurs at very low concentrations. So low, in fact, that it would be undetectable were it not for the special processing which the juice undergoes.

The intriguing possibility suggested by this discovery is that azetidine-2-carboxylic acid occurs widely in plants as a metabolic mistake—as the result of imprecise selection of a substrate by an enzyme. It might be that azetidine-2-carboxylic acid is an example of "metabolic noise".

Physiological sciences have advanced to their present state of knowledge, largely by the use of a simple mechanistic outlook as expressed by the law of mass action. The "law of the minimum" or "limiting factor hypothesis" was the first, and until recently, the major exception to this way of regarding complex processes but the self-regulating and interlocking pathways of metabolism are now coming to provide an even more serious exception. The investigation of self-regulating systems in living organisms by the "cause and proportional effect" design of experiment is inadequate because the integrated system may not respond at all to a stimulus, or respond in the reverse manner to that expected on the basis of a simple logic.

From the relative simplicity of the feedback loop, through the repressor/inducer/operator system of bacterial protein synthesis to the incompletely understood complexity of hormonal action, life processes show the capacity for self-regulation. By gathering together in one volume examples of the different kinds of control mechanism we can appreciate more fully the intricacy of metabolism and perhaps become more aware of the dangers of trusting a hypothesis when it accommodates the facts. As Whitehead said "Seek simplicity, then distrust it". Perhaps this book will help to make us more cautious in our interpretation of physiological data.

The review papers in this book were read at the Phytochemical Society Symposium held at Rutherford College of the University of Kent at Canterbury, and at the Sittingbourne Laboratories of Shell Research Limited. Thanks are due to many people in these two organizations for their hospitality and help with the arrangements for the meetings and to Shell Research Limited for financial support. The Phytochemical Society is grateful to the speakers for providing such up-to-date reviews of the topics and for their prompt co-operation during the editing.

Finally, the editor wishes to record his thanks to Mrs Margaret Ogle and Mrs Marilyn Jury for efficient secretarial help, both in the organization of the meeting and with the processing of the manuscripts, and to the staff of Academic Press for expert assistance in preparing the book for publication.

November 1972

B. V. MILBORROW

Contents

CHAPTER 1

Metabolic Control in Higher Plants
D. D. Davies

CHAPTER 2

Some Aspects of the Regulation of Amino Acid
Biosynthesis in Bacteria
H. Tristram

CHAPTER 3

Amino Acid Biosynthesis and its Control in Plants

B. J. Miflin

CHAPTER 4

The Regulation of Ribosomal RNA Synthesis

J. Ingle

CHAPTER 5

Plant Cell Cultures:
Their Potential for Metabolic Studies

H. E. Street

CHAPTER 6

Ethylene and Protein Synthesis

D. J. Osborne

CHAPTER 7

Biosynthesis of Gibberellins

C. A. West

CHAPTER 8

Stereochemical Aspects of Enzyme Action

J. W. Cornforth

CHAPTER 9

The Control of Fatty Acid Biosynthesis in Plants

B. Sedgwick

CHAPTER 10

Hormones and Carbohydrate Metabolism in Germinating Cereal Grains

D. E. Briggs

CHAPTER 11

The Synthesis of Chloroplast Enzymes

J. W. Bradbeer

CHAPTER 12

Regulatory Mechanisms in the Photocontrol of
Flavonoid Biosynthesis

H. Smith

CHAPTER 13

The Non-protein Amino Acids of Plants:
Concepts of Biosynthetic Control

L. Fowden

CHAPTER 1

Metabolic Control in Higher Plants

D. D. DAVIES

*School of Biological Sciences, University of East Anglia,
Norwich, Norfolk, England*

I. INTRODUCTION

The common inheritance of living organisms is reflected in their biochemistry. Thus the study of intermediary metabolism has shown that the major metabolic pathways are common to all living organisms. In the mid-1950s, the study of metabolic control took a major step forward with the discovery that many metabolic sequences are controlled by their end products in a manner analogous to the negative feedback circuits of electronics (Umbarger, 1956; Yates and Pardee, 1956). Detailed study of feedback systems has shown considerable variation from species to species and it is now apparent that whilst we can reasonably anticipate the existence of, say the Krebs cycle in a given species, we cannot assume the presence of particular control mechanisms.

Fortunately we can define basic patterns of control, and the principles involved in the control of single metabolic pathways, branched pathways and cyclical process have been recognized (see Stadtman, 1970 for review). Other contributors to this symposium will be concerned with these principles and with specific examples of allosteric enzymes.

II. METABOLIC INTERLOCK

My intention is to discuss the integration of metabolic control. One aspect of this has been termed metabolic interlock (Jensen, 1969) and can be illustrated by reference to the biosynthesis of serine in plants. Slaughter and Davies

(1968) have shown that serine inhibits the first step in the specific metabolic sequence leading to its synthesis. Subsequently Slaughter (1970) has shown that

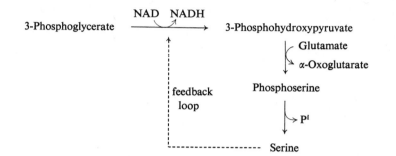

3-phosphoglycerate dehydrogenase is inhibited by L-methionine and thus we have a metabolic interlock between the sequence leading to serine biosynthesis and the end product (methionine) of another sequence in which the methyl group of methionine is indirectly derived from the hydroxymethyl group of serine.

Our understanding of the interlocking of metabolic processes is only just beginning, but considerable attention has been paid to the energy charge of the adenine nucleotides and to the redox state of the pyridine nucleotides. I will briefly review the limited amount of work which has been done with plants, but will concentrate on two other aspects: (i) the role of pH in control and (ii) the role of pyridoxal phosphate and pyridoxamine phosphate in the control of nitrogen metabolism.

III. Energy Charge

The role of adenine nucleotides in the regulation of carbohydrate metabolism was discussed by Krebs in 1964. Subsequently Atkinson and Walton (1967) developed the concept of energy charge—a parameter intended to indicate quantitatively the energy state of the cell. The energy charge is defined as $[(ATP) + 0.5\,(ADP)]/[(ATP) + (ADP) + (AMP)]$ and the work of Atkinson has shown that the activities of many enzymes are regulated by energy charge as shown in Fig. 1.

Work with plants has not contradicted the generalization proposed by Atkinson, but little evidence has been obtained in support of the concept. Thus in animals and bacteria, phosphofructokinase is inhibited by ATP and citrate and this inhibition is modulated by ADP, AMP and inorganic phosphate. The inhibition of phosphofructokinase by ATP and citrate has been confirmed for the carrot (*Daucus carota*) enzyme (Dennis and Coultate, 1966) for the enzyme of corn (*Zea mays*) scutellum (Garrard and Humphreys, 1968) and for the pea seed enzyme (Kelly and Turner, 1969) which is also inhibited by phos-

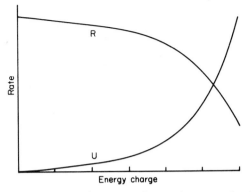

FIG. 1. Generalized response to the energy charge of enzymes involved in regulation of ATP-regenerating (R) and ATP-utilizing (U) sequences (After Alkinson and Walton, 1967).

phoenolpyruvate, though at low concentrations of ATP the enzyme is stimulated by phosphoenolpyruvate. However ADP and AMP were relatively ineffective in modulating the response to ATP.

The activity of NAD specific isocitrate dehydrogenase of bacteria and fungi is modulated by AMP and the activity of the animal enzyme is modulated by ADP. However, detailed studies of plant isocitrate dehydrogenase (Cox and Davies, 1967; Dennis and Coultate, 1967) have shown that the enzyme responds neither to single adenine nucleotides nor to energy charge.

However the activities of a number of plant enzymes are modulated by adenine nucleotides, for example the decarboxylation of α-oxoglutarate by preparations from pea (*Pisum sativum*) and cauliflower (*Brassica oleracea*) mitochondria (Davies and Kenworthy, 1970; Wedding and Black, 1971) is stimulated by AMP. Another example is nitrate reductase whose activity is modulated by ADP (Nelson and Ilan, 1969; Eaglesham and Hewitt, 1971). Finally, attention should be drawn to the work of Bomsel and Pradet (1968) and Pradet and Bomsel (1969) who have examined the levels of adenine nucleotides in wheat (*Triticum sativum*) and lettuce (*Lactuca sativa*) in the presence and absence of oxygen. They noted wide variations in the ratios ATP/ADP and ATP/AMP while the energy charge remained relatively constant. These results are considered to be a confirmation of the energy charge hypothesis of Atkinson. It therefore seems necessary to wait further experimental data before evaluating the extent to which the Atkinson hypothesis is applicable to plants.

IV. CONTROL INVOLVING THE RATIO NAD/NADH

Krebs (1969) has pointed out that equilibrium enzymes may play an important role in regulation since they determine the concentrations of substrates for non-equilibrium regulatory enzymes. Thus lactate dehydrogenase maintains the reactants pyruvate, lactate, NAD and NADH at equilibrium and an increase in the ratio NAD/NADH will tend to increase gluconeogenesis by increasing the concentration of pyruvate. A related equilibrium situation exists in potato (*Solanum tuberosum*) tubers where under anaerobic conditions lactate accumulates until pyruvate decarboxylase is activated. Eventually ethanol is formed and the lactate concentration declines. In such a situation it is difficult to see the biological significance of product inhibition unless the

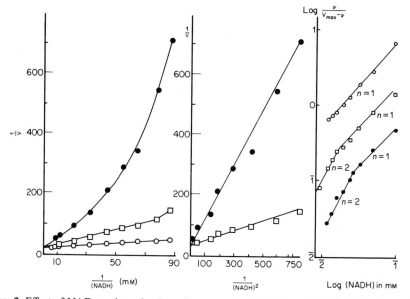

Fig. 2. Effect of NAD on the reduction of pyruvate by NADH at pH 6·1. MES buffer pH 6·1, 33 mM; pyruvate 0·2 mM; ● NAD 3 mM; □ NAD 1 mM; ○ Control (Davies and Davies, 1972).

inhibition is sufficiently great to convert an equilibrium situation into a non-equilibrium situation. The inhibition of lactate dehydrogenase by NAD may or may not have physiological significance, but some evidence for the allosteric binding of NAD (Fig. 2) tends to support the physiological significance of the inhibition.

Isocitrate dehydrogenase has been considered as a control site in plant mitochondria (Laties, 1967). However the enzyme does not respond to energy charge and it has been suggested (Cox and Davies, 1967; Dennis and Coultate, 1967) that the NAD/NADH ratio may be an important regulatory factor. Some evidence for the allosteric binding of NADH has been obtained with crude preparations of isocitrate dehydrogenase (Fig. 3) but purified preparations show competitive inhibitions by NADH. (Cox and Davies, 1967; Duggleby and Dennis, 1970). In considering the physiological significance of such competitive inhibition, Atkinson (1968) has pointed out that surprisingly large changes in enzyme activity can result from small changes in relative

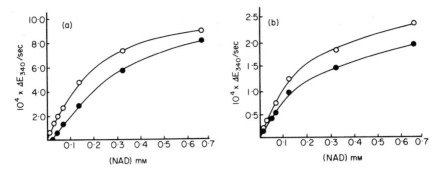

FIG. 3. Effect of NAD$^+$ concentration on the rate of reaction in the presence and absence of NADH with (a) a crude extract (b) a partially purified isocitrate dehydrogenase. ● no NADH ▲ 0·073 mM-NADH. Buffer HEPES pH 7·6 0·05 mD-Isocitrate 2 mM. MnSO$_4$ 1 mM (After Cox and Davies, 1967).

affinities (Fig. 4). Control of the type implicit in this figure requires only changes of affinities for substrates already bound to enzymes and consequently the development of such control should be much easier than the evolution of a wholly new property such as the formation of an allosteric site.

Finally, the role of NADH as an allosteric effector of a number of enzymes should be mentioned. In particular it has been shown to inhibit NADP specific malic enzyme (Sanwal and Smando, 1969) and NADP specific glucose-6-phosphate dehydrogenase (Sanwal 1970) in a number of bacteria.

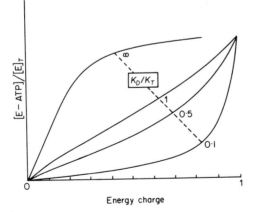

FIG. 4. Calculated % saturation, as a function of energy charge, of an enzymic ATP-binding site, assuming a total adenylate pool (ATP + ADP + AMP) of 5 mM, Michaelis constant for ATP, Kt, of 0·2 mM, and various Michaelis constants for ADP, Kd, as specified by the Kd/Kt ratios identifying the curves.

V. Interaction Between Effectors

The possible interaction between the NAD/NADH ratio and the enzymes involved in generating the NADP/NADPH ratio, indicated in the previous section, is one example of widespread interlocking of metabolic control. Atkinson (1968) has pointed to the way in which end-product inhibition interacts with energy charge control (Fig. 5). Veech *et al.* (1970) have discussed the

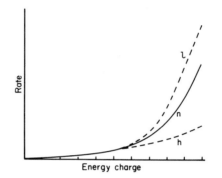

Fig. 5. Generalized interaction between energy charge and end-product concentration in the control of a regulatory enzyme in a biosynthetic sequence. The curves correspond to low (l), normal (n), and high (h) concentrations of the end-product modifier (After Atkinson, 1968).

interlock between ATP/ADP and NAD/NADH ratios. Thus the ratio $(ATP)/(ADP)(P_i)$ measured in rat liver under various nutritional states was found to be close to the value calculated from the concentrations of substrates and equilibrium constants of the enzymes lactate dehydrogenase, glyceraldehyde phosphate dehydrogenase and 3-phosphoglycerate kinase which are assumed to function close to equilibrium (Veech *et al.*, 1970). This implies that the redox state of the NAD couple in the cytoplasm is linked to and partially controlled by, the phosphorylation state of the adenine nucleotides.

In the next section, the effect of pH on metabolic control will be discussed. Since the equilibrium constants of reactions involving pyridine nucleotides are pH dependent and since most enzyme substrates are ionized the controlling effect of pH will be manifest in a wide range of control systems.

VI. Control of and by pH

The effect of pH on an enzyme catalysed reaction is, in idealized form, represented by a bell-shaped curve. Neglecting for the time being the difficulties inherent in interpreting pH data (see Dixon and Webb, 1964) it is clear that, in general, enzymes exhibit pH optima and that these optima vary from one enzyme to another. Unless we assume a microclimate for each enzyme, it follows that many of the enzymes functioning in the cell must be operating away from their pH optima, so that the rates of the reaction they catalyse will be sensitive to small changes in pH. At a fixed pH the rates of many metabolic processes are interlocked and regulated. If the pH of the cell varies we may anticipate considerable changes in the relative balances between metabolic pathways. For example, the effect of change of pH on the relative activities of the Embden–Meyerhof pathway and the pentose–phosphate pathway in red blood cells is shown in Table I. Analysis of the situation in red blood cells

TABLE I

The pH dependence of the Embden–Meyerhof and the oxidative pentose–phosphate pathway in erythrocytes of man (Albrecht *et al.*, 1971)

| | % Glucose metabolized via | |
pH	Embden–Meyerhof	Pentose–phosphate
6·9	55	45
7·4	11	89
8·2	2	98

suggests that below pH 6·8 control of glycolysis is exerted by hexokinase and phosphofructokinase; from pH 6·8–7·4, phosphofructokinase is largely responsible for control; from pH 7·6–8·3 hexokinase, phosphofructokinase and triose phosphate dehydrogenase jointly control the rate of glycolysis (Albrecht *et al.*, 1971).

Another example is the effect of pH on the formation of acetoin and acetolactate by extracts of yeast. The acetoin forming system has an acid pH optimum whereas the acetolactate forming system has an optimum at pH 7·2, thus the route from pyruvate to acetoin or to acetolactate is controlled by pH (Fig. 6).

To maintain the complex balance between metabolic pathways, the pH of the cytoplasm must be controlled. One way of achieving this would involve carboxylation and decarboxylation reactions

$$R + CO_2 \xrightarrow{\text{carboxylation}} R - COO^- \xrightarrow{\text{decarboxylation}} CO_2 + R$$

Such a system would be self regulating provided that the enzymes involved have appropriate pH optima. The pH of the cytoplasm of plant cells is not known with any certainty, though the acid lability of NADH and particularly NADPH, suggests that the pH is above pH 7—say pH 7·2. A self-adjusting system of carboxylating and decarboxylating enzymes would then require the properties shown in Fig. 7.

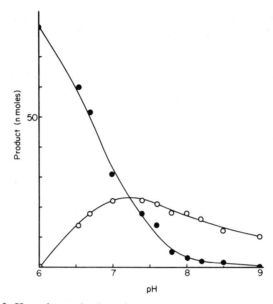

FIG. 6. Effect of pH on the production of acetoin (●) and acetolactate (○) by extracts of *Saccharomyces cerevisiae* (redrawn from the data of Magee and Robichon-Szulmajster, 1967).

In this system the two pH curves are for simplicity assumed to intersect at the normal pH of the cytoplasm. If a disturbance causes the pH to decrease, then the activity of the decarboxylating enzyme will increase whilst the activity of the carboxylating enzyme will decrease and the system will tend to adjust the pH back to the original value. The converse changes will, of course, occur if the pH of the cytoplasm increases. The sensitivity of the system will depend upon the slopes of the pH *vs* activity curves, but the basic requirements are that carboxylating enzymes should have alkaline optima whilst decarboxylating enzymes should have acidic optima.

A survey of the literature (Table II) shows that a number of carboxylating and decarboxylating enzymes have the anticipated pH optima. However, published pH versus activity curves are frequently difficult to interpret since they usually represent a maximum velocity (i.e. substrate saturation) at a particular pH but there is seldom any assurance that substrate saturation has been maintained throughout the pH range. Furthermore, the pH optimum under conditions of substrate saturation may be significantly different from the optimum with lower and presumably more "physiological" concentrations of substrate. This may explain the fact that the pH optima for isocitrate dehydrogenase and malic enzyme appear to be too alkaline for the general hypothesis. Consider for example, the case of NAD specific isocitrate dehydrogenase from pea seedlings (Cox and Davies, 1969). The affinity of the enzyme

TABLE II

pH Optima of carboxylating and decarboxylating enzymes

E.C. No.	Enzyme	Source	Optimum pH	References
1.1.1.41	Isocitrate dehydrogenase (NAD specific)	Pea	7·6 (Isocitrate 1 mM) 6·9 (Isocitrate 50 μM)	Cox and Davies, 1969
1.1.1.40	Malic enzyme (NADP specific)	Wheat germ	7·3	Harary et al., 1953
1.1.1.39	Malic enzyme (NAD specific)	Cauliflower (*Brassica oleracea*)	6·7–6·9	Macrae, 1971a
4.1.1.2	Oxalate decarboxylase	*Collybia velutipes* (fungus)	2·5–4·0	Shimazono and Hayaishi, 1957
4.1.1.15	Glutamate decarboxylase	Barley	5·9	Fowden, 1954
	Glutamate decarboxylase	Tulip (*Tulipa*)	5·8	
4.1.1.11	Aspartate decarboxylase	*Alcaligenes faecalis* (bacterium)	5·0	Novogrodsky and Meister, 1964
4.1.1.–	Leucine decarboxylase	Red algae	4·2–6·0	Hartmann, 1972
4.1.1.1	Pyruvate decarboxylase	Wheat germ	6·0	Davies et al., 1972
6.4.1.2	Acetyl-CoA carboxylase	Wheat germ	9	Hatch and Stumpf, 1961
6.4.1.1	Pyruvate carboxylase	Yeast	8·3	Ruiz-Amil et al., 1965
4.1.1.31	Phosphoenolpyruvate carboxylase	Peanut (*Arachis hypogea*)	7·9–8·3	Maruyama and Lane, 1962
6.4.1.3	Propionyl-CoA carboxylase	Pig heart	8·5	Kaziro et al., 1961
4.1.1.–	Ribulosediphosphate carboxylase	Maize	7·8	Andrews and Hatch, 1971

D. D. DAVIES

TABLE III

Effect of pH on the affinity (S) of
pea NAD-specific isocitrate dehydrogenase for isocitrate and on
the Hill number (n) (Cox and
Davies, 1969)

pH	$S\ 0.5\ (\mu M)$	n
6·4	8	1·0
7·0	33	1·6
7·6	96	2·6
8·0	310	2·8

TABLE IV

Effect of isocitrate concentration on the pH optimum of isocitrate
dehydrogenase (Cox and Davies, 1969)

Isocitrate concentration (mM)	1·0	0·5	0·15	0·05
pH optimum	7·6	7·5	7·3	6·9

for isocitrate is markedly affected by pH (Table III) and the pH optimum
decreases as the substrate concentration decreases (Table IV). Thus if the
concentration of isocitrate in mitochondria is of the order 0·05 mM, the pH
versus activity curve is consistent with a role for isocitrate dehydrogenase in
the buffering system described by Fig. 7. Similarly, the malic enzyme of liver

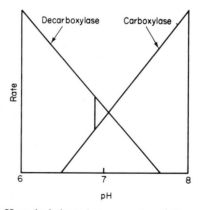

FIG. 7. Effect of pH on the balance between carboxylation and decarboxylation.

shows an alkaline pH optimum at high malate concentrations but the optimum moves towards the acid region at low malate concentrations (Fig. 8). The affinity for malate shows an approximately 10-fold increase for each pH unit (Table V). Consequently, if malic enzyme functions at low substrate concentrations, the pH profile would be compatible with Fig. 7, and with assigning a role in the regulation of pH.

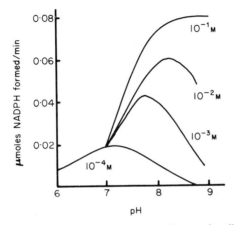

FIG. 8. Influence of malate concentration on the pH optimum of malic enzyme. The assay solution contained: $MnSO_4$ 1 μmole, NADP 0·2 μmole malic enzyme 3 μg, buffer 100 μmoles (pH 7–9, Tris pH 6–7, histidine) sodium malate at concentrations indicated, final volume 1 ml, temp 23°C (redrawn from Rutter and Lardy, 1958).

TABLE V

Effect of pH on the affinity
of malic enzyme for malate
(Rutter and Lardy, 1958)

pH	K_m (mM)
8·5	3·3
7·5	0·39
6·5	0·04

A. CARBOXYLATION REACTIONS

Excess cation uptake by plants leads to an increase in organic acid production. The rate of CO_2 fixation is of the order required to account for the observed synthesis of organic acids (Hiatt and Hendricks, 1967). The enzyme implicated in the carboxylation is phosphoenolpyruvate carboxylase. The pH optimum

of the maize enzyme is 8·5 (Wong and Davies, 1970) and the steep slope of the
pH activity curve (Fig. 9) suggests that the pH change induced by excess cation
uptake would produce a large increase in the rate of CO_2 fixation. Thus an
increase of pH from 7·2–7·6 produces a 7 to 8-fold increase in the rate of
carboxylation. The carboxylase is inhibited by malate and oxaloacetate (Ting,
1968) and though it is not clear to what extent the inhibition by malate is due

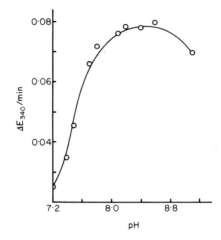

FIG. 9. pH Activity curve for phosphoenolpyruvate carboxylase of maize seedlings. The
enzyme was assayed by coupling with malate dehydrogenase (after Wong and Davies, 1970).

to chelation of Mg^{2+} (Wong and Davies, 1970) there is nevertheless end
product inhibition which reinforces control by pH.

B. DECARBOXYLATION REACTIONS

Loss of organic acids is associated with excess uptake of anions and also with
the ripening process in fruits. A number of enzymes may be involved in these
decarboxylations and the NAD-specific malic enzyme of cauliflower provides
a model system (Macrae, 1971a, b). The enzyme is associated with the mitochondria and shows a pH optimum at 6·8 with a sharp drop in activity at
pH values above 7·0: a change of one pH unit produces a 6-fold change in
activity. The enzyme competes with malate dehydrogenase for malate and the
effect of a low pH on the metabolism of malate is to direct malate in the direction of decarboxylation (Table VI). These results can be explained in terms of
the pH activity profiles of malic enzyme and malate dehydrogenase. Additional
to the controlling effect of pH on malic enzyme is the activating effect of CoA.
The concentration of free CoA may be expected to be inversely related to the
concentration of pyruvate, hence low concentrations of pyruvate will stimulate
malate decarboxylation and high concentrations of pyruvate will remove the

TABLE VI

The effect of pH on the products of malate oxidation (Macrae, 1971b)

pH	Pyruvate formed (μmoles)	Oxaloacetate formed (μmoles)	Ratio pyruvate/oxaloacetate
7·93	0·04	0·39	0·1
7·44	0·23	0·20	1·15
7·01	0·58	0·04	14·5
6·69	0·54	0·02	27·0

The reaction mixture contained 0·3 M mannitol, 10 mM KCl, 5 mM $MgCl_2$, 5 mM KH_2PO_4, 10 mM TES, 2·25 mg bovine serum albumin, 0·8 mM sodium arsenite, 1·33 mM ADP, 10 mM L-malate and mitochondria in a final volume of 3·0 ml.

stimulation by converting the CoA into acetyl-CoA. The malate decarboxylating system of apples is thought to involve NADP-specific malic enzyme and pyruvic decarboxylase (Hulme et al., 1963). The pH optimum of pyruvate decarboxylase· is clearly acidic and the significance of this for metabolic control is discussed in the next section.

1. Pyruvate Decarboxylase

Under aerobic conditions pyruvate is oxidatively decarboxylated via the Krebs cycle. For reasons of efficiency, pyruvic decarboxylase, which catalyses the decarboxylation of pyruvate to form acetaldehyde, should not function under aerobic conditions. The pH vs activity profile of pyruvic decarboxylase

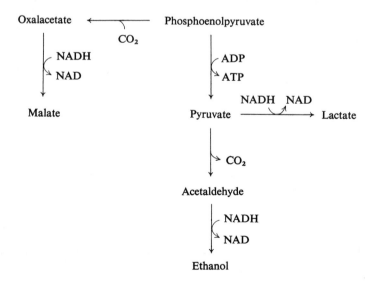

from wheat germ shows an optimum at pH 6, and little or no activity at pH 7·2 (Fig. 10). If 7·2 is the pH of the cells cytoplasm, the pyruvate decarboxylase will not be active.

However, under anaerobic conditions most plants produce ethanol and it is therefore proposed that the initial stage of anaerobiosis involves the production of acid which will lower the pH and "switch on" pyruvate decarboxylase. Two systems are obvious candidates for this role—phosphoenolpyruvate carboxylase in association with malate dehydrogenase forming one system, lactate dehydrogenase being the other. We have investigated the properties of potato lactate dehydrogenase for the proposed role in anaerobiosis (Davies and Davies, 1972). At alkaline pH the enzyme is only slightly inhibited by ATP, but as the pH falls ATP becomes a strong inhibitor of lactate dehydrogenase (Fig. 11). The inhibition by ATP at alkaline pH values appears to be competitive with respect to NADH but at acidic pH values sigmoid kinetics appear and the Hill number for ATP is 2 (Fig. 12).

These kinetics are consistent with the suggested role for lactate dehydrogenase and the following sequence is proposed. On transfer from air to nitrogen the pH is assumed to be 7·2 and lactate dehydrogenase functions at its maximum rate. The production of lactate leads to a fall in pH and, to prevent an overproduction of acid, the enzyme then functions as an ATP controlled pH-stat. The acidic environment produced by lactate switches on pyruvate decarboxylase and the partitioning of pyruvate between lactate and ethanol production

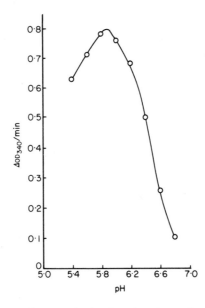

FIG. 10. pH activity curve for pyruvic decarboxylase from wheat germ. The enzyme was assayed by coupling with alcohol dehydrogenase (After Davies *et al.*, 1972).

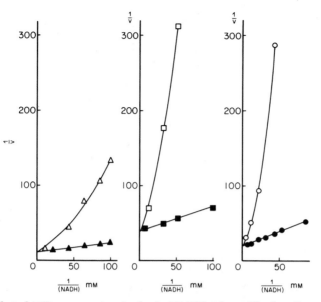

FIG. 11. Effect of ATP on pyruvate reduction by NADH at three pH values. Pyruvate 0·2 mM.
△ pH 7·1 ▲ pH 7·1 + ATP (0·2 mM) TES buffer (33 mM); □ pH 6·5 ■ pH 6·5 + ATP (0·2 mM)
TES buffer (33 mM); ○ pH 6·1 ● pH 6·1 + ATP (0·1 mM) MES buffer (33 mM).

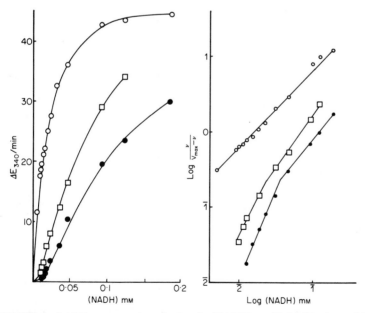

FIG. 12. Effect of ATP on pyruvate reduction by NADH at pH 6·1 (Davies and Davies,
1972). MES buffer (33 mM), Pyruvate (0·2 mM). ○ Control; □ ATP (0·05 mM); ● ATP
(0·1 mM).

can then be interpreted in terms of the pH versus activity profiles of lactate dehydrogenase and pyruvate decarboxylase.

VII. Control by Pyridoxal and Pyridoxamine Phosphates

Higher plants compete for limited supplies of nitrogen and the availability of nitrate is reflected by the level of nitrate reductase which is a substrate-inducible enzyme and also subject to regulation by endogenous metabolites (for review see Hewitt, 1970). The product of nitrate reduction is ammonia which enters into organic combination, largely by the reductive amination of α-oxoglutarate. The α-oxoglutarate is derived from carbohydrates via the Krebs cycle and consequently we might expect some degree of interlock in the regulation of carbohydrate and nitrogen metabolism.

The nitrogen status of the plant could be reflected by the ratio pyridoxamine phosphate/pyridoxal phosphate and coordination between the enzymes of carbohydrate and nitrogen metabolism could be achieved if the enzymes activities were regulated by either pyridoxal phosphate or pyridoxamine phosphate.

The inhibition of nitrate reductase by pyridoxamine phosphate has been reported by Sims et al., 1968. Duckweed (Lemna minor) contains two nitrate reductases—the NADH enzyme is not inhibited by pyridoxamine phosphate but is inhibited by ammonia—the NADPH enzyme is inhibited by pyridoxamine phosphate. Under the conditions employed by Sims et al. (1968), 50% inhibition of nitrate reductase was observed in the presence of 1 mM pyridoxamine phosphate and 0·1 mM pyridoxamine phosphate produced 20% inhibition. Whether or not these inhibitions have physiological significance remains to be established but inhibition by pyridoxamine phosphate is consistent with the proposed mechanism of regulation.

Pyridoxal phosphate is known to inhibit a number of enzymes and particularly to inhibit enzymes of carbohydrate metabolism (Domschke and Domagk, 1970). In a number of cases inhibition has been shown to result from the formation of a Schiff base between pyridoxal phosphate and the terminal amino group of specific lysyl residues (Anderson et al., 1966; Rippa et al., 1967). The physiological significance of the inhibition by pyridoxal phosphate depends upon its specificity. My colleague Artur Teixeira and I have examined the effect of pyridoxal phosphate on glutamate dehydrogenase from pea seedlings. The inhibition is time dependent; at pH 7·5 and 25°C, pyridoxal phosphate produces the maximum inhibition of glutamate dehydrogenase after incubation for between 5 and 10 min (Fig. 13). The inhibition is sensitive to pH—requiring pH values above 6 and below 8 (Fig. 14). The inhibition is non-competitive with respect to α-oxoglutarate (Fig. 15). Incubating the inhibited enzyme with one of its substrates relieves the inhibition (Fig. 16). We have investigated the specificity of the inhibition produced by pyridoxal

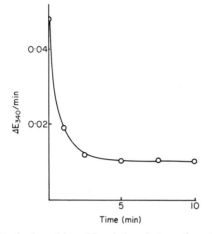

FIG. 13. Effect of incubation with pyridoxal phosphate on the activity of glutamate dehydrogenase. Pyridoxal phosphate 0·2 mM, α-oxoglutarate 2 mM, NH₄Cl 133 mM, NADH 0·2 mM, tris buffer pH 7·5.

phosphate on a range of enzymes prepared from pea seedlings and preliminary results suggest that glutamate dehydrogenase is particularly sensitive and that enzymes of glycolysis and the pentose phosphate pathway are, in general, subject to inhibition. Other enzymes examined showed little or no inhibition.

Assuming that the NAD glutamate dehydrogenase of pea seedlings functions in the direction of reductive amination then the observed inhibition is consistent with the proposed mechanism of regulation.

A low ratio pyridoxamine phosphate/pyridoxal phosphate reflects a low

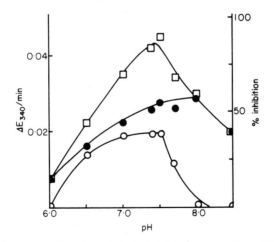

FIG. 14. Effect of pH on the inhibition of glutamate dehydrogenase by pyridoxal phosphate. □ Control activity; ● Activity with pyridoxal phosphate (0·02 mM); ○ % inhibition.

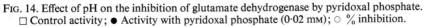

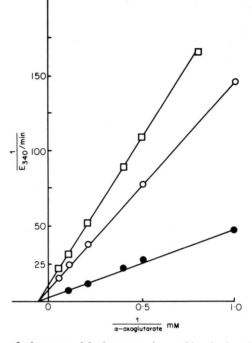

FIG. 15. Inhibition of glutamate dehydrogenase by pyridoxal phosphate. ● Control; ○ Pyridoxal phosphate 0·1 mM; □ Pyridoxal phosphate 0·2 mM.

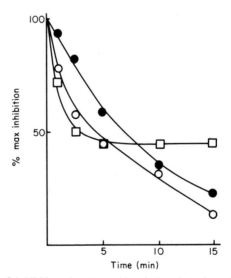

FIG. 16. Reversal of inhibition due to pyridoxal-phosphate by substrates of glutamate dehydrogenase. ● NADH 0·2 mM; ○ NH₄Cl 133 mM; □ α-oxoglutarate 2·0 mM.

nitrogen status; the inhibition of glutamate dehydrogenase and enzymes of glycolysis by pyridoxal phosphate is consistent with the lack of ammonia. A high ratio reflects a high nitrogen status and inhibition of the ammonia generating system by pyridoxamine phosphate would be consistent with the proposed regulation.

On balance, I believe there is a presumptive case for involving the ratio pyridoxamine phosphate/pyridoxal phosphate in metabolic regulation. However, speculation in the area of metabolic control is all too easy. Critical experiments are required, so back to the bench!

REFERENCES

Albrecht, V., Roigas, H., Schultze, M., Jacobasch, G. and Rapoport, S. (1971). *Europ. J. Biochem.* **20**, 44.

Anderson, B. M., Anderson, C. D. and Churchich, J. E. (1966). *Biochemistry* **5**, 2893.

Andrews, T. J. and Hatch, M. D. (1971). *Phytochemistry* **10**, 9.

Atkinson, D. E. (1968). *Biochemistry* **7**, 4030.

Atkinson, D. E. and Walton, G. M. (1967). *J. biol. Chem.* **242**, 3239.

Bomsel, J. L. and Pradet, A. (1968). *Biochim. Biophys. Acta* **162**, 230.

Cox, G. F. and Davies, D. D. (1967). *Biochem. J.* **105**, 729.

Cox, G. F. and Davies, D. D. (1969). *Biochem. J.* **113**, 813.

Davies, D. D. and Davies, S. (1972). In preparation.

Davies, D. D. and Kenworthy, P. (1970). *J. exp. Bot.* **21**, 247.

Davies, D. D., Grego, S. and Kenworthy, P. (1972). In preparation.

Dennis, D. T. and Coultate, T. P. (1966). *Biochem. biophys. Res. Commun.* **25**, 187.

Dennis, D. T. and Coultate, T. P. (1967). *Life Sci.* **6**, 2353.

Dixon, M. and Webb, E. C. (1964). "Enzymes." Longmans Green, London.

Domschke, W. and Domagk, G. F. (1970). *Hoppe-Seyler's Z. Physiol. Chem.* **350**, 1111.

Duggleby, R. G. and Dennis, D. T. (1970). *J. biol. Chem.* **245**, 3751.

Eaglesham, A. R. J. and Hewitt, E. J. (1971). *FEBS Letters* **16**, 315.

Fowden, L. (1954). *J. exp. Bot.* **5**, 28.

Garrard, L. A. and Humphreys, T. E. (1968). *Phytochemistry* **7**, 1949.

Harary, I., Korey, S. R. and Ochoa, S. (1953). *J. biol. Chem.* **203**, 595.

Hartman, T. (1972). *Phytochemistry* **11**, 1327.

Hatch, M. D. and Stumpf, P. K. (1961). *J. biol. Chem.* **236**, 2879.

Hewitt, E. J. (1970). *In* "Nitrogen Nutrition of the Plant" (E. A. Kirkby, ed.). Agricultural Chemistry Symposium, Univ. of Leeds.

Hiatt, A. J. and Hendricks, S. B. (1967). *Z. Pflanzenphysiol.* **56**, 220.

Hulme, A. C., Jones, J. D. and Wooltorton, L. S. C. (1963). *Proc. R. Soc.* B **158**, 514.

Jensen, R. A. (1969). *J. biol. Chem.* **244**, 2816.

Kaziro, Y., Ochoa, S., Warner, R. C. and Chen, J. (1961). *J. biol. Chem.* **236**, 1917.

Kelly, G. J. and Turner, J. F. (1969). *Biochem. J.* **115**, 481.

Krebs, H. A. (1964). *Proc. R. Soc.* B **159**, 545.

Krebs, H. A. (1969). *In* "Current Topics in Cellular Regulation" (B. L. Horecker and E. R. Stadtman, eds), Vol. VI, p. 49. Academic Press, New York and London.

Laties, G. C. (1967). *Phytochemistry* **6**, 181.

Macrae, A. R. (1971a). *Biochem. J.* **122**, 495.

Macrae, A. R. (1971b). *Phytochemistry* **10**, 1453.

Magee, P. T. and de Robichon-Szulmajster, H. (1967). *Europ. J. Biochem.* **3**, 502.

Maruyama, M. and Lane, M. D. (1962). *Biochem. biophys. Acta* **65**, 207.

Nelson, N. and Ilan, I. (1969). *Plant Cell Physiol.* (*Tokyo*) **10**, 143.

Novogrodsky, A. and Meister, A. (1964). *J. biol. Chem.* **239**, 879

Pradet, A. and Bomsel, J. L. (1969). *XI International Botanical Congress Abstracts* p. 173.

Rippa, M., Spanio, L. and Pontremoli, S. (1967) *Archs Biochem. Biophys* **118**, 48.

Ruiz-Amil, M., de Torrontegui, G., Palacian, E., Catalina, L. and Losada, M. (1965). *J. biol. Chem.* **240**, 3485.

Rutter, W. and Lardy, H. A. (1958). *J. biol. Chem.* **233**, 374.

Sanwal, B. D. (1970). *J. biol. Chem.* **245**, 1626.

Sanwal, B. D. and Smando, W. R. (1969). *J. biol. Chem.* **244**, 1817.

Shimazono, H. and Hayaishi, G. (1957). *J. biol. Chem.* **227**, 151.

Sims, A. P., Folkes, B. and Bussey, A. H. (1968). *In* "Recent Aspects of Nitrogen Metabolism in Plants" (E. J. Hewitt and C. V. Cutting, eds), Academic Press, London and New York.

Slaughter, J. C. (1970). *FEBS Letters* **7**, 245.

Slaughter, J. C. and Davies, D. D. (1968). *Biochem. J.* **109**, 749.

Stadtman, E. R. (1970). *In* "The Enzymes" (P. D. Boyer, ed.), 3rd Edition, Vol. 1, p. 397. Academic Press, New York and London.

Ting, I. P. (1968). *Pl. Physiol.* **43**, 1919.

Umbarger, H. E. (1956). *Science, N. Y.* **123**, 848.

Veech, R. L., Raijman, L. and Krebs, H. A. (1970). *Biochem. J.* **117**, 499.

Wedding, R. T. and Black, M. K. (1971). *J. biol. Chem.* **246**, 1638.

Wong, K. and Davies, D. D. (1970). Unpublished.

Yates, R. A. and Pardee, A. B. (1956). *J. biol. Chem.* **221**, 757.

CHAPTER 2

Some Aspects of the Regulation of Amino Acid Biosynthesis in Bacteria

H. TRISTRAM

Department of Botany and Microbiology, University College, London, England

I. INTRODUCTION

Many "non-protein" amino acids possess growth-inhibitory properties towards microorganisms, and occasionally towards other organisms. Some of these compounds have been obtained by chemical syntheses designed to yield products showing a close structural similarity to one of the normal protein amino acids; others occur as natural products, especially in higher plants and fungi. The distribution and possible modes of biosynthesis of some of the "non-protein" amino acids of higher plants has been treated elsewhere (Fowden, 1970; Fowden, p. 323 of this volume). In order to retain biological activity the structural modifications tolerated among the synthetic amino acid analogues are restricted to changes which retain the overall shape and ionic characteristics of the corresponding protein amino acid (Richmond, 1962).

A number of synthetic analogues, and latterly, higher plant non-protein amino acids, bearing a close structural relationship to one or more of the common protein amino acids have proved useful tools in the study of the metabolic control of amino acid biosynthesis in bacteria. The mode of action of some of these compounds, especially those formed by some higher plants, and the contribution of these studies to our understanding of metabolic control of amino acid biosynthesis will be discussed.

II. Proline Analogues

A. GROWTH INHIBITION

Several imino acids, all structurally related to proline(1) are growth-inhibitory to *Escherichia coli* and *Salmonella typhimurium*, inhibition being specifically annulled by proline (Smith *et al.*, 1962; Fowden and Richmond, 1963; Fowden *et al.*, 1963; Unger and DeMoss, 1966; Rowland and Tristram, 1972). Toxic compounds include L-azetidine-2-carboxylic acid (2), the lower homologue of proline and *cis*-3,4-methano-L-proline (3), both higher plant products, the former isolated from some Liliaceae and Leguminosae (Fowden, 1955; Sung and Fowden, 1969) and the latter from *Aesculus parviflora* (Fowden *et al.*, 1969). *Cis*-4-hydroxy-L-proline, found in free form in leaves of the sandal-wood tree (*Santalum album*), is slightly growth-inhibitory towards *E. coli*, whereas *trans*-4-hydroxy-L-proline which is widely distributed in bound form in plant proteins, is without effect on bacterial growth (I. Rowland, unpublished experiments). Proline analogues of synthetic origin which are highly growth-inhibitory include 3,4-dehydro-DL-proline(4) and thiazolidine-4-carboxylic acid. However L-pipecolic acid(5) and L-4,5-dehydropipecolic acid, both higher plants products, show no growth inhibitory activity towards bacteria. A number of the analogues mentioned above are also toxic to some higher plants and animals (Fowden *et al.*, 1967).

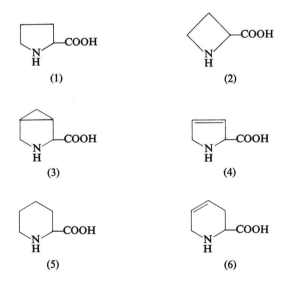

The toxicity of azetidine-2-carboxylic acid and 3,4-dehydroproline may be ascribed, at least in part, to their ability to stoichiometrically replace proline residues in proteins, with the possibility of formation of biologically impaired enzymes (Fowden *et al.*, 1967). Incorporation of both analogues into proteins of *E. coli, S. typhimurium* and mung bean (*Phaseolus aureus*) was demonstrated by conventional paper chromatographic methods (Fowden and Richmond, 1963; Fowden *et al.*, 1963; Hussain, 1968). A very sensitive method for the demonstration of incorporation of an analogue into protein has been exploited in the author's laboratory. The ability of an analogue to support protein

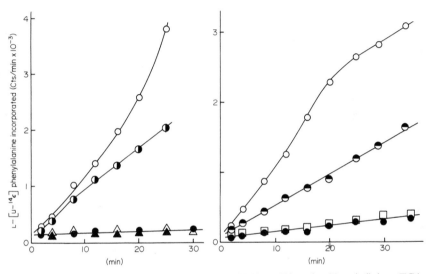

FIG. 1. Incorporation of L-[U-¹⁴C]phenylalanine (0·825 μCi/μmole; 20 μg/ml) into TCA-insoluble material by *E. coli* 55-1 in the presence (○) and absence (●) of L-proline (20 μg/ml) or 3,4-dehydro-DL-proline, (◑); L-azetidine-2-carboxylic acid, (◕); L-pipecolic acid (□); *cis*- (▲); or *trans*-3,4-methano-L-proline, (△); all at 20 μg/ml. The graphs represent two separate experiments (unpublished results of I. Rowland).

synthesis is determined in a starved bacterial strain, auxotrophic for the corresponding protein amino acid, by following incorporation of a labelled unrelated amino acid. In this way it was shown (Fig. 1) that 3,4-dehydroproline and azetidine-2-carboxylic both supported protein synthesis in a proline auxotroph of *E. coli*, whereas pipecolic acid did not (I. Rowland, unpublished experiments); neither *cis*- nor *trans*-3,4-methanoprolines supported protein synthesis in either *E. coli* or *S. typhimurium* (Rowland and Tristram, 1972).

The effectiveness of azetidine-2-carboxylic acid and 3,4-dehydroproline in promoting synthesis of biologically impaired protein in bacteria is enhanced by their ability to mimic proline in operating control mechanisms shutting off proline synthesis, so starving the cell of endogenous proline.

B. PROLINE BIOSYNTHESIS

Elucidation of the pathway of proline biosynthesis was achieved by use of proline auxotrophs of *E. coli* (Umbarger and Davis, 1962) and by isotopic labelling studies (Roberts *et al.*, 1955). Some doubt remains about the details of the earlier steps of the pathway, due largely to the failure to obtain *in vitro* systems capable of forming proline from glutamate, but it is generally agreed that glutamate is converted to glutamyl-γ-semialdehyde which is cyclized

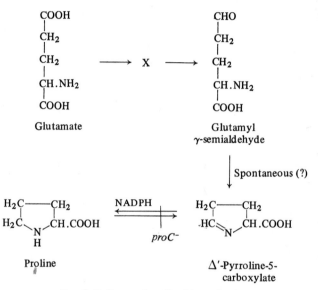

FIG. 2. Pathway of proline biosynthesis.

(probably non-enzymically) to Δ'-pyrroline-5-carboxylate. The latter is converted to proline by a NADPH-dependent Δ'-pyrroline-5-carboxylate reductase which has been partially purified from a variety of organisms. The pathway, as presently understood, is represented in Fig. 2. The postulated intermediate, designated X in Fig. 2, is possibly γ-glutamyl phosphate (cf. formation of β-aspartyl phosphate), or may be an adenylylated derivative of glutamate. Evidence for the existence of compound X is largely circumstantial, including the observation that formation of Δ'-pyrroline-5-carboxylate from glutamate by washed suspensions of *E. coli* is AMP-dependent (Strecker, 1957) and the conversion of glutamate to the semialdehyde involves two genes, *proA* and *proB*, in *E. coli* and *S. typhimurium* (Taylor, 1970; Sanderson, 1970). Owing to the lack of a suitable *in vitro* system, the early steps of proline biosynthesis in bacteria are usually studied by following the accumulation of Δ'-pyrroline-5-carboxylate from glutamate in washed suspensions lacking a functional Δ'-pyrroline-5-carboxylate reductase due to a mutation in the *proC* gene.

C. CONTROL OF PROLINE BIOSYNTHESIS

That proline biosynthesis is under tight metabolic control in *E. coli* (Roberts *et al.*, 1955) and *S. typhimurium* (Gaudie, 1969) is evident from application of the so-called isotopic competition method. Synthesis of Δ'-pyrroline-5-carboxylate reductase is not subject to repression in either *E. coli* (Baich and Pierson, 1965) or *S. typhimurium* (I. Rowland, unpublished experiments) but the early enzymes* are weakly repressible by proline (Tristram and Thurston, 1966). Lack of marked repression of proline biosynthetic enzymes suggests that proline synthesis is controlled mainly by endproduct inhibition of an early (probably the first) enzyme of the pathway. That the formation of Δ'-pyrroline-5-carboxylate from glutamate is indeed inhibited by proline was first demonstrated by Strecker (1957) and has since been confirmed in *E. coli* (Baich and Pierson, 1965; Tristram and Thurston, 1966) and *S. typhimurium* (Rowland and Tristram, 1972).

A number of proline analogues, including azetidine-2-carboxylic acid and 3,4-dehydroproline mimic proline by their ability to inhibit Δ'-pyrroline-5-carboxylate formation by *proC*⁻ strains of *E. coli* (Tristram and Thurston, 1966) and *S. typhimurium* (I. Rowland, unpublished experiments). Although proline itself and 3,4-dehydroproline are potent feedback inhibitors of proline biosynthesis, earlier experiments suggested that azetidine-2-carboxylic acid was far less effective. More recently, powerful inhibition of Δ'-pyrroline-5-carboxylate formation by azetidine-2-carboxylic acid has been demonstrated in *E. coli* (Baich and Smith, 1968; I. Rowland, unpublished experiments). By contrast, pipecolic acid and 4,5-dehydropipecolic acid, both possessing 6-membered ring structures, were inactive as false feedback inhibitors (Tristram and Thurston, 1966).

As already noted, *cis*-3,4-methanoproline is a potent inhibitor of the growth of *E. coli* and *S. typhimurium*; this analogue also strongly inhibits Δ'-pyrroline-5-carboxylate formation in both bacterial species. The diastereo isomer *trans*-3,4-methanoproline, whilst also inhibiting growth and proline biosynthesis, is a much less effective inhibitor than the *cis* form (Fig. 3). The affinity of *cis*-3,4-methanoproline for the bacterial proline permease was also higher than that of the *trans* isomer (Rowland and Tristram, 1972).

It is of interest to compare the molecular conformations of *cis*- and *trans*-3,4-methanoprolines and pipecolic acid. The latter compound, in contrast to 3,4-methanoproline, has been shown to be devoid of growth-inhibitory activity (Fowden and Richmond, 1963); it lacks affinity for proline permease (Tristram and Neale, 1968) and does not mimic proline as a feedback inhibitor of proline biosynthesis (Tristram and Thurston, 1966). Both isomers of 3,4-methanoproline possess the "boat" conformation (Fujimoto *et al.*, 1971), whereas

---

* In view of the difficulties involved in identifying the first reaction specific to proline synthesis (see text) the term "early enzymes" is used to denote the enzyme(s) involved in conversion of glutamate to Δ'-pyrroline-5-carboxylate.

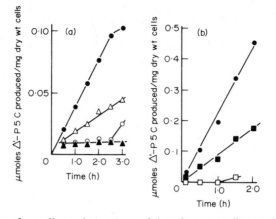

Fig. 3. The effect of L-proline and *cis*- or *trans*-3,4-methano-L-proline on the production of Δ'-pyrroline-5-carboxylic acid by (a) *E. coli Vpro*-1 and (b) *S. typhimurium proC*110. Each flask contained 0·05 M potassium phosphate buffer (pH 7·6), 0·01 M L-glutamate, 0·015 M lactate, 0·005 M AMP and feedback inhibitors, as indicated. Cell concentration, 2·4 mg dry wt/ml. Control, ●; 10^{-3} M L-proline, ○; $8·3 \times 10^{-4}$ M L-proline, □; $8·3 \times 10^{-5}$ M *cis* isomer, ■; $1·25 \times 10^{-4}$ M *cis* isomer, ▲; $1·25 \times 10^{-4}$ M *trans* isomer, △ (after Rowland and Tristram, 1972).

pipecolic acid exists in the "chair" conformation (R. Abraham and R. C. Sheppard, personal communication). If *cis*-3,4-methanoproline (Fig. 4a) can interact with the binding site of proline permease and with the allosteric site of the feedback-inhibited enzyme of the proline biosynthetic pathway,

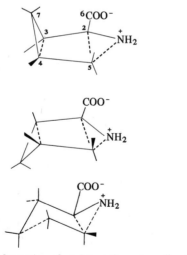

Fig. 4. Conformation of *cis*-3,4-methano-L-proline (a), *trans*-3,4-methano-L-proline (b) and L-pipecolic acid (c). From data of Fujimoto *et al.* (1971) and R. C. Sheppard (personal communication).

it is possible that interaction between the *trans* isomer (Fig. 4b) and the same sites is impaired by steric hindrance due to the presence of the downward projecting (as drawn in Fig. 4) cyclopropyl ring. Similar considerations may explain the failure of pipecolic acid (Fig. 4c) to interact with the same sites, though the greater size of the 6-membered ring of pipecolic acid compared with the 5-membered pyrrolidine ring of proline and the methanoprolines may be a factor in preventing interaction. Further, 6-membered rings have substituent groups occupying axial and equatorial positions, whereas in 5-membered rings the substituent groups are approximately midway between axial and equatorial orientations (R. C. Sheppard, personal communication).

It is relatively easy to obtain microbial mutants resistant to the growth-inhibitory effects of analogues. Such mutants, which may be resistant for one of a variety of reasons (Fowden *et al.*, 1967), have proved very useful in the study of metabolic regulation of microbial biosynthesis (Tristram, 1968; Umbarger, 1971). 3,4-Dehydroproline-resistant mutants were isolated from *E. coli* (Baich and Pierson, 1965; Tristram and Neale, 1968) and *S. typhimurium* (Gaudie, 1969); many excreted considerable quantities of proline, indicating a loss of metabolic control. The proline biosynthetic pathway of one such mutant was no longer subject to feedback inhibition by proline (Baich and Pierson, 1965), presumably due to a mutationally altered allosteric site. It is probable that most proline-excreting analogue-resistant mutants will possess similar phenotypes since, as stated above, prevention of overproduction of proline is due largely to the efficiency of feedback inhibition rather than to enzyme repression. Proline-excreting mutants resistant to *cis*-3,4-methano-proline have recently been obtained in *E. coli* (I. Rowland, unpublished experiments) and these are currently being mapped. The information should allow the recognition of the *pro* gene responsible for coding the polypeptide chain carrying the allosteric (feedback) site.

III. PHENYLALANINE ANALOGUES

Another group of compounds, mainly isolated from higher plants, behave as antagonists of aromatic amino acids in a number of bacterial systems. These compounds, apart from being β-substituted alanines, bear little obvious structural resemblance to the aromatic amino acids. Their distribution within the plant kingdom and possible mode of biosynthesis has been reviewed elsewhere (Fowden, 1970; Fowden, p. 323 of this volume). Many display a high affinity for bacterial aromatic amino acid permeases. *E. coli* and *S. typhimurium* each possess a general aromatic amino acid permease and three other aromatic permeases, specific for either phenylalanine, tyrosine or trypto-phan (Ames, 1964; Ames and Roth, 1968; Brown, 1970). The general aromatic amino acid permease displays a broad specificity, transporting a wide variety of aromatic analogues (Willshaw and Tristram, 1972). These results will not be discussed in detail; attention will be concentrated mainly on those compounds of higher plant origin. The results described come from unpublished experiments of my colleague, Mrs. G. Willshaw.

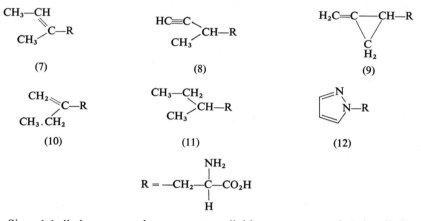

Since labelled compounds were not available, a measure of their affinity for the aromatic permeases was obtained by determining the degree of inhibition of accumulation of aromatic amino acids in *E. coli* cells. 2-Amino-4-methylhex-4-enoic acid (AMHA)(7), at 250-fold excess, caused 65 % inhibition of the general aromatic amino acid permease. Other structurally-related compounds, such as 2-amino-4-methylhex-5-ynoic acid (EL-2)(8), β-(methyl-enecyclopropyl)-alanine (hypoglycin A)(9) and 2-amino-4-ethylpent-4-enoic acid (ethallylglycine)(10), the latter not a natural compound, caused significant though never greater than about 30 % inhibition (Table I). 2-Amino-4-methyl-hexanoic acid (homoisoleucine)(11) was a poor substrate of the general aromatic amino acid permease. Table I (lines 1 and 2) show that the specificity of the phenylalanine-specific and tyrosine-specific permeases is not absolute; tyrosine caused some inhibition of the phenylalanine-specific permease (and vice versa). The tyrosine-specific permease was inhibited only weakly by AMHA and hypoglycin A, but the former compound was a good substrate of the phenylalanine specific permease, as was ethallylglycine (Table I).

Regulation of aromatic amino acid biosynthesis in bacteria involves both repression and feedback inhibition. In *E. coli* the first reaction specific to aromatic amino acid synthesis is mediated by 3-deoxy-D-*arabino*-heptulosonic acid-7-phosphate (DAHP) synthetase which exists in three isoenzymic forms, each specified by a separate gene on the single bacterial chromosome and each repressible and feedback-inhibited by one of the aromatic amino acids. The isoenzymes can be physically separated from each other and mutant *E. coli* strains, each capable of producing only one of the isoenzymes, have been obtained. In *E. coli* regulation of DAHP synthetase activity appears to be the major control point of the communal portion of the aromatic biosynthetic pathway (Pittard and Gibson, 1970).

TABLE I

Inhibition[a] of phenylalanine-sensitive DAHP synthetase and accumulation of L-[U-14C]phenylalanine or L-[U-14C]tyrosine by phenylalanine, tyrosine and analogues[b]

Inhibitor	DAHP synthetase[c]	Phenylalanine-specific permease[e]	Tyrosine-specific permease[e]	General aromatic permease[f]	
				Phe	Tyr
L-Phenylalanine	94	95	20	—	—
L-Tyrosine	12	38	92	—	65
DL-2-Amino-4-methylhex-4-enoic acid [AMHA]	6[d]	95	13	67	11
DL-2-Amino-4-ethylpent-4-enoic acid [ethallylglycine]	68	82	22	32	—
DL-2-Amino-4-methylhexanoic acid [homoisoleucine]	0	53	6	24	20
L-2-Amino-4-methylhex-5-ynoic acid [EL-2]	0	32	—	24	—
β-(Methylenecyclopropyl)-L-alanine [hypoglycin A]	25	25	25	25	—
L-Pyrazol-1-ylalanine	82	26	0	23	—

[a] Inhibition expressed as % of initial rate in absence of analogue.
[b] Unpublished experiments of Mrs G. Willshaw.
[c] Inhibitor concentration: 10^{-3} M (L-form) (see also footnote d).
[d] AMHA concentration: 5×10^{-4} M (L-form).
[e] Accumulation of 2×10^{-6} M L-[U-14C]phenylalanine (50 μCi/μmole) or L-[U-14C]tyrosine (50 μCi/μmole) measured in the presence and absence of 2×10^{-4} M inhibitor (L-form).
[f] Accumulation of 2×10^{-7} M L-[U-14C]phenylalanine (50 μCi/μmole) or L-[U-14C]tyrosine (50 μCi/μmole) measured in the presence and absence of 5×10^{-5} M inhibitor (L-form).

A number of higher plant non-protein amino acids were tested for ability to mimic phenylalanine in feedback-inhibiting the phenylalanine-sensitive DAHP synthetase of *E. coli* AB3259 (*aroG*$^+$ *aroF*$^-$ *aroH*$^-$), a strain lacking the tyrosine- and tryptophan-sensitive synthetases. The specificity of the allosteric site of the synthetase differed appreciably from that of the amino acid binding site of the aromatic permease (Table I). DAHP synthetase (phenylalanine-sensitive) was inhibited 82% by 10^{-3} M pyrazol-1-ylalanine (12), whereas this compound possessed little affinity for the general aromatic amino acid permease. (The specific permeases have not been tested.) Conversely AMHA, which had a high affinity for both the general aromatic and phenyl-alanine-specific permeases, did not inhibit the phenylalanine-specific DAHP synthetase significantly. Smith *et al.* (1964) had previously reported the failure of a synthetic preparation of AMHA to inhibit phenylalanine-sensitive DAHP synthetase. Ethallylglycine, whilst having little affinity for the general aromatic amino acid permease, possessed a high affinity for the phenylalanine-specific transport system and caused about 70% inhibition of DAHP synthetase (phenylalanine-sensitive).

AMHA inhibited growth of *Leuconostoc dextranicum*, the inhibition being readily reversed by phenylalanine (Edelson *et al.*, 1959). This compound was also a substrate, as measured by the ATP-pyrophosphate exchange reaction, for the phenylalanyl-tRNA synthetase of mung bean (*Phaseolus aureus*) (Smith and Fowden, 1968) and five species of *Aesculus*, including *A. californica* (Fowden *et al.*, 1970; Anderson and Fowden, 1970). Although AMHA is produced by *A. californica* and the phenylalanyl-tRNA synthetase of this species promotes ATP-pyrophosphate exchange in the presence of AMHA, the analogue is not incorporated into *Aesculus* protein. It must be assumed that either the aminoacyladenylate can not be transferred to tRNA$_{phe}$ by the *Aesculus* enzyme, or alternatively AMHA is localized in the plant cell at sites metabolically unavailable to the phenylalanyl-tRNA synthetase (Anderson and Fowden, 1970). However, when tested by the hydroxamate method, neither AMHA nor ethallylglycine were substrates of the phenylalanyl-tRNA synthetase of *E. coli* (Conway *et al.*, 1962).

The recognition of this group of compounds, lacking ring structures, as phenylalanine analogues, is of particular interest since it suggests that interaction of phenylalanine with bacterial permeases, some feedback inhibitor sites and some phenylalanyl-tRNA synthetases does not require the presence of an intact aromatic ring. The minimum structure beyond the β-carbon of the alanyl side chain which is required for activity is several carbon atoms with a coplanar configuration. The planar methylbutenyl group of AMHA, with two methyl groups, both in *cis* configuration, imparts properties of a phenylalanine antagonist. Nevertheless, as Edelson *et al.* (1959) have pointed out, there is evidence of a structural requirement other than coplanarity for a molecule to act as a phenylalanine analogue since growth inhibition of *L. dextranicum* by methallylglycine was reversed by leucine, but not by phenylalanine. The growth-inhibitory activity of 2-amino-4-methylhexanoic acid (homoisoleucine), the saturated form of AMHA, towards *E. coli* 9723 and *L. dextranicum* was reversed by leucine but not by phenylalanine (Edelson *et al.*, 1959). Significantly, planarity of the substituent groups of homoisoleucine would not be realized.

The acetylenic amino acid 2-amino-4-methylhex-5-ynoic acid (EL-2) only slightly inhibited phenylalanine transport by either the general aromatic permease or phenylalanine-specific permease and had no detectable activity as a feedback inhibitor of DAHP synthetase (G. Willshaw, unpublished experiments).

IV. ENZYME REPRESSION

A. JACOB-MONOD MODEL

In 1961 Jacob and Monod (1961a, b) formulated their model of genetic control of enzyme induction and repression. The model, based on an intensive genetic and biochemical study of lactose utilization by *E. coli*, is now well known and will not be described in detail. Briefly the model envisages the synthesis of messenger RNA (mRNA) initiated near a specific region of the DNA termed the operator. An operator is adjacent to, and coordinates the expression of a group of contiguous genes specifying the primary structure of some or all the enzymes of a metabolic pathway. The group of structural genes, together with its associated operator, constitutes an operon. Also present in the DNA is a genetic element known as the regulator gene which produces a cytoplasmic determinant, the repressor protein. The repressor has the capacity of reversibly combining with the appropriate operator region of DNA or with the controlling metabolite (or effector) (Bourgeois, 1971). In enzyme induction the inducing substance (or sometimes a metabolic derivative of it) acts as effector; in enzyme repression the effector is the repressing metabolite, usually the final endproduct of a pathway or possibly, more frequently, a derivative of the final endproduct. Further investigation has necessitated the introduction of another genetic element concerned with genic expression, namely the promoter. This is visualized as a site of initiation of transcription, the function of the operator being solely to interact with the repressor (in inducible systems) or the repressor-effector complex (in repressible systems). The promoter is thought to the located outside the operator, on the side distal from the structural genes of the operon though there is evidence that the group of genes specifying the enzymes of tryptophan biosynthesis may contain two promoters, one adjacent to the operator and a second promoter situated within the operon, between two of the structural genes (Martin, 1969).

In repressible systems the repressor protein is thought to interact with the effector and the product of this interaction reversibly complexes with the operator. When the operator is blocked by the repressor-effector complex, expression of the structural genes does not occur and the corresponding enzymes are not formed (repressed state). If the intracellular concentration of effector falls, the repressor-effector complex dissociates and the repressor reverts to a form in which it can no longer complex with operator. Freeing of the operator allows the structural genes to be transcribed, the resulting mRNA being translated to yield proteins specified by the operon (Fig. 5).

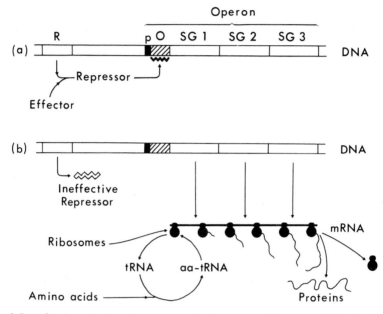

FIG. 5. Postulated mechanism of gene expression in a repressible system. (a) In the presence of repressor metabolite (effector). The product of the regulator gene (R) complexes with the effector and associates with the operator (O). Structural genes (SG1, 2 and 3) are not expressed. (b) In the absence of effector the repressor is unable to associate with the operator, the structural genes are transcribed into a single (polycistonic) mRNA which is translated into polypeptide chains. The roles of ancillary factors are omitted (reprinted from Tristram, 1968).

As in inducible systems, in which mutation may lead to constitutive synthesis of all the proteins of an operon, so repressible systems may, as a result of mutation, become constitutive. Such non-repressible mutants, no longer subject to repression by the final product of a pathway, may arise as a result of mutation in the operator, which loses its affinity for the repressor-effector complex. Alternatively, non-repressible mutations may occur in the regulator gene which produces an altered repressor which either fails to complex with the effector, or forms a repressor-effector complex no longer having affinity for the operator (Jacob and Monod, 1961a, b).

Attention has been drawn, especially by Umbarger (1969a, 1971) to the dangers of uncritical and indiscriminate application of this model. These criticisms are even more cogent when the model is invoked to explain regulatory phenomena in eukaryotes. The study of regulation of several amino acid biosynthetic pathways in bacteria, frequently utilizing amino acid analogues as tools, has led to the conclusion that in many instances the Jacob-Monod model in its simplest form may not adequately account for the observations, some of which are described below.

B. HISTIDINE BIOSYNTHESIS

In *S. typhimurium* the intracellular level of histidine controls its own synthesis by coordinate repression of all the enzymes of the pathway and feedback inhibition of phosphoribosyl-ATP pyrophosphorylase, the first enzyme specific to histidine synthesis. As will become apparent, control of repression by histidine is indirect, being mediated not by histidine itself, but by a histidine derivatives. The nine genes specifying the ten enzymes of the pathway are clustered on the bacterial chromosome and a genetic region possessing the properties of an operator has been identified at one end of the cluster, adjacent to the gene *hisG* which specifies the first enzyme of the pathway (Loper *et al.*, 1964; Ames *et al.*, 1967; Hartman *et al.*, 1971).

The toxic histidine analogue, β-thiazol-2-ylalanine mimics histidine in causing inhibition of phosphoribosyl-ATP pyrophosphorylase, so starving the cell of histidine and curtailing protein synthesis. A series of mutants selected for resistance to the growth-inhibitory effect of the analogue were insensitive to feedback inhibition by histidine or the analogue; these mutations mapped in *hisG*, the gene specifying the pyrophosphorylase (Loper *et al.*, 1964; Ames *et al.*, 1967). Binding of histidine to the phosphoribosyl-ATP pyrophosphorylase of wild type *S. typhimurium* leads to inhibition of enzyme activity accompanied by a conformational change in the protein; the enzyme of a feedback-resistant strain was still able to bind histidine, but no conformational change was observed (Blasi *et al.*, 1971).

Under appropriate experimental conditions growth of *S. typhimurium* is inhibited by β-1,2,4-triazol-3-ylalanine. This analogue is incorporated into bacterial protein. The synthesis of histidinol phosphate phosphatase and imidazoleacetol phosphate transaminase and, by inference, the remaining histidine biosynthetic enzymes, is repressed by the analogue (Levin and Hartman, 1963). α-Methyl-histidine, another toxic analogue of histidine, curtails protein synthesis by preventing formation of histidinyl-tRNA. In *E. coli* cells treated with this analogue the histidine biosynthetic enzymes were present in greater than normal amounts, i.e. synthesis of the enzymes was derepressed (Schlesinger and Magasanik, 1964). The pathway of histidine synthesis, points of control and action of histidine analogues are summarized in Fig. 6.

The concomitant inhibition of histidinyl-tRNA formation and derepression of histidine biosynthetic enzymes by α-methylhistidine first directed attention

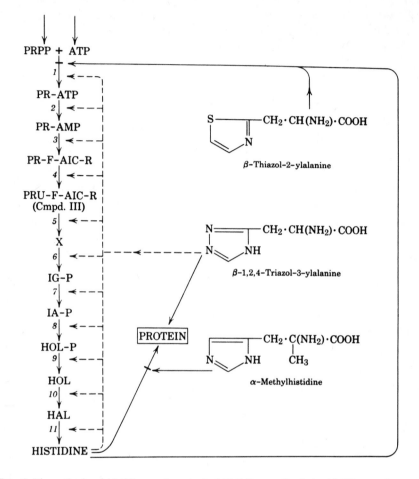

FIG. 6. Biosynthesis of histidine and control of histidine synthesis by histidine and some analogues (for references, see text). → indicates enzyme repression; ↔ indicates reactions inhibited by histidine or analogues. Abbreviations: PRPP: phosphoribosylpyrophosphate; PR-ATP: N-1-(5'-phosphoribosyl)-ATP. PR-AMP: N-1-(5'-phosphoribosyl)-AMP. PR-F-AIC-R: N-(5'-phosphoribosyl-formimino)-5-amino-1-(5''-phosphoribosyl)-4-imidazole-carboxamide. PRU-F-AIC-R: N-(5'-phospho-1'-ribulosyl-formimino)-5-amino-1-(5''-phosphoribosyl)-4-imidazole-carboxamide. X: unidentified intermediate (see Loper *et al.*, 1964). IG-P: imidazole glycerol phosphate. IA-P: imidazole acetol phosphate. HOL-P: histidinol phosphate. HOL: histidinol. HAL: histidinal [postulated enzyme-bound intermediate involved in conversion of histidinol to histidine (see Loper *et al.*, 1964)] (after Fowden *et al.*, 1967).

to the idea that the signal for repression of the enzymes of amino acid biosynthetic pathways might be, not the free amino acid itself, but the corresponding aminoacyl-tRNA. In the case of histidine synthesis this suggestion has
received added weight from the study of *S. typhimurium* mutants selected for
resistance to *β*-1,2,4-triazol-3-ylalanine. These mutants were derepressed,
producing high levels of histidine biosynthetic enzymes, even in the presence of
high concentrations of histidine. On the basis of genetic analysis the mutants
were divisible into several classes. *HisO* mutations were located at one ex-
tremity of the cluster of structural genes, adjacent to *hisG*; from dominance
studies it was concluded that *hisO* had the properties of an operator gene.
Other derepressed mutants mapped in ths *hisR, S, T, U* and *W* genes which
were neither linked to the histidine gene cluster, nor to each other. *HisS* mutants
possessed altered histidinyl-tRNA synthetases with decreased affinities for
histidine, compared with the enzyme of wild type strains; this gene is the
structural gene for histidinyl-tRNA synthetase. Mutations in *hisR* resulted in a
reduction of the cellular level of tRNA to about 50 % of the normal (wild type)
level (Roth *et al.*, 1966a, b; Roth and Ames, 1966; Silbert *et al.*, 1966; Anton,
1968; Fink and Roth, 1968). The *hisR* gene is thought to specify the structure
of histidine-specific tRNA (Silbert *et al.*, 1966), a suggestion substantiated by
the observation that the introduction of an extra copy of *hisR* leads to a large
increase in cellular content of tRNA$_{his}$ (Brenner and Ames, 1972).

The *hisT* gene codes for a protein (Chang *et al.*, 1971) which is concerned in
formation of pseudouridine in the anticodon loop of tRNA$_{his}$ (Brenner and
Ames, 1972). Available published evidence for the function of this protein is
not very compelling. Even less convincing is the claim that *hisW* also codes for
a tRNA maturation enzyme. At the time of writing the function of *hisU* appears
to be unknown. Only one tRNA$_{his}$ species can be detected in *S. typhimurium*,
so disposing of the earlier view that *hisR, W* and *U* may code for differing species
of histidine-specific tRNA. Although the tRNA$_{his}$ of a *hisT* mutant displays
slightly different chromatographic properties when compared with normal
tRNA$_{his}$, no differences are detectable between the latter and tRNA from
hisR, W or *U* mutants (Brenner and Ames, 1972).

Thus no repressor protein or gene corresponding to the regulator gene of
the Jacob and Monod model have yet been identified as part of the mechanism
regulating histidine synthesis in *S. typhimurium*. Nevertheless the structural
genes specifying the enzymes of the pathway are clustered on the chromosome
and an operator-like region has been identified at one end of the cluster. Available evidence suggests that the level of histidinyl-tRNA rather than free histidine is the signal for repression, but the nature of the protein, if any, with which
it complexes, is unknown. It seems likely that the major part of the histidinyl-
tRNA synthetase and tRNA$_{his}$ of *S. typhimurium* are complexed *in vivo*; this
conclusion led to the suggestion that this complex is itself the repressor (de
Lorenzo and Ames, 1970). This hypothesis predicts that the complex should
display a histidine-dependent binding to the operator (*hisO*) gene. There is

some evidence that phosphoribosyl-ATP pyrophosphorylase may be involved in repression of the histidine pathway (see Section IV, E).

C. ROLE OF AMINOACYL-tRNA SYNTHETASES

Reference has already been made to observations which implicated histidinyl-tRNA in repression of the histidine biosynthetic enzymes (see Section IV, B). Further evidence for the involvement of aminoacyl-tRNAs in repression of the enzymes of other amino acid biosynthetic pathways has been sought mainly from studies with amino acid analogues or mutants possessing altered amino-acyl-tRNA synthetases, the mutants often being obtained by selection for resistance to the growth-inhibitory effects of analogues. The results obtained in different laboratories have sometimes been contradictory and will not be reviewed in detail, but rather the principal types of experimental approach will be outlined, together with a summary of the main conclusions reached and an indication of difficulties involved in interpretation of the observations.

If, in order to operate a repression mechanism, an amino acid must be transferred to tRNA, a correlation between ability of an analogue to be charged to tRNA and ability to repress enzyme synthesis should be demonstrable. However attempts to demonstrate such a correlation have frequently failed.

Mutants producing an altered aminoacyl-tRNA synthetase with lowered affinity for a particular amino acid frequently have derepressed levels of enzymes involved in the synthesis of that amino acid (e.g. *hisS* mutations; see Section IVB). Such mutants fail to fully charge the relevant tRNA under normal conditions of growth, since elevated amino acid levels are required to saturate the impaired aminoacyl-tRNA synthetase. Nevertheless these mutants can usually be repressed by addition of relatively high concentrations of the relevant amino acid to the culture medium. Strains having temperature-sensitive aminoacyl-tRNA synthetases which fail to fully aminoacylate tRNAs at the restrictive temperature, have proved particularly useful. These strains are usually derepressed at the restrictive temperature (when tRNA is uncharged, or poorly charged), but have normal enzyme levels at the permissive temperature (when tRNA is near-fully charged). By comparing normal strains with mutants possessing altered aminoacyl-tRNA synthetases, attempts have been made to demonstrate a correlation between the extent of charging of tRNA and the degree of derepression of enzyme synthesis.

Using one or more of these approaches aminoacyl-tRNAs were implicated in repression of enzymes involved in the synthesis of histidine (Roth and Ames, 1966; Stulberg *et al.*, 1969) and the branched-chain amino acids (Alexander *et al.*, 1971; Blatt and Umbarger, 1970; Dwyer and Umbarger, 1968; Eidlic and Neidhardt, 1965; Freundlich, 1967; Freundlich and Trela, 1969; McLaughlin *et al.*, 1969; Szentirmai *et al.*, 1968; Trela and Freundlich, 1969; Williams and Freundlich, 1969). Contrary to these results, study of other pathways has failed to implicate aminoacyl-tRNAs in repression of enzymes involved in biosynthesis of tyrosine (Ravel *et al.*, 1965; Schlesinger and Nester, 1969), phenylalanine (Neidhardt, 1966), arginine (Hirshfield *et al.*, 1968) and methionine (Gross and Rowbury, 1969, 1971). The position regarding control of synthesis of enzymes of the tryptophan pathway is particularly confused. Studying the activity of tryptophan analogues as effectors of repression in *E. coli* mutants having altered tryptophanyl-tRNA synthetases (*trpS* mutants) and in strains with normal synthetases, Doolittle and Yanofsky (1968) and Mosteller and Yanofsky (1971) concluded there was no correlation between ability of an analogue to repress and its activity as a substrate of tryptophanyl-tRNA synthetase. Other workers have isolated *trpS* mutants with severely impaired tryptophanyl-tRNA synthetases and elevated levels of tryptophan biosynthetic enzymes. However some of these mutants could not be as strongly repressed, nor as highly derepressed as the wild type strain (Hiraga *et al.*, 1967; Kano *et al.*, 1968; Ito *et al.*, 1969a, b).

Studies of this type, especially when involving use of analogues, frequently raise difficulties of interpretation which do not always receive sufficient attention. As Calvo and Fink (1971) have pointed out, many mutations in aminoacyl-tRNA synthetase structural genes, whilst leading to impaired aminoacylation of tRNA, may not alter the role of the protein in generating a repression signal. Further, when a mutation leading to production of an altered aminoacyl-tRNA synthetase is accompanied by derepressed levels of enzymes of the appropriate biosynthetic pathway, it is important to establish that the phenotype is not the result of multiple mutations. Failure to observe a correlation between % tRNA charged and degree of repression does not exclude the possibility of participation of a minor tRNA species in repression (Hirshfield *et al.*, 1968; Leisinger and Vogel, 1969). Similarly, failure to observe a correlation between activity of an analogue as an effector of repression and its transfer to tRNA is difficult to interpret. Since the nature of the repression signal is not precisely known, it may only be necessary for an analogue to bind to an aminoacyl-tRNA synthetase to effect repression (Calvo and Fink, 1971).

Apparent repression by an analogue which is not incorporated into protein (with the resultant possibility of synthesis of inactive protein) may be due to prevention of derepression by inhibition of general protein synthesis by the analogue. Such a situation could arise if the analogue inhibits synthesis of the corresponding protein amino acid by, for example, false feedback inhibition.

One of the main problems in demonstrating enzyme repression by an amino acid analogue is the difficulty of distinguishing between true repression and the continued synthesis of catalytically inactive protein due to incorporation of the analogue into enzyme molecules. The demonstration that an analogue does not impair synthesis, in active form, of several metabolically unrelated enzymes is frequently relied on as evidence that the analogue, as a result of incorporation, does not impair biological function. This is not a compelling argument since a particular analogue, on incorporation into protein, frequently leads to formation of some enzymes which are biologically non-functional whereas other enzymes, containing residues of the same analogue, are fully active (Fowden *et al.*, 1967). Diminution in the amount of mRNA corresponding to a particular gene (or genes) has been used as a measure of repression at the level of transcription (Mosteller and Yanofsky, 1971). This approach avoids the possibility of conclusions being influenced by formation of "false" protein, but Martin (1969) has pointed out need for caution in accepting quantitative estimates of mRNA obtained by DNA-RNA hybridization techniques.

That formation of biologically-impaired "false" protein did not account for apparent repression of histidine biosynthetic enzymes by β-1,2,4-triazol-3-ylalanine (see Section IV, B) was shown by the demonstration that histidinol phosphate phosphatase and imidazole-acetol phosphate transaminase, the two enzymes studied, were synthesized in an enzymically-active state in cells of a non-repressible mutant exposed to the analogue. However in a normally repressible strain incubated under conditions otherwise suitable for enzyme synthesis, addition of analogue prevented the development of enzyme activity (Levin and Hartman, 1963).

Arginine is known to repress, in a non-coordinate fashion, all eight arginine biosynthetic enzymes. The *bona fide* repression, by the arginine analogue canavanine, of arginine biosynthetic enzymes has been elegantly demonstrated by Faanes and Rogers (1968). If a strain of *E. coli* K-12, auxotrophic for arginine, was starved of arginine for a short period, the subsequent addition of the amino acid resulted in a burst of ornithine transcarbamylase synthesis lasting 3–4 min. These observations, together with observations on other bacterial strains, were interpreted as representing the synthesis of ornithine transcarbamylase mRNA during the period of arginine starvation (derepression conditions) which was subsequently translated following re-establishment of conditions suitable for active protein synthesis resulting from restoration of arginine. The short burst of enzyme synthesis resulted from the decay of the short-lived mRNA and further synthesis of ornithine transcarbamylase was prevented by the repressive effect of the added arginine. Using this technique Faanes and Rogers (1968) were able to show that the presence of canavanine during the period of starvation for a required amino acid seriously curtailed the amount of ornithine transcarbamylase synthesis following addition of the required amino acid and the re-establishment of conditions for protein synthesis. Since canavanine was effectively present only during the period of mRNA synthesis and not during the subsequent translation of mRNA into protein, the suppression of ornithine transcarbamylase synthesis could not be attributed to formation of enzymically-inactive protein, but to prevention of ornithine transcarbamylase mRNA synthesis; in other words, canavanine was shown to repress at least one of the arginine biosynthetic enzymes.

A study of a different kind, not subject to some of the objections mentioned above, has implicated threonyl-tRNA synthetase in repression of threonine biosynthetic enzymes. The antibiotic borrelidin inhibits the threonyl-tRNA synthetases of a variety of organisms, preventing the transfer of threonine to $tRNA_{thr}$ (for references see Nass et al., 1971). In E. coli K-12 the first reaction specific to threonine, methionine and lysine synthesis is catalysed by aspartokinase, which exists in three isoenzymic forms, one repressible by lysine, another repressible by methionine, the third repressible by threonine plus isoleucine. One of two isoenzymic forms of homoserine dehydrogenase is also repressed by threonine plus isoleucine, the other by methionine. Methionine-sensitive aspartokinase and homoserine dehydrogenase activity is very low in strain K-12 and absent from strain B. If cells of E. coli K-12 or B (in which the lysine-sensitive aspartokinase had been repressed) were exposed to borrelidin, aspartokinase and homoserine dehydrogenase were derepressed about 5-fold, suggesting the involvement of threonyl-tRNA in repression (Nass et al., 1969).

D. TRANSLATIONAL CONTROL

The model of Jacob and Monod (1961a, b) assumed that control of protein biosynthesis is exerted at the level of transcription. This concept has received strong support from the demonstration that mRNA corresponding to the lactose and galactose operons can be detected in E. coli only when cells are exposed to inducer (Attardi et al., 1963; Naono et al., 1965). Further, mRNAs corresponding to the tryptophan or histidine operons are detectable only under derepressed conditions, that is, only when enzymes of tryptophan or histidine biosynthesis are being actively synthesized (Martin, 1963; Imamoto et al., 1965). Epstein and Beckwith (1968) have pointed out that the fact that lac repressor protein binds to lac operator DNA and is displaced from it on addition of inducer, as postulated by the model, is strong evidence for control at the transcriptional level. In recent years there has accumulated a growing body of evidence suggesting that control of synthesis of amino acid biosynthetic enzymes may be more complex than the model suggests. Firstly, control of mRNA translation may be involved in the regulation of synthesis of amino acid biosynthetic enzymes; secondly, evidence has been presented that the first enzyme specific to a biosynthetic pathway may be implicated in expression of the structural genes of the pathway. Consideration of earlier studies suggesting regulation at the translational level led to the formulation of several models to account for such a mode of control (Cline and Bock, 1966; Vogel and Vogel, 1967; Roth et al., 1966b). In the present account only more recent contributions will be discussed.

Lavallé and de Hauwer (1970) measured, by a hybridization technique, the amount of mRNA corresponding to the tryptophan operon in an E. coli K-12 derivative in which the tryptophan biosynthetic enzymes were normally repressible ($trpR^+$) and in a non-repressible mutant ($trpR^-$). Addition of tryptophan to a culture of the $trpR^+$ strain growing in minimal (glucose-minerals) medium resulted in immediate establishment of a new rate of transcription of the tryptophan genes characteristic of repressed conditions (about 50 % of the rate observed in cells grown in minimal medium lacking tryptophan) and thereafter the level of mRNA remained constant. However the rate of *translation* of the tryptophan mRNA, measured by the appearance of tryptophan synthetase activity, was decreased by a factor of *ca* 15. During the next several generation times the rate of translation progressively increased until it reached that characteristic of a repressed culture and from then on remained constant. In other words, following repression by addition of tryptophan, these was a disproportionate fall in rate of translation compared with the fall in rate of transcription; further a new steady state was finally established after a relatively long period during which the rate of translation increased to a constant value characteristic of a repressed culture. When *trpR*+ cells were grown in a rich medium containing all the amino acids except tryptophan, the synthesis of enzymes of tryptophan biosynthesis was derepressed. Addition of tryptophan to such a culture led to the same pattern of behaviour with respect to mRNA synthesis, namely, the rate of transcription was reduced by a factor of *ca* 3. However the rate of translation was reduced by a factor of *ca* 150; in fact synthesis of tryptophan synthetase almost ceased. Again, over a period of several generations, the rate of translation gradually increased to that characteristic of repressed cells. On the contrary, during release of repression by removal of tryptophan, a change in rate of tryptophan synthetase synthesis was accounted for by the observed change in rate of mRNA synthesis. That is, shortage of tryptophan immediately led to increased efficiency of translation of the few mRNA molecules available.

Although it was not possible to measure levels of mRNA, a similar phenomenon was demonstrated during repression of synthesis of ornithine transcarbamylase and acetyl ornithinase, two of the enzymes of arginine biosynthesis (Lavallé, 1970).

The results of these investigations showed that control is exerted not only at the transcriptional level (synthesis of mRNA), as predicted by the Jacob-Monod model, but also at the level of translation. It was postulated that a hypothetical molecule existed in high concentrations in cells under conditions of derepression which, in the presence of metabolite repressor (tryptophan or arginine in the examples studied), led to a severe restriction in translation of existing mRNAs. On repression by addition of tryptophan or arginine this hypothetical entity was no longer synthesized and existing molecules were diluted out by continued growth and division of the cells, resulting in a progressive increase of the rate of translation until the rate characteristic of the repressed state was achieved. The nature of the postulated entity is not known, but it was pointed out (Lavallé and de Hauwer, 1970; Lavallé, 1970) that the two requirements of the hypothesis, namely that regulation of formation must be under repressive control by tryptophan (or arginine) and it must possess a recognition site for tryptophan (or arginine) severely restricts the possibilities. Only the biosynthetic enzymes responsible for synthesis of tryptophan (or arginine) are likely candidates among molecules presently known. What seems certain is that translation is prevented by a diffusible, cytoplasmic entity, since the genes specifying the enzymes of the arginine biosynthetic pathway are not clustered on the bacterial chromosome and hence, unlike the tryptophan operon, which is transcribed into a single polycistronic mRNA, are presumably transcribed into a number of discrete messengers. Nevertheless the translation of at least two of these mRNA species (corresponding to ornithine transcarbamylase and acetyl ornithinase), and by inference, the messengers specifying the remaining arginine enzymes, was curtailed.

E. ROLE OF FEEDBACK-INHIBITED ENZYMES

The early observations of Cohen and Jacob (1959) and Moyed (1960) provided the first indication that the first enzyme of a biosynthetic pathway may be involved in the regulation of synthesis of the enzymes of that pathway. In *E. coli* the conversion of chorismate to tryptophan involves five reactions catalysed by a group of enzymes specified by a cluster of five contiguous genes on the bacterial chromosome. The gene cluster possesses many of the properties of an operon; several pieces of evidence point to the presence of an operator region at one end of the cluster. The relationship between genes, gene products and reactions is, however, complex. Control of the pathway is effected by repression of all the enzymes, together with feedback inhibition of the first two reactions of the pathway by tryptophan (Pittard and Gibson, 1970). Several growth-inhibitory analogues of tryptophan, including 5-methyltryptophan, mimic the natural amino acid in effecting endproduct inhibition of the early reactions of tryptophan biosynthesis. A mutant selected for resistance to 5-methyltryptophan was pleiotropic; the conversion of shikimate-5-phosphate to anthranilate was less sensitive to inhibition by the analogue or tryptophan and also produced increased amounts of the anthranilate-producing enzyme (Moyed, 1960). These observations were confirmed and extended by Somerville and Yanofsky (1965) who isolated 5-methyltryptophan-resistant mutants, the anthranilate synthetases of which were insensitive to tryptophan. The mutations mapped in *trpE*, the gene specifying one component of the anthranilate synthetase-anthranilate phosphoribosyl transferase complex and many were pleiotropic, since the tryptophan enzymes were insensitive to repression by tryptophan. Other mutations in the same gene (*trpE*), leading to loss of anthranilate synthetase activity, were also altered in the pattern of repression of tryptophan biosynthetic enzymes.

Experiments of Kovach et al. (1969) indicate a role for phosphoribosyl-ATP pyrophosphorylase, the enzyme specified by the *hisG* gene and the first enzyme specific to histidine biosynthesis, in repression of enzymes of the histidine operon. β-1,2,4-Triazol-3-ylalanine repressed the aminotransferase specified by the *hisC* gene only if the feedback inhibitor site of the first enzyme was intact. In mutants in which the site was altered by mutation, such that the mutant enzyme was insensitive to feedback inhibition, triazolealanine was unable to effect repression of the aminotransferase. These observations suggest a role of the feedback-inhibited enzyme in repression, but interpretation is complicated by the observation that the histidine biosynthetic enzymes in the same feedback-resistant strains were still repressed by histidine. Further, in two *hisG* mutants in which the feedback sites were still intact, but in which the catalytic sites were impaired, histidine still repressed synthesis of the aminotransferase whereas triazolealanine repressed synthesis in one mutant (*hisG*70) but not in the other (*hisG*52).

More recently these workers showed, by column chromatography on Sephadex G-100 and by a nitrocellulose filter binding technique, an affinity of phosphoribosyl-ATP pyrophosphorylase for histidinyl-tRNA. On the other hand other aminoacyl-tRNAs were bound by the enzyme, though binding was progressively inhibited by increasing Mg^{2+} concentration whereas binding of histidinyl-tRNA increased to a maximum at 10 mm Mg^{2+} and then decreased. Stripped tRNA competed with histidinyl-tRNA for binding (Kovach *et al.*, 1970).

Although phenylalanyl-tRNA has not yet been firmly implicated in the repression of the enzymes of phenylalanine biosynthesis, formation of a complex between phenylalanyl-tRNA and the phenylalanine-sensitive DAHP synthetase of *E. coli* has been detected by gel filtration on Sephadex G-75 (Duda *et al.*, 1968). The binding of phenylalanyl-tRNA to the enzyme protein was much stronger than was that of valyl-tRNA and it was suggested that the interaction may fulfill a role in repression of synthesis of the phenylalanine biosynthetic enzymes.

The significance of the experiments of Kovach and his coworkers and of Duda and his collaborators is difficult to assess at present. It would be informative to know more about the binding of the relevant aminoacyl-tRNAs to the phosphoribosyl-ATP pyrophosphorylase and DAHP synthetase in, for example, feedback resistant strains and also more about the nature of the sites to which the aminoacyl-tRNAs are bound. Further, both groups of workers obtained evidence that (uncharged) $tRNA_{his}$ or $tRNA_{phe}$ bind to the relevant enzyme, so it is essential to know more about the relative affinities of charged and uncharged tRNAs for the enzyme binding sites. Another difficulty of interpretation arises from a consideration of other studies of interactions between nucleic acids and proteins. For instance, the *lac* repressor protein interacts strongly with the *lac* operator region of DNA from *E. coli* (Bourgeois, 1971), but also interacts, with varying degrees of affinity, with other DNase sequences (Lin and Riggs, 1970; Riggs *et al.*, 1972). These observations raise the question of the specificity of the interactions observed by Kovach *et al.* (1970) and Duda *et al.* (1968), though both groups claim specificity, at least with respect to the tRNAs involved in complex formation. Kovach and his coworkers also showed that binding of histidinyl-tRNA was correlated with specific enzymic activity during purification of the pyrophosphorylase and that the enzyme was not contaminated with histidinyl-tRNA synthetase, suggesting the binding observed was specific for the first enzyme of histidine biosynthesis.

In *S. typhimurium* the enzymes concerned with the biosynthesis of isoleucine and valine are under multivalent repressive control. The simultaneous presence of a superabundance of isoleucine, valine, leucine and pantothenate is required for repression; limitation in supply of any one of the amino acids leads to derepression of enzyme synthesis (Umbarger, 1969b). Threonine deaminase, the first enzyme specific to isoleucine synthesis, is subject to feedback inhibition by isoleucine, the inhibition being antagonized by valine. Hatfield and Burns (1970) have produced evidence that the enzyme plays a role in regulation of synthesis of the isoleucine and valine enzymes. The threonine deaminase of *S. typhimurium* is tetrameric; each identical monomer contains a sulphydryl group which forms an –S–S– bridge with the sulphydryl group of another monomer. The resulting dimers exist in equilibrium with a tetramer which binds two pyridoxal-5'-phosphate molecules. The coenzyme-containing tetramer is enzymically inactive (immature enzyme), but is converted to active (mature) enzyme in the presence of any one of the ligands for which the enzyme possesses binding sites (isoleucine, valine or threonine). The process of maturation is blocked in the presence of isoleucine plus valine or isoleucine plus threonine (for references, see Hatfield and Burns, 1970). More recently it was shown that the immature tetramer specifically binds leucyl-tRNA, whereas the mature, catalytically active enzyme does not. It was suggested that the tetramer containing bound leucyl-tRNA acts directly as a repressor, possibly by binding to DNA (Hatfield and Burns, 1970). Repression of synthesis of isoleucine and valine enzymes would be effected by the simultaneous availability of isoleucine, valine and leucyl-tRNA. Derepression would occur under conditions of isoleucine or valine limitation since available aporepressor (immature enzyme) would be depleted by enzyme maturation. Limitation of leucine supply would result in reduced levels of leucyl-tRNA and derepression would result from failure to convert the aporepressor to active (holo-) repressor. The main features of this model are illustrated in Fig. 7.

In *E. coli* three of the five enzymes specific to valine and isoleucine synthesis are derepressed under conditions of valine limitation. Addition of valine to such cells repressed enzyme formation. Addition of *threo*-α-amino-β-chlorobutyrate also prevented development of enzyme activity, whereas α-aminobutyrate could not replace valine in preventing enzyme synthesis. α-Amino-β-chlorobutyrate, but not α-aminobutyrate was transferred to tRNA (Freundlich, 1967). Though not unequivocal (Umbarger, 1969b), the evidence was interpreted as indicating repression by α-amino-β-chlorobutyrate and that charging of $tRNA_{val}$ by valine (or analogue) was a necessary step for repression (Freundlich, 1967). Hatfield and Burns (1970) reported that α-amino-β-chlorobutyrate (but not α-aminobutyrate) was also able to promote the conversion of immature tetrameric threonine deaminase to catalytically active (mature) enzyme.

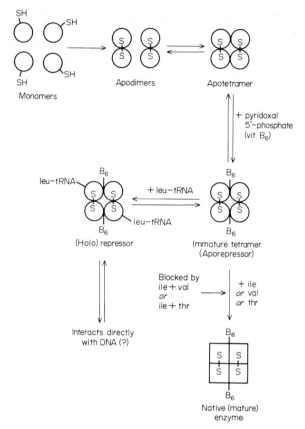

FIG. 7. A model for multivalent repression, by threonine deaminase, of valine and isoleucine biosynthetic enzymes of *S. typhimurium* (after Hatfield and Burns, 1970).

Despite the appeal of this model it fails to take into account the fact that valine and isoleucine must probably be transferred to their cognate tRNAs before participating in the multivalent repression of isoleucine and valine biosynthetic enzymes (Umbarger, 1969b). It would be of great interest to know whether the maturation of threonine deaminase can be promoted by the appropriate aminoacyl-tRNAs rather than by the free amino acids.

V. CONCLUDING REMARKS

The latter part of this review has been devoted to outlining some of the difficulties and anomalies in the study of control of amino acid biosynthesis in bacteria. An exhaustive review of the considerable literature on this subject has not been attempted, but rather an attempt has been made to indicate the dangers of too facile an acceptance of the Jacob-Monod model, when applied to amino acid biosynthetic systems. Many observations, frequently reviewed (Calvo and Fink, 1971; Martin, 1969; Pittard and Gibson, 1970; Umbarger, 1969a, b), fulfill the requirements of the original model. For example, clearly defined gene clusters with a DNA region having the properties of an operator have been identified as specifying the enzymes concerned with synthesis of valine and isoleucine, leucine, histidine and tryptophan. However, the genes specifying the enzymes of arginine and methionine synthesis are widely scattered on the bacterial chromosome. Although the scattered genes specifying arginine and methionine biosynthetic enzymes may be associated with individual operators, only in one instance has such an operator been demonstrated (Jacoby and Gorini, 1969). The control of methionine biosynthesis is clearly complex and not yet fully understood; neither operators nor a repressor gene in the sense specified by the Jacob-Monod model have been described (Smith, 1971). Genes analogous to the regulator (repressor protein-forming) gene of the Jacob-Monod model have been identified with reasonable certainty only in the case of leucine (*leuR*), arginine (*argR*) and possibly tryptophan (*trpR*) and tyrosine (*tyrR*) biosynthesis; recently isolation of the supposed repressor protein product of the *argR* gene has been claimed (Udaka, 1970). As Umbarger (1971) has pointed out; demonstration of a protein component necessary for the generation of a repression signal is no longer sufficient evidence to postulate that the molecule is a repressor protein. The criterion for identification of a repressor protein must now be a specific binding to a DNA region having the properties of an operator.

The degree of involvement of aminoacyl-tRNAs and their synthetases in the generation of repression signals is still far from clear and the evidence, in many cases, is equivocal. Nevertheless there is strong evidence implicating aminoacyl-tRNA synthetases, either directly or indirectly, in repression of enzymes responsible for synthesis of histidine, the branched-chain amino acids and possibly threonine.

Growing evidence suggests that control at the translational level may be of importance in regulation of amino acid biosynthesis, and may be superimposed on a transcriptional control. Recent suggestions that the first enzyme of a pathway may be involved in control of synthesis of enzymes of the pathway is of considerable interest, especially in view of the demonstration that, in a few systems at least, the first enzyme can specifically complex with aminoacyl-tRNA.

In presenting their novel model for repression of some of the isoleucine and valine biosynthetic enzymes by threonine deaminase, Hatfield and Burns (1970) pointed out that "it is not surprising that the cell should have evolved a regulatory mechanism wherein a single protein could function in both a regulatory and a catalytic capacity".

The observations presented above do not exclude the usefulness of the Jacob-Monod model in studies of amino acid biosynthesis. In its simplest form the model has proved inadequate, but may need only comparatively minor modification to accommodate many observations which currently are difficult to rationalize. Amino acid analogues, owing to their capacity for mimicking the metabolic activities of normal protein amino acids, have been very useful tools, not yet fully exploited, for the study of control mechanisms, and specially for obtaining many different classes of non-repressible and other regulatory mutants.

REFERENCES

Alexander, R. R., Calvo, J. M. and Freundlich, M. (1971). *J. Bacteriol.* **106**, 213–220.

Ames, G. F. (1964). *Archs Biochem. Biophys.* **104**, 1–18.

Ames, G. F. and Roth, J. R. (1968). *J. Bacteriol.* **96**, 1742–1749.

Ames, B. N., Goldberger, R. F., Hartman, F. E., Martin, R. G. and Roth, J. R. (1967). *In* "Regulation of Nucleic Acid and Protein Biosynthesis" (V. V. Konigsberger and L. Bosch, eds), pp. 272–287. B.B.A. Library, Vol. 10. Elsevier, Amsterdam.

Anderson, J. W. and Fowden, L. (1970). *Biochem. J.* **119**, 677–690.

Antón, D. N. (1968). *J. molec. Biol.* **33**, 533–546.

Attardi, G., Naono, S., Rouvière, J., Jacob, F. and Gros, F. (1963). *Cold Spring Harb. Symp. quant. Biol.* **28**, 363–372.

Baich, A. and Pierson, D. J. (1965). *Biochim. biophys. Acta* **104**, 397–404.

Baich, A. and Smith, F. I. (1968). *Experientia* **24**, 1107.

Blasi, F., Aloj, S. M. and Goldberger, R. F. (1971). *Biochemistry* **10**, 1409–1417.

Blatt, J. M. and Umbarger, H. E. (1970). *Bacteriol. Proc.* p. 136.

Bourgeois, S. (1971). *Curr. Top. Cell. Regn.* **4**, 39–75.

Brenner, M. and Ames, B. N. (1972). *J. biol. Chem.* **247**, 1080–1088.

Brown, K. D. (1970). *J. Bacteriol.* **104**, 177–188.

Calvo, J. M. and Fink, G. R. (1971). *A. Rev. Biochem.* **40**, 943–968.

Chang, G. W., Roth, J. R. and Ames, B. N. (1971). *J. Bacteriol.* **108**, 410–414.

Cline, A. L. and Bock, R. M. (1966). *Cold Spring Harb. Symp. quant. Biol.* **31**, 321–333.

Cohen, G. and Jacob, F. (1959). *C.r. hebd. Séanc, Acad. Sci., Paris* **248**, 3490–3492.

Conway, T. W., Lansford, E. M. and Shive, W. (1962). *J. biol. Chem.* **237**, 2850–2854.

Doolittle, W. F. and Yanofsky, C. (1968). *J. Bacteriol.* **95**, 1283–1294.

Duda, E., Staub, M., Venetianer, P. and Dénes, G. (1968). *Biochem. biophys. Res. Commun.* **32**, 992–997.

Dwyer, S. B. and Umbarger, H. E. (1968). *J. Bacteriol.* **95**, 1680–1684.

Edelson, J., Skinner, C. G., Ravel, J. M. and Shive, W. (1959). *J. Am. chem. Soc.* **81**, 5150–5153.

Eidlic, L. and Neidhardt, F. C. (1965). *Proc. natn. Acad. Sci. U.S.A.* **53**, 539–543.

Epstein, W. and Beckwith, J. (1968). *A. Rev. Biochem.* **37**, 411–436.

Faanes, R. and Rogers, P. (1968). *J. Bacteriol.* **96**, 409–420.

Fink, G. R. and Roth, J. R. (1968). *J. molec. Biol.* **33**, 547–557.

Fowden, L. (1955). *Nature, Lond.* **176**, 347–348.

Fowden, L. (1970). *Prog. Phytochem.* **2**, 203–266.

Fowden, L. and Richmond, M. H. (1963). *Biochim. biophys. Acta* **71**, 459–461.

Fowden, L., Neale, S. and Tristram, H. (1963). *Nature, Lond.* **199**, 35–38.

Fowden, L., Lewis, D. and Tristram, H. (1967). *Adv. Enzymol.* **29**, 89–163.

Fowden, L., Smith, A., Millington, D. S. and Sheppard, R. C. (1969). *Phytochemistry* **8**, 437–443.

Fowden, L., Anderson, J. W. and Smith, A. (1970) *Phytochemistry* **9**, 2349–2357.

Freundlich, M. (1967). *Science, N.Y.* **157**, 823–825.
Freundlich, M. and Trela, J. M. (1969). *J. Bacteriol.* **99**, 101–106.
Fujimoto, Y., Irreverre, F., Karle, J. M., Karle, I. L. and Witkop, B. (1971). *J. Am. chem. Soc.* **93**, 3471–3477.
Gaudie, D. (1969). Ph.D. Thesis, Univ. of London.
Gross, T. S. and Rowbury, R. J. (1969). *Biochim. biophys. Acta* **184**, 233–236.
Gross, T. S. and Rowbury, R. J. (1971). *J. gen. Microbiol.* **65**, 5–21.
Hartman, P. E., Hartman, Z. and Stahl, R. C. (1971). *Adv. Genet.* **16**, 1–34.
Hatfield, G. W. and Burns, R. O. (1970). *Proc. natn. Acad. Sci. U.S.A.* **66**, 1027–1035.
Hiraga, S., Ito, K., Hamada, K. and Yura, T. (1967). *Biochem. biophys. Res. Commun.* **26**, 522–527.
Hirshfield, I. N., DeDeken, R., Horn, P. C., Hopwood, D. A. and Maas, W. K. (1968). *J. molec. Biol.* **35**, 83–93.
Hussain, A. (1968). Ph.D. Thesis, Univ. of London.
Imamoto, F., Morikawa, N., Sato, K., Mishima, S. and Nishimura, T. (1965). *J. molec. Biol.* **13**, 157–168.
Ito, K., Hiraga, S. and Yura, T. (1969a). *J. Bacteriol.* **99**, 279–286.
Ito, K., Hiraga, S. and Yura, T. (1969b). *Genetics* **61**, 521–538.
Jacob, F. and Monod, J. (1961a). *J. molec. Biol.* **3**, 318–356.
Jacob, F. and Monod, J. (1961b). *Cold Spring Harb. Symp. quant. Biol.* **26**, 193–209.
Jacoby, G. A. and Gorini, L. (1969). *J. molec. Biol.* **39**, 73–87.
Kano, Y., Matsushiro, A. and Shimura, Y. (1968). *Molec. gen. Genet.* **102**, 15–26.
Kovach, J. S., Phang, J. M., Ference, M. and Goldberger, R. F. (1969). *Proc. natn. Acad. Sci. U.S.A.* **63**, 481–488.
Kovach, J. S., Phang, J. M., Blasi, F., Barton, R. W., Ballesteros-Olmo, A. and Goldberger, R. F. (1970). *J. Bacteriol.*, **104**, 787–792.
Lavallé, R. (1970). *J. molec. Biol.* **51**, 449–451.
Lavallé, R. and de Hauwer, G. (1970). *J. molec. Biol.* **51**, 435–447.
Leisinger, T. and Vogel, H. J. (1969). *Biochim. biophys. Acta* **182**, 572–574.
Levin, A. P. and Hartman, P. E. (1963). *J. Bacteriol.* **86**, 820–828.
Lin, S.-Y. and Riggs, A. D. (1970). *Nature, Lond.* **228**, 1184–1186.
Loper, J. C., Grabnar, M., Stahl, R. C., Hartman, Z. and Hartman, P. E. (1964). *Brookhaven Symp. Biol.* **17**, 15–50.
Lorenzo, F. de, and Ames, B. N. (1970). *J. biol. Chem.* **245**, 1710–1716.
McLaughlin, C. S., Magee, P. T. and Hartwell, L. H. (1969). *J. Bacteriol.* **100**, 579–584.
Martin, R. G. (1963). *Cold Spring Harb. Symp. quant. Biol.* **28**, 357–361.
Martin, R. G. (1969). *A. Rev. Genet.* **3**, 181–216.
Moyed, H. S. (1960). *J. biol. Chem.* **235**, 1098–1102.
Mosteller, R. D. and Yanofsky, C. (1971). *J. Bacteriol.* **105**, 268–275.
Naono, S., Rouvière, J. and Gros, F. (1965). *Biochem. biophys. Res. Commun.* **18**, 664–674.
Nass, G., Poralla, K. and Zähner, H. (1969). *Biochem. biophys. Res. Commun.* **34**, 84–91.
Nass, G., Poralla, K. and Zähner, H. (1971). *Naturwissenshaften* **58**, 603–610.
Neidhardt, F. C. (1966). *Bacteriol. Rev.* **30**, 701–719.
Pittard, J. and Gibson, F. (1970). *Curr. Top. Cell. Regn.* **2**, 29–63.
Ravel, J. M., White, M. N. and Shive, W. (1965). *Biochem. biophys. Res. Commun.* **20**, 352–359.
Richmond, M. H. (1962). *Bacteriol. Rev.* **26**, 398–420.
Riggs, A. D., Lin, S. and Wells, R. D. (1972). *Proc. natn. Acad. Sci. U.S.A.* **69**, 761–764.

Roberts, R. B., Abelson, P. H., Cowie, D. B., Bolton, E. T. and Britten, R. J. (1955). "Studies of Biosynthesis in *Escherichia coli*." Publs. Carnegie Instn. No. 607.

Roth, J. R. and Ames, B. N. (1966). *J. molec. Biol.* **22**, 325–334.

Roth, J. R., Antón, D. N. and Hartman, P. E. (1966a). *J. molec. Biol.* **22**, 305–323.

Roth, J. R., Silbert, D. F., Fink, G. R., Voll, M. J., Antón, D. N., Hartman, P. E. and Ames, B. N. (1966b). *Cold Spring Harb. Symp. quant. Biol.* **31**, 383–392.

Rowland, I. and Tristram, H. (1972). *Chem.-Biol. Interactions* **4**, 377–388.

Sanderson, K. E. (1970). *Bacteriol. Rev.* **34**, 176–193.

Schlesinger, S. and Magasanik, B. (1964). *J. molec. Biol.* **9**, 670–682.

Schlesinger, S. and Nester, E. W. (1969). *J. Bacteriol.* **100**, 167–175.

Silbert, D. F., Fink, G. R. and Ames, B. N. (1966). *J. molec. Biol.* **22**, 335–347.

Smith, D. A. (1971). *Adv. Genet.* **16**, 141–165.

Smith, I. K. and Fowden, L. (1968). *Phytochemistry* **7**, 1065–1075.

Smith, L. C., Ravel, J. M., Skinner, C. G. and Shive, W. (1962). *Archs Biochem. Biophys.* **99**, 60–64.

Smith, L. C., Ravel, J. M., Lax, S. R. and Shive, W. (1964). *Archs Biochem. Biophys.* **105**, 424–430.

Somerville, R. L. and Yanofsky, C. (1965). *J. molec. Biol.* **11**, 747–759.

Strecker, H. J. (1957). *J. biol. Chem.* **225**, 825–834.

Stulberg, M. P., Isham, K. R. and Stevens, A. (1969). *Biochim. biophys. Acta* **186**, 297–304.

Sung, M.-L. and Fowden, L. (1969). *Phytochemistry* **8**, 2095–2096.

Szentirmai, A., Szentirmai, M. and Umbarger, H. E. (1968). *J. Bacteriol.* **95**, 1672–1679.

Taylor, A. L. (1970). *Bacteriol. Rev.* **34**, 155–175.

Trela, J. M. and Freundlich, M. (1969). *J. Bacteriol.* **99**, 107–112.

Tristram, H. (1968). *Sci. Prog., Oxford* **56**, 449–477.

Tristram, H. and Neale, S. (1968). *J. gen. Microbiol.* **50**, 121–137.

Tristram, H. and Thurston, C. F. (1966). *Nature, Lond.* **212**, 74–75.

Udaka, S. (1970). *Nature, Lond.* **228**, 336–338.

Umbarger, H. E. (1969a). *A. Rev. Biochem.* **38**, 323–370.

Umbarger, H. E. (1969b). *Curr. Top. Cell. Regn.* **1**, 57–76.

Umbarger, H. E. (1971). *Adv. Genet.* **16**, 119–140.

Umbarger, H. E. and Davis, B. D. (1962). *In* "The Bacteria" (I. C. Gunsalus and R. Y. Stanier, eds), Vol. 3, pp. 167–251. Academic Press, New York.

Unger, L. and DeMoss, R. D. (1966). *J. Bacteriol.* **91**, 1556–1563.

Vogel, H. J. and Vogel, R. H. (1967). *A. Rev. Biochem.* **36**, 519–538.

Williams, L. and Freundlich, M. (1969). *Biochim. biophys. Acta* **186**, 305–316.

Willshaw, G. and Tristram, H. (1972). *Biochem. J.* **127**, 71P.

CHAPTER 3

Amino Acid Biosynthesis and its Control in Plants

B. J. MIFLIN

*Department of Plant Science, University of Newcastle upon Tyne,
Northumberland, England*

I. BIOSYNTHETIC PATHWAYS

The pathways of synthesis of the protein amino acids have been reviewed previously by Davies (1968) and Fowden (1967). This brief introduction into the biosynthetic routes will be concerned primarily with results published subsequently to the earlier reviews. No attempt has been made to mention all papers relevant to the subject. The quotation of any reference for a given step does not imply that this was the initial observation, rather, where several reports of an enzyme have been made, usually the latest reference found has been given. Figures 1–5 contain flow diagrams for the main pathways. In the diagrams those steps for which enzymic evidence is available are given in bold arrows and the reference letter included by the arrow. The pathways are based on those established in micro-organisms as described in détail by Meister (1965) and Greenberg (1969).

A. ASPARTATE FAMILY

Little progress has been made in studying the reactions involved in the synthesis of the aspartate family of amino acids (Fig. 1) particularly those

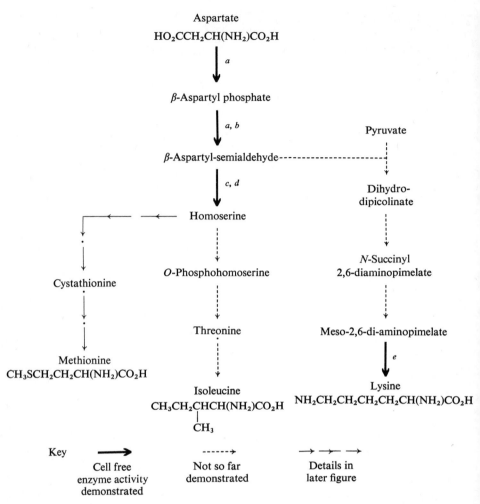

Fig. 1. The synthetic pathway of the aspartate family of amino acids. The letters against the arrows of this and subsequent figures relate to the references dealing with work on that enzyme. *a*, Bryan *et al.* (1970); *b*, Sasoka (1961); *c*, Bryan (1969); *d*, Sasoka and Inagaki (1960); *e*, Shimura and Vogel (1966).

leading to lysine production and those between homoserine and threonine. The activity of the first, key enzyme aspartokinase has only recently been demonstrated in plants (Bryan *et al.*, 1970; Cheshire and B. J. Miflin, unpublished). Labelling studies showing the incorporation of [^{14}C]diaminopimelic acid into lysine (Finlayson and McConnell, 1960), the presence in plants of *meso*-2,6-diaminopimelic acid decarboxylase (Shimura and Vogel, 1966) and the specific isotope distribution in the piperidine ring of anabasine after [4-^{14}C] aspartate [2-^{14}C]succinate and [2-^{14}C]lysine feeding (Griffith and Griffith, 1966), all point to the diaminopimelate pathway as the chief route of lysine biosynthesis in higher plants. However, it appears that under conditions of nitrogen starvation (Medvedev and Kretovich, 1966) or in certain species, e.g. the sensitive plant (*Mimosa pudica*) (Notation and Spenser, 1964), there is suggestive evidence that the fungal, aminoadipate, pathway may be operating. Direct [^{14}C]aspartate incorporation experiments (Dunham and Bryan, 1971) and isotope dilution studies (Dougall and Fulton, 1967b) suggest that the indicated pathway to threonine via homoserine is operating.

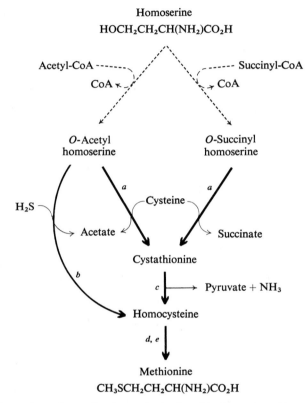

FIG. 2. Alternative pathways for the synthesis of methionine. *a, b, c,* Giovanelli and Mudd (1966, 1967, 1971, respectively); *d,* Burton and Sakami (1969); *e,* Dodd and Cossins (1970).

Recent studies have shown that a number of alternate pathways are available for methionine synthesis (Fig. 2). Firstly a trans-sulphuration of either *O*-acetylhomoserine or *O*-succinylhomoserine to cystathionine (Giovanelli and Mudd, 1966) followed by β-elimination of pyruvate and ammonia from cystathionine to give homocysteine (Giovanelli and Mudd, 1971) can occur. The second possibility is the direct sulphuration of *O*-acetylhomoserine to homocysteine (Giovanelli and Mudd, 1967). Nothing is known of the formation of the *O*-alkylhomoserines and of the relative physiological importance of the

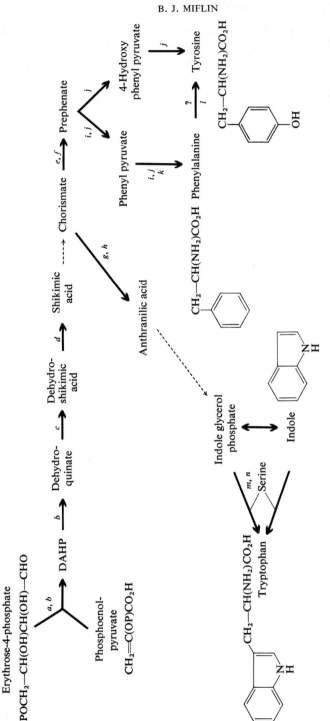

Fig. 3. The pathway for the synthesis of aromatic amino acids. *a*, Minamikawa (1967); *b*, Nandy and Ganguli (1961); *c*, Balinsky *et al.* (1971); *d*, Sanderson (1966); *e*, Cotton and Gibson (1968); *f*, Gilchrist *et al.* (1972); *g*, Belser *et al.* (1971); *h*, Weber and Bock (1970); *i*, Gamborg and Keeley (1966); *j*, Gamborg and Simpson (1964); *k*, Redkina *et al.* (1969); *l*, Nair and Vining (1965); *m*, Chen and Boll (1971); *n*, Delmer and Mills (1968b).

different pathways. Methylation of homocysteine to methionine can take place using either the mono- or tri-glutamate forms of methyltetrahydrofolic acid (Dodd and Cossins, 1970; Burton and Sakami, 1969).

B. AROMATIC AMINO ACIDS

Much recent work has been done on the biosynthesis of the aromatic amino acids. Careful studies with labelled intermediates have confirmed that trypto-phan, tyrosine and phenylalanine are all synthesized via the shikimic acid pathway (Fig. 3) (Delmer and Mills, 1968a). Also, virtually all of the key enzymic steps of the pathway have been demonstrated in cell-free extracts. One of the most interesting observations from the evolutionary point of view is that tryptophan synthetase in plants, as in bacteria but not fungi, is a 2 component enzyme consisting of an A and a B protein. Separately, these proteins carry out partial reactions, but together they catalyse the synthesis of tryptophan from indole glycerol phosphate (Delmer and Mills, 1968b; Chen and Boll, 1971). In particular there is evidence that the hybrid synthetase formed from *Escherichia coli* A protein and tobacco (*Nicotiana tabacum*) B protein is enzymically active and also that the B proteins from the two sources have immunological sites in common (Delmer and Mills, 1968b). Little further evidence has become available, on the possibility of plants generally being able to hydroxylate phenylalanine to tyrosine, since this topic was discussed by Davies (1968).

C. GLUTAMATE FAMILY

Enzymatic evidence for the involvement of acetylated intermediates in the synthesis of amino acids from glutamate is cited in Fig. 4. Also, since both [^{14}C]glutamate and [^{14}C]N-acetylglutamate donate carbon to proline when fed to intact tissue (Morris et al., 1969) and isotope dilution studies indicate that N-acetylglutamic semialdehyde and N-acetylornithine are intermediates in arginine synthesis (Dougall and Fulton, 1967a), it is likely that the acetylated pathway is normally operating in green plants. In the acetylated pathway the utilization of a transacetylase, transferring the acetyl group from acetylornithine to glutamate, would ensure that the pathway did not involve the net breakdown of acetyl-CoA. Although proline can readily be formed from ornithine (e.g. see Mazelis and Fowden, 1969) it is probable that proline normally is synthesized via glutamic semialdehyde. Ornithine gives rise to arginine via the ornithine-urea cycle, all the enzymes of which have been found in plants (Fig. 5). However, it is unlikely that it performs in the same way in plants as it does in animals. One interesting enzyme, found in *Chlorella*, is arginine desimidase which splits ammonia from arginine to give citrulline thus bypassing the arginase step, which appears to be missing in this organism (Shafer and Thompson, 1968).

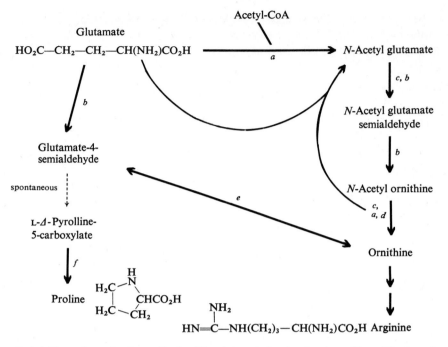

FIG. 4. The pathway for the synthesis of the glutamate family of amino acids. *a*, Morris and Thompson (1968); *b*, Morris *et al.* (1969); *c*, Hoare and Hoare (1966); *d*, Staub and Denes (1966); *e*, Mazelis and Fowden (1969); *f*, Noguchi *et al.* (1966).

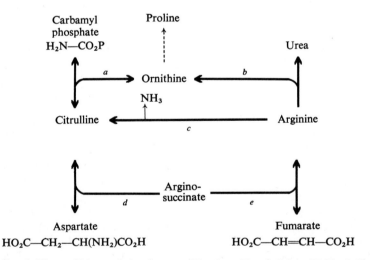

FIG. 5. The ornithine cycle in plants. *a*, Kleczkowski and Cohen (1964); *b*, Notation and Spenser (1964); *c*, Shafer and Thompson (1968); *d*, Shargool (1971); *e*, Rosenthal and Naylor (1969).

D. BRANCHED CHAIN AMINO ACIDS

The pathway to the branched chain amino acids (Fig. 6) has been understood for some time, although recent studies have involved a considerable degree of purification and characterization of threonine deaminase (Sharma and Mazumder, 1970) and a demonstration that acetolactate synthetase also forms acetohydroxybutyrate (Miflin, 1971). The synthesis of the remaining amino acids, with the exception of histidine and cysteine, has already been well documented.

E. OTHERS

Little is known of histidine biosynthesis except that histidinol (Dougall and Fulton, 1967b) and imidazolglycerol phosphate would appear to be intermediates (Seigel and Gentile, 1966; Davies, 1971). Cysteine can be formed by plant enzymes according to the following reaction (Giovanelli and Mudd, 1967, 1968; Morris and Thompson, 1968) although the formation of O-acetylserine has not been demonstrated.

$$O\text{-Acetylserine} + H_2S \rightarrow Cysteine + Acetate$$

Serine is relatively inactive in this reaction.

II. CONTROL OF AMINO ACID BIOSYNTHESIS

Two questions can be asked. Does the plant regulate the synthesis of its amino acids? If so, how is this regulation achieved? There are now many observations from different sources and obtained using different techniques to show that the *in vivo* synthesis of amino acids is under metabolic control. All of the experiments individually are open to alternative explanations, but together they provide convincing evidence for this contention.

A. EVIDENCE FOR CONTROL *In vivo*

The oldest information on this point, though not originally considered in this context, probably comes from studies of the growth inhibitory effects of single amino acids and the relief of these effects by other amino acids (for a review see Street, 1966). The classic studies of Gladstone (1939) and Umbarger (1961) have shown that at least some of these growth inhibitions are related to feedback control mechanisms. Briefly, the system works as outlined in Fig. 7, such that if the reaction A $\rightarrow$ B is shut off either by D or E acting as a feedback regulator then an external excess of just one of the amino acids will inhibit growth due to the shortage of the other. When both are supplied growth returns

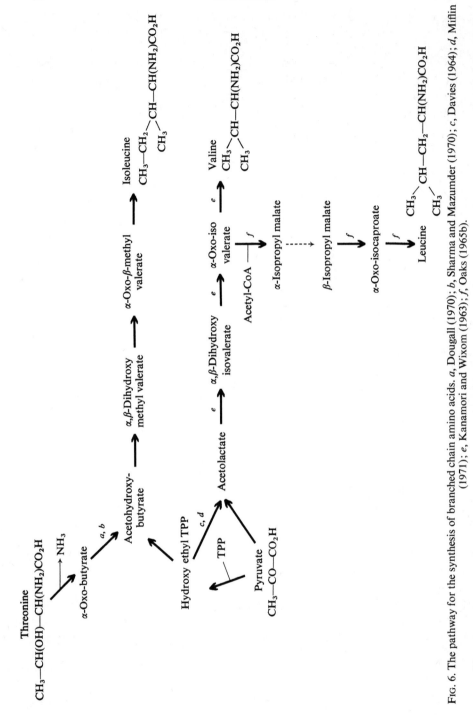

Fig. 6. The pathway for the synthesis of branched chain amino acids. *a*, Dougall (1970); *b*, Sharma and Mazumder (1970); *c*, Davies (1964); *d*, Miflin (1971); *e*, Kanamori and Wixom (1963); *f*, Oaks (1965b).

to normal. If growth inhibition and relief patterns can be shown to exist between terminal amino acids of the same family, then this would suggest that a regulating mechanism is operating. Studies in the author's laboratory (Miflin, 1969a, see also Table III) with barley (*Hordeum vulgare*) and by Børstlap (1970) with duckweed (*Spirodela*) have shown such effects exist for the branched chain amino acids. The terminal amino acids of the aspartate pathway also show the same kind of relationships in affecting the growth of rice (*Oryza sativa*) callus tissue (Furuhashi and Yatazawa, 1970) and liverwort (*Marchantia*) (Dunham and Bryan, 1969).

Nitrogen supply	Synthetic pathway	Amino acids present	Growth
NO$_3$	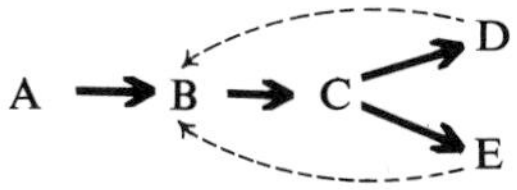 A → B → C ⟨ D E	D, E	Normal
NO$_3$ + D	A	D	Inhibited
NO$_3$ + D + E	A	D, E	Normal

FIG. 7. A theoretical test for possible feedback of amino acid biosynthesis in branched pathways.

More precise evidence for feedback regulation stems from the work of Oaks (1965a) in which a general precursor, uniformly labelled with [^{14}C] (e.g. [^{14}C]acetate or [^{14}C]glucose) is fed to tissues in the presence and absence of excess, exogenously supplied, non-radioactive amino acids. The incorporation of the precursor into both the soluble and protein pools of the amino acids is then measured. It is important that both pools are measured so that any effects due to isotope dilution are not confused with regulation—a good example of how the results are analysed is given by Oaks *et al.* (1970). Although all internal pools of the amino acids appear to be swamped by the exogenous amino acids the criticism could be levelled that the system is artificially contrived and thus differs from that in which amino acids accumulate naturally at the end of a synthetic pathway. Fletcher and Beevers (1971) have carried out experiments which attempt to achieve this latter situation. In these they have blocked protein synthesis with cycloheximide, thus preventing the utilization of the amino acids and causing the accumulation of amino acids. They have then looked at the effect of this decreased utilization on the synthesis of amino acids from [^{14}C]acetate. Their results show good evidence for feedback regulation and correlate well with other findings, as can be seen in Table I which summarizes the results of the various approaches. These studies do not distinguish between feedback inhibition of enzyme activity and feedback repression of enzyme synthesis as the biochemical mechanism operative in the control. This can only be done by isolating and studying the key enzymes.

B. J. MIFLIN

TABLE I

Summary of *in vivo* evidence for feedback regulation in plants

	Growth inhibition[a]	[14C] Precursor feeding[b]	Cyclohex- imide inhibition studies[c]	Amino- triazole inhibition[d]
Aspartate family				
Asparagine				
Aspartate		−		
Lysine	+	+	+	
Threonine	+	+	+	
Methionine	+			
Glutamate family				
Glutamine				
Glutamate		−	−	
Proline	+	+	+	
Arginine	+	+	+	
Branched chain amino acids				
Isoleucine		+	+	
Leucine	+	+	+	
Valine	+	+		
Aromatic amino acids				
Tryptophan		+		
Tyrosine	+			
Phenylalanine	?			
Histidine				+
Others				
Glycine		−		
Serine		−	−	
Alanine		−		

This table is based on the work reported in the following references:
[a] Miflin, 1969, Dunham and Bryan, 1969, Furuhashi and Yatazawa, 1970.
[b] Oaks, 1965a, Oaks *et al.*, 1970.
[c] Fletcher and Beevers, 1971.
[d] Davies, 1971.
A + sign indicates the presence and a − sign indicates the absence of evidence for feedback control of synthesis.

B. EVIDENCE FOR REPRESSION

Several enzymes have been investigated to see if the levels of extractable activity present in cells is affected by the addition of amino acids in whose synthesis they are involved. All of these experiments have shown that, under these conditions, the levels of the enzymes are not lowered (Table II). Weber and Bock (1968) rightly point out that this type of experiment does not prove

TABLE II

Studies which have shown an absence of repression of enzymes involved in
the synthesis of the carbon skeletons of amino acids

Enzyme	Metabolites tested	Reference
Threonine deaminase	Threonine, Leucine, Isoleucine, Valine	Dougall, 1970
Acetolactate synthetase	Leucine, Isoleucine, Valine	Miflin and Cave, 1972
Tryptophan synthetase	Tryptophan	Widholm, 1971
Anthranilate synthetase	Tryptophan	Widholm, 1971
Chorismate mutase	Tyrosine, Phenylalanine	Chu and Widholm, 1972
3-deoxy-D-arabino-heptulosonic-7-phosphate synthetase	Tyrosine, Phenylalanine	Weber and Bock, 1968

the complete absence of repression since the enzymes may be already maximally repressed before the amino acids are supplied exogenously. However,
even if this were so, the results show that repression is far from absolute and
that, in studies where indirect evidence indicates that the levels of exogenous
amino acids turn off the internal synthesis, repression is not the explanation for
the evidence of *in vivo* control presented in the previous section.

C. FEEDBACK MODIFICATION OF ENZYME ACTIVITY

In contrast to repression there are numerous examples showing that the
activity of isolated enzymes is allosterically regulated by the presence of amino
acids. This modification is usually negative (i.e. activity is inhibited) but, particularly where interlocking pathways are concerned, it may be positive. Some
of the relevant evidence will be reviewed for each amino acid family.

1. *Branched Chain Amino Acids*

The enzyme catalysing the first unique step in isoleucine biosynthesis is
threonine deaminase. This enzyme is inhibited by isoleucine (Sharma and
Mazumder, 1970; Dougall, 1970) and the inhibition is relieved by the addition
of valine. Under conditions that, *in vivo*, might lead to an imbalance of the
branched chain amino acid due to the underproduction of isoleucine (viz. a
low level of threonine and a very low level of isoleucine) valine stimulates the
activity of threonine deaminase (Sharma and Mazumder, 1970). D,L-norvaline
and L-aspartate have also been reported to activate the enzyme (Blekhman
et al., 1971).

Acetohydroxyacid synthetase, which forms both acetolactate and acetohydroxybutyrate, is the first enzyme unique to leucine, isoleucine and valine

synthesis and can be considered as the first enzyme in a triple branched pathway. This enzyme thus provides a good example for considering the mechanisms that the plant has evolved for the control of such branched pathways and how they compare with those described in bacteria (see Umbarger, 1969). The enzyme has been extracted from barley and partially purified (Miflin, 1969b; 1971). Its activity is inhibited by the end product amino acids, much in accordance with the predictions that can be made from the growth inhibition studies referred to previously (Table III). Thus leucine, valine and isoleucine, in

TABLE III

Correlation of inhibition of growth and acetolactate synthetase in barley
by amino acids

Amino acid added	Concentration in growth medium (mM)	Inhibition of growth (%)	Inhibition of enzyme (%)
Leucine	2	43	55
Valine	2	49	37
Isoleucine	2	9	16
Valine + leucine	2 + 0·2	65	81
Leucine + valine + isoleucine	2 + 0·2 + 0·2	11	76

In the enzyme studies the total concentration of the amino acids was 1 mM in each case. From Miflin 1969a and 1971.

decreasing order of effectiveness, are all inhibitory when applied singly. The results also suggested that there was some interaction between leucine and valine. This was tested and a synergistic effect of leucine and valine on growth inhibition was found (Miflin, 1969a). In the presence of all three amino acids growth is restored but the enzyme remains inhibited. The increased effectiveness of leucine and valine together suggests a co-operative feedback mechanism. The generally accepted definition is that of Stadtman (1966) who defines co-operative inhibition as occurring when "an *excess* of any of the ultimate end products causes a partial inhibition of the first enzymic step, whereas the simultaneous *excess* of two or more of the end products results in a greater inhibition than the sum of the fractional inhibitions caused by each independently" (my italics). The results of an experiment designed to test this definition is shown in Fig. 8. The upper curves of Fig. 8a show the inhibition caused by increasing concentrations of various amino acids and Fig. 8b shows the difference between the amino acids, added separately, and together. It can be seen clearly in this latter graph that the inhibition by leucine and valine cannot be classed as "co-operative" by Stadtman's definition since there is only synergistic inhibition at levels of leucine and valine well below "excess".

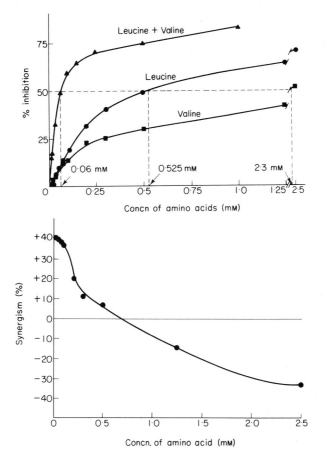

FIG. 8. The inhibition of acetolactate synthetase by leucine and valine. (a) The percentage of inhibition caused by increasing concentrations of amino acid present as either leucine, valine, or leucine plus valine. (b) Variation in the synergistic inhibition of leucine and valine with differing concentrations. The percentage synergism is calculated as the percentage of inhibition caused by leucine and valine together – (percentage of inhibition caused by leucine + percentage inhibition caused by valine). The amino acid concentration of each amino acid is given. Taken from Miflin (1971).

However, on theoretical grounds it seems illogical and inappropriate to define terms in reference to situations arising in the presence of excess levels of end product, since the purpose of feedback regulation is to prevent such excess levels occurring. What is important is the situation operating during the phase of inhibition when maximum changes in reaction rate occur with minimum changes in modifier concentration. The author has suggested, therefore, that the relative efficiencies of various inhibitors (or, where appropriate, stimulators), supplied singly or in combination, may be most meaningfully expressed in terms of the concentration of inhibitor required to give 50% inhibition ($I_{50\%}$) (Miflin, 1971). Co-operative inhibition is defined as occurring when the total concentration of a combination of inhibitors required to cause 50% inhibition is less than that for the most effective single inhibitor. Since 50% inhibition is brought about by leucine and valine together at approximately 1/10th the concentration of leucine needed (see the top graph Fig. 8a), then leucine and valine are acting as co-operative inhibitors. The $I_{50\%}$ value for valine is 2·3 mM. Leucine and isoleucine also inhibit in a co-operative fashion. Similar patterns for the inhibition of acetolactate synthetase by the branched chain amino acids have been observed for a range of higher plants (Miflin and Cave, 1972). These patterns differ from those in bacteria in that, in bacteria, only valine inhibits acetolactate synthetase activity and that multivalent control is achieved through repression of enzyme synthesis (Umbarger, 1969).

The final regulatory step in the pathway is isopropylmalate synthetase, which is the first enzyme unique to leucine biosynthesis, and which is inhibited by leucine (Oaks, 1965b). A summary of these feedback loops is given in Fig. 9.

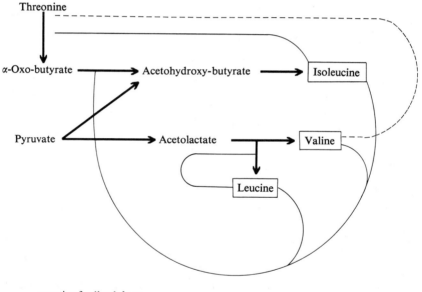

—— negative feedback loop
--- positive feedback loop

Fig. 9. Feedback control loops in branched chain amino acid biosynthesis.

2. Aromatic Amino Acids

Partial knowledge of the regulation of this pathway has been obtained (Fig. 10). Tryptophan regulates its own synthesis by feedback inhibition of anthranilate synthetase (Belser et al., 1971; Weber and Bock, 1969). The activity of chorismate mutase is modified by tryptophan, tyrosine and phenyl-alanine in algae (Weber and Bock, 1969), peas (*Pisum sativum*) (Cotton and Gibson, 1968) and mung beans (*Phaseolus aureus*) (Gilchrist *et al.*, 1972). Tyrosine and phenylalanine inhibit the enzyme and tryptophan activates it in the absence of the other two; it also relieves the inhibition caused by their presence. The interaction between phenylalanine and tyrosine is less than additive in algae (Weber and Bock, 1969); this precludes the possibility that two isoenzymes exist, each separately regulated by one of the pair. In mung

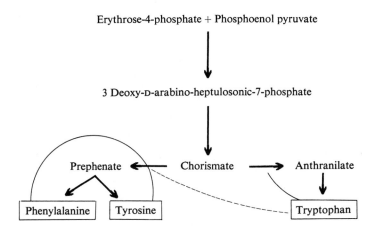

FIG. 10. Feedback control loops in aromatic amino acid biosynthesis.

beans Gilchrist *et al.* (1972) have shown that two isoenzymes exist, only one of which is subject to regulation by both amino acids. In this system the possibility of any co-operation between phenylalanine and tyrosine is so far unresolved, since the two together have only been tried at concentrations 10–100 times greater than that needed for one of them to cause maximum inhibition.

In contrast to the situation in bacteria there does not appear to be evidence of a regulatory step after prephenate (Gamborg and Keeley, 1966). 3-Deoxy-D-arabinoheptulosonic-7-phosphate synthetase is a regulatory enzyme in bacteria, and is also inhibited by tyrosine and, to a lesser extent, phenylalanine in blue-green algae and *Euglena* (Weber and Bock, 1968).

3. *Glutamate Family*

So far only the acetylated pathway has been studied in any detail at the enzymic level and two of the steps are subject to feedback regulation (Fig. 11). In particular, the acetylase and transacetylase are controlled in *Chlorella* and radish (*Raphanus sativus*) as indicated in the diagram, although the stimulation by ammonia is greater in the radish enzyme (Morris and Thompson, 1971).

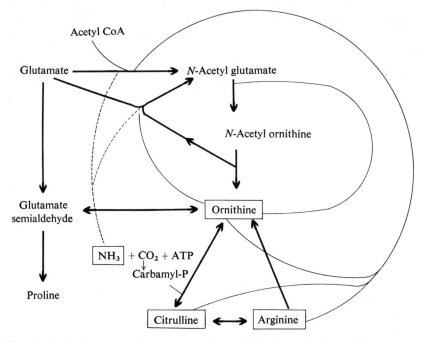

FIG. 11. Feedback control loops in the synthesis of the glutamate family of amino acids.

The significance of this stimulation is not yet fully known but could be related to the synthesis of carbamyl phosphate and its subsequent condensation with ornithine, an eventual product of the enzyme. In blue-green algae the N-acetyl-glutamate kinase has been demonstrated and is inhibited by ornithine (Hoare and Hoare, 1966). So far there is no evidence as to whether both regulatory steps are operating in the same organism. It is notable that, although Oaks et al. (1970) have shown that excess proline shuts off its own synthesis, proline is ineffective as a regulator of the acetylated pathway. No information on the regulation of the non-acetylated route has been found.

4. *Aspartate Family*

Growth inhibition studies with rice callus tissue (Furuhashi and Yatazawa, 1970) and with *Marchantia* (Dunham and Bryan, 1969, 1971) suggest that lysine and threonine co-operate synergistically to inhibit growth. Neither is particularly effective on its own. This inhibition is relieved by the further addition of methionine or homoserine. These results suggest that lysine and threonine are co-operative feedback regulators of aspartokinase and, in shutting off this enzyme, they also shut off methionine production. This is further supported by results showing that the levels of free methionine are negligible in the presence of lysine and threonine (Furuhashi and Yatazawa, 1970) and that the synthesis of the aspartate family of amino acids is blocked under these conditions (Table IV). So far results with isolated aspartokinase are preliminary and show that the enzyme is inhibited by lysine and, to a lesser extent, threonine (Bryan et al., 1970). The nature of any co-operative interaction remains to be elucidated. The only other enzyme in the aspartate pathway that has been shown

TABLE IV

Correlations of inhibition of growth and amino acid synthesis in *Marchantia* by amino acids of the aspartate family

Amino acid additions	Growth rate mm/h	Amount of synthesis from [^{14}C] aspartate (cts/min $\times$ 10^{-3})		
		Lysine	Methionine	Threonine
Control	10	178	243	125
Lysine + threonine	2	31	111	61
Lysine + threonine + methionine	6	31	97	71

Calculated from the data of Dunham and Bryan (1971).

to be feedback-regulated is homoserine dehydrogenase (Bryan, 1969). This enzyme is inhibited by threonine and, somewhat surprisingly, by serine, aspartic acid and cysteine. The nature of the inhibition by the amino acids differs according to whether NADPH or NADH is the cofactor and Bryan has suggested that only threonine acts solely at a specific regulatory site on the enzyme distinct from the catalytic site.

III. SUMMARY

Davies (1968), amongst others (e.g. Vogel, 1965), has pointed out previously that the pathways of amino acid synthesis in plants most closely resemble those found in bacteria. Further evidence has become available to support this contention, particularly with respect to the nature of tryptophan synthetase and the role of acetylated intermediates in cysteine and ornithine biosynthesis. In contrast, the control mechanism operating in specific pathways show a number of differences between the two groups.

The most striking difference is the apparent absence of enzyme repression in higher plants. This lack is unlikely to be due to a general absence of the phenomenon of induction and repression since there is evidence that large variations in the extractable levels of enzymes, in response to the presence and absence of substrates, do occur (e.g. nitrate and nitrite reductase—Filner *et al.*, 1969). However, the underlying mechanisms are not clear and appear to require continual protein turnover, involving *de novo* enzyme synthesis, even under repressive conditions (Zielke and Filner, 1971). The rates of protein synthesis and degradation in duckweed (*Lemna*) have recently been calculated and a rate constant obtained, 0.014 h^{-1} for synthesis, corresponding to an average halflife for a protein of 7 days (Trevawas, 1972). The occurrence of this continual protein turnover in plants, which is usually absent in bacteria, involves a continuous requirement for new amino acid biosynthesis, since the amino acids from the degraded protein are further broken down and not returned to the synthetic pool (e.g. Bidwell *et al.*, 1964). Under these conditions the relatively more coarse and long-term regulation afforded by repression and induction would seem to be less appropriate than the finer and more immediate regulation afforded by feedback modification of enzyme activity. Alternative or additional reasons for the loss of repressive mechanisms during evolution may be put forward. For example, induction and repression may be inoperative at certain stages of the cell cycle (Knutsen, 1965); these stages may be short in relation to the generation time of rapidly dividing bacterial cells but long in relation to the life span of maturing and differentiated plant cells. Thus the usefulness of the system would be severely limited.

The mechanisms of feedback modification so far described for plant systems are capable, not only of turning pathways on and off, but also of maintaining a necessary balance between the various end products to meet the needs of the cell in any given situation. This is particularly well demonstrated in the branched chain amino acid pathway where simple feedback inhibition, co-operative feedback inhibition and feedback activation all occur. Feedback modification loops may also occur between pathways (e.g. see Slaughter, 1970). Having made these points it must be stressed that regulatory abilities of plant cells vary considerably, particularly with age. For example, proline synthesis becomes progressively more refractory to end product inhibition with increasing maturity (Oaks *et al.*, 1970). This may be due, either to an *in vivo* desensitization of the enzymes, or to compartmentalization of the end product away from the enzymes that synthesize it.

One final point is the agronomic significance of these control mechanisms. In many crop plants the levels of some of the nutritionally essential amino acids (from the animal standpoint) are low. Since these amino acids are also those whose synthesis is closely regulated, it is important to know, and perhaps to alter, the characteristics of this regulation.

ACKNOWLEDGEMENTS

The author is grateful to the A.R.C. for financial support of his work presented in this paper and to Dr J. M. Widholm for providing copies of his manuscripts prior to publication.

REFERENCES

Balinsky, D., Dennis, A. W. and Cleland, W. (1971). *Biochemistry N. Y.* **10**, 1947.
Belser, W. L., Murphy, J. B., Delmer, D. P. and Mills, S. E. (1971). *Biochim. biophys. Acta* **237**, 1.
Bidwell, R. G. S., Barr, R. A. and Steward, F. C. (1964). *Nature, Lond.* **203**, 367.
Blekhman, G. I., Kagan, Z. S. and Kretovich (1971). *Biokhimiya* **36**, 1050.
Borstlap, A. C. (1970). *Acta. bot. Neerl.* **19**, 211.
Bryan, J. K. (1969). *Biochim. biophys. Acta* **171**, 205.
Bryan, P. A., Cawley, R. D., Brunner, C. E. and Bryan, J. K. (1970). *Biochem. biophys. Res. Commun.* **41**, 1211.
Burton, E. G. and Sakami, W. (1969). *Biochem. biophys. Res. Commun.* **36**, 228.
Chen, J. and Boll, W. G. (1971). *Can. J. Bot.* **49**, 1155.
Chu, M. and Widholm, J. M. (1972). *Physiol. Pl.* **26**, 24.
Cotton, R. G. H. and Gibson, F. (1968). *Biochim. biophys. Acta* **156**, 187.
Davies, D. D. (1968). *In* "Nitrogen Metabolism in Plants" (E. J. Hewitt and C. V. Cutting, eds), p. 125. Academic Press, London and New York.
Davies, M. E. (1964). *Pl. Physiol. Lancaster* **39**, 53.
Davies, M. E. (1971). *Phytochemistry* **10**, 783.
Delmer, D. P. and Mills, S. E. (1968a). *Pl. Physiol., Lancaster* **43**, 81.
Delmer, D. P. and Mills, S. E. (1968b). *Biochim. biophys. Acta* **167**, 431.
Dodd, W. A. and Cossins, E. A. (1970). *Biochim. biophys. Acta* **201**, 461.
Dougall, D. K. (1970). *Phytochemistry* **9**, 959.
Dougall, D. K. and Fulton, M. M. (1967a). *Pl. Physiol., Lancaster* **42**, 387.
Dougall, D. K. and Fulton, M. M. (1967b). *Pl. Physiol., Lancaster* **42**, 941.
Dunham, V. L. and Bryan, P. K. (1969). *Pl. Physiol., Lancaster* **44**, 1601.
Dunham, V. L. and Bryan, P. K. (1971). *Pl. Physiol., Lancaster* **47**, 91.
Filner, P., Wray, J. L. and Varner, J. E. (1969). *Science, N. Y.* **165**, 358.
Finlayson, A. J. and McConnell, W. B. (1960). *Biochim. biophys. Acta* **45**, 622.
Fletcher, J. S. and Beevers, H. (1971). *Pl. Physiol., Lancaster* **48**, 261.
Fowden, L. (1967). *A. Rev. Pl. Physiol.* **18**, 85.
Furuhashi, K. and Yatazawa, M. (1970). *Pl. Cell Physiol.* **11**, 569.
Gamborg, O. L. (1965). *Can. J. Biochem.* **43**, 723.
Gamborg, O. L. and Keeley, F. W. (1966). *Biochim. biophys. Acta.* **115**, 65.
Gamborg, O. L. and Simpson, F. J. (1964). *Can. J. Biochem.* **42**, 583.
Gilchrist, D. G., Woodin, T. S., Johnson, M. S. and Kosuge, T. (1972). *Pl. Physiol., Lancaster*, **49**, 52.
Giovanelli, J. and Mudd, S. H. (1966). *Biochem. biophys. Res. Comm.* **25**, 366.
Giovanelli, J. and Mudd, S. H. (1967). *Biochem. biophys. Res. Comm.* **27**, 150.
Giovanelli, J. and Mudd, S. H. (1968). *Biochem. biophys. Res. Comm.* **31**, 275.
Giovanelli, J. and Mudd, S. H. (1971). *Biochim. biophys. Acta* **227**, 654.
Gladstone, G. P. (1939), *Br. J. exp. Path.* **20**, 189.
Greenberg, D. M. (1969). "Metabolic Pathways" Vol. III. Academic Press, New York and London.
Griffith, T. and Griffith, G. D. (1966). *Phytochemistry* **5**, 1175.
Hoare, D. S. and Hoare, S. L. (1966). *J. Bact.* **92**, 375.
Kanamori, M. and Wixom, R. L. (1963). *J. biol. Chem.* **238**, 998.
Kleczkowski, K. and Cohen, P. P. (1964). *Archs Biochem. Biophys.* **107**, 271.
Knutsen, G. (1965). *Biochim. biophys. Acta* **103**, 495.
Mazelis, M. and Fowden, L. (1969). *Phytochemistry* **8**, 801.
Medvedev, A. I. and Kretovich, W. L. (1966). *Biokhimiya* **34**, 659.

Meister, A. (1965). "Biochemistry of the amino acids." Academic Press, New York and London.
Miflin, B. J. (1969a). *J. exp. Bot.* **20**, 810.
Miflin, B. J. (1969b). *Phytochemistry* **8**, 2271.
Miflin, B. J. (1971). *Archs Biophys. Biochem.* **146**, 542.
Miflin, B. J. and Cave, P. R. (1972). *J. exp. Bot.* **23**, 511.
Minamikawa, T. (1967). *Pl. Cell Physiol.* **8**, 695.
Morris, C. J. and Thompson, J. F. (1968). *Biochem. biophys. Res. Commun.* **31**, 281.
Morris, C. J. and Thompson, J. F. (1971). *Pl. Physiol., Lancaster* **47** (Suppl.), abs. 101.
Morris, C. J., Thompson, J. F. and Johnson, C. M. (1969). *Pl. Physiol., Lancaster* **44**, 1023.
Nair, P. M. and Vining, L. C. (1965). *Phytochemistry* **4**, 401.
Nandy, M. and Ganguli, N. C. (1961). *Biochem. biophys. Acta* **48**, 608.
Noguchi, M., Koiwai, A. and Tamaki, E. (1966). *Agric. biol. Chem.* **30**, 452.
Notation, A. D. and Spenser, I. D. (1964). *Can. J. Biochem.* **42**, 1803.
Oaks, A. (1965a). *Pl. Physiol., Lancaster* **40**, 149.
Oaks, A. (1965b). *Biochim. biophys. Acta.* **111**, 79.
Oaks, A., Mitchell, D. J., Barnard, R. A. and Johnson, F. J. (1970). *Can. J. Bot.* **48**, 2249.
Redkina, T. V., Uspenskaya, Zh. V. and Kretovich, W. L. (1969). *Biokhimiya* **34**, 247.
Rosenthal, G. A. and Naylor, A. W. (1969). *Biochem. J.* **112**, 415.
Sanderson, G. W. (1966). *Biochem. J.* **98**, 248.
Sasoka, K. (1961). *Pl. Cell Physiol.* **2**, 231.
Sasoka, K. and Inagaki, H. (1960). *Mem. Res. Inst. Food Sci., Kyoto Univ.* **21**, 12.
Seigel, J. N. and Gentile, A. C. (1966). *Pl. Physiol., Lancaster* **41**, 670.
Shafer, J. and Thompson, J. F. (1968). *Phytochemistry* **7**, 391.
Shargool, P. D. (1971). *Phytochemistry* **10**, 2029.
Sharma, R. K. and Mazumder, R. (1970). *J. biol. Chem.* **245**, 3008.
Shimura, Y. and Vogel, H. J. (1966). *Biochim. biophys. Acta.* **118**, 396.
Slaughter, J. C. (1970). *FEBS Letters* **7**, 245.
Splitoesser, W. E. (1969). *Phytochemistry* **8**, 735.
Stadtman, E. R. (1966). *Adv. Enzymol.* **28**, 41.
Staub, M. and Denes, G. (1966). *Biochim. biophys. Acta.* **128**, 82.
Street, H. E. (1966). *In* "Cells and Tissues in Culture" (Willmer, E. N. ed.), Vol. 3. Academic Press, London and New York.
Trevawas, A. (1972). *Pl. Physiol., Lancaster,* **49**, 40.
Umbarger, H. E. (1961). *Cold Spring Harb. Symp. quant. Biol.* **26**, 301.
Umbarger, H. E. (1969). *A. Rev. Biochem.* **38**, 323.
Vogel, H. J. (1965). *In* "Evolving Genes and Proteins" (V. Bryson and H. J. Vogel, eds), p. 25. Academic Press, New York and London.
Weber, H. L. and Bock, A. (1968). *Arch. Mikrobiol.* **61**, 159.
Weber, H. L. and Bock, A. (1969). *Arch. Mikrobiol.* **66**, 250.
Weber, H. L. and Bock, A. (1970). *Europ. J. Biochem.* **16**, 244.
Widholm, J. M. (1971). *Pl. Physiol. Lancaster* **25**, 75.
Zielke, H. R. and Filner, P. (1971). *J. biol. Chem.* **246**, 1772.

CHAPTER 4

The Regulation of Ribosomal RNA Synthesis

J. INGLE

Department of Botany, University of Edinburgh, Edinburgh, Scotland

II. Transcription and Processing of rRNA 69
III. The rRNA Gene 76
    A. The Number of Copies of the Gene 76
    B. The Possibility of Gene Amplification 79
    C. Utilization of the Genes 85
References 90

I. Introduction

Although one of the most interesting features of the plant cell is that it contains at least 3 distinct types of ribosome, the 80s cytoplasmic ribosome, the 70s chloroplast ribosome and the mitochondrial ribosome, each containing its own discrete ribosomal RNAs (rRNA), I want to limit this presentation to a discussion of cytoplasmic rRNA, and to concentrate on certain aspects of regulation of synthesis.

II. Transcription and Processing of rRNA

It was originally demonstrated by Scherrer and Darnell (1962), in HeLa cells that the initial transcription product in the nucleolus of the rRNA genes was a large molecule, $4 \cdot 1 \times 10^6$ daltons, much larger than the final stable rRNAs with molecular weights of $1 \cdot 75 \times 10^6$ and $0 \cdot 7 \times 10^6$. The details of the processing of this molecule into the final rRNAs have been reviewed recently by Burdon (1971) and Maden (1971) and are summarized in Fig. 1. The exact arrangement of components within the initial $4 \cdot 1 \times 10^6$ dalton transcription unit is not clear, but results from several types of experiments indicate that this molecule contains both of the final rRNA sequences. The oligonucleotide fingerprints obtained by pancreatic RNase digestion of the 1·75, 0·7, 4·1 and $3 \cdot 1 \times 10^6$ components showed that both the 4·1 and $3 \cdot 1 \times 10^6$ RNA digests contained all the oligonucleotides present in the 1·75 and 0·7 × 10⁶ rRNAs, together with others not present in rRNA (Jeanteur *et al.*, 1968). The 2′-O-methylation pattern of the 4·1 × 10⁶ molecule was identical to that of 1·75 plus 0·7 × 10⁶ rRNAs (Wagner *et al.*, 1967). The 1·75 and 0·7 × 10⁶ rRNAs competed with the 4·1 × 10⁶ precursor for hybridization with DNA by 35 and 13 % respectively, and together competed with only 50 % of the precursor

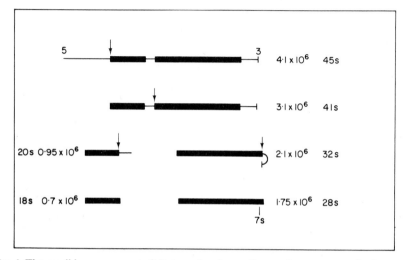

Fig. 1. The possible arrangement of ribosomal and non-ribosomal sequences, and subsequent processing of HeLa precursor rRNA. Ribosomal-RNA sequences are indicated by the heavy lines, and cleavage points by arrows.

(Jeanteur and Attardi, 1969). The fact that the 3·1 × 10⁶ molecule still contains both the 1·75 and 0·7 × 10⁶ rRNA sequences suggests that the 1·0 × 10⁶ daltons of non-rRNA removed during the first stage of processing is located at either one or both ends of the molecule. For simplicity it is shown at the 5′ (initiation) end of the precursor molecule. Experiments using cordycepin (3′-deoxyadenosine), which causes premature termination of transcription of the precursor, resulted in the accumulation of incomplete precursor molecules, which yielded some 0·7 × 10⁶ rRNA but no 1·75 × 10⁶ rRNA (Siev *et al.*, 1969). This suggests that the 0·7 × 10⁶ rRNA sequence is closer than the 1·75 × 10⁶ sequence to the initiation (5′) end of the precursor molecule. Methylation of the 4·1 × 10⁶ precursor, which largely involves methylation at the 2′-O position of ribose, is tightly coupled to transcription of the precursor, since it appears that only *nascent* precursor molecules are methylated by the nucleolar methylases (Weinberg *et al.*, 1967). The initial transcription unit is also associated with proteins, and processing of the precursor rRNA probably occurs within *nascent* ribosomal particles (Warner and Soeiro, 1967).

The initial product is cleaved to yield the 3·1 × 10⁶ molecule which, since it still contains both the 0·7 and 1·75 × 10⁶ rRNA sequences, must have lost only non-ribosomal RNA. The second cleavage then produces the 0·95 and 2·1 × 10⁶ components, containing only the 0·7 or 1·75 × 10⁶ rRNA respectively. Further removal of non-ribosomal RNA yields the stable rRNAs. The The non-ribosomal RNA removed during processing of the 4·1 × 10⁶ precursor molecule appears to be essentially unmethylated, and is probably rapidly degraded since components corresponding to the non-ribosomal sequences do not appear, even transiently, within the nucleus. A further important point is that the third RNA component of the ribosome, the 5s RNA, is not part of this initial transcription unit, and is synthesized quite independently of the high molecular weight rRNAs.

A similar situation occurs in all the eukaryotes studied, in that there is a large, polycistronic precursor molecule containing one sequence of each of the rRNAs. However, in cold-blooded animals, up to and including reptiles, and in higher plants, the size of the initial transcription unit is much less than that in HeLa. (Table I). Consequently, whereas mammals, rodents, marsupials and birds conserve only 50 to 60% of the transcription unit, lower organisms retain about 80% of the initial precursor rRNA. In pea (*Pisum sativum*) roots the initial transcription product is 2·5 × 10⁶ daltons, (Fig. 2) and this appears to produce the 0·7 × 10⁶ rRNA and a 1·4 × 10⁶ product, which in turn yields the

TABLE I

Size of the ribosomal transcription unit

	Mol wt × 10⁻⁶		% Precursor conserved
	Transcription unit	rRNAs	
Mammal (HeLa and fibroblasts)	4·4	1·75 + 0·70, 2·45	54
Rodent (mouse)	4·19	1·70 + 0·65, 2·35	56
Marsupial (potoroo)	4·19	1·70 + 0·65, 2·35	56
Bird (chicken)	3·92	1·61 + 0·63, 2·24	57
Reptile (iguana)	2·74	1·51 + 0·62, 2·13	78
Amphibian (frog)	2·76	1·58 + 0·61, 2·19	79
Fish (trout)	2·70	1·55 + 0·65, 2·20	81
Insect (*Drosophila*)	2·85	1·40 + 0·65, 2·05	72
Plants			
tobacco (*Nicotiana tabacum*)	2·76	1·29 + 0·66, 1·95	71
pea (*Pisum sativum*)	2·3	1·3 + 0·7, 2·0	87
artichoke (*Helianthus tuberosus*)	2·3	1·3 + 0·7, 2·0	87
carrot (*Daucus carota*)	2·8–2·2	1·3 + 0·7, 2·0	71–91
mung bean (*Phaseolus aureus*)	2·9–2·5	1·3 + 0·7, 2·0	69–80

Data taken from Birnstiel *et al.* (1971) except for pea and artichoke (Rogers *et al.*, 1970), carrot (Leaver and Key, 1970) and mung bean (Grierson and Loening, 1972).

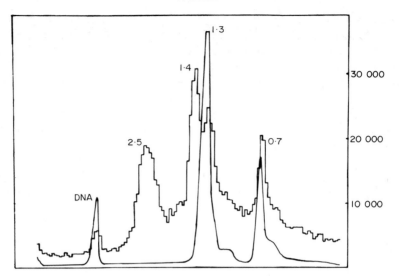

FIG. 2. Gel fractionation of total nucleic acid from pea roots. Roots were incubated in [³²P]orthophosphate (160 μCi/ml) for 1 h and total nucleic acid was prepared and fractionated on a 2·4% polyacrylamide gel for 4 h at 50V. The continuous scan represents the A265 nm, and the histogram the radioactivity per 0·5 mm slice.

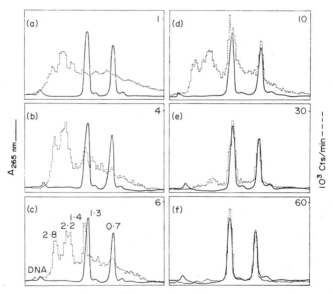

Electrophoretic mobility

FIG. 3. Gel fractionation of total nucleic from carrot-root discs. Carrot discs were incubated with [³²P]orthophosphate (80 μCi/ml) for 10, 20, 40, 60, 120 and 240 min (a to f respectively), total nucleic acid was prepared and fractionated by gel electrophoresis (——) A265 nm; (————) radioactivity per 0·5 mm slice (Leaver and Key, 1970).

1.3×10^6 rRNA. Analysis of the RNA synthesized in carrot (*Daucus carota*) discs showed, after 10 min incorporation of radioactive precursor, a background of polydisperse RNA plus two large precursor molecules of 2.8 and 2.2×10^6 daltons (Fig. 3). After 20 min incubation these two molecules constituted the bulk of the newly synthesized RNA. The 1.4×10^6 dalton precursor was

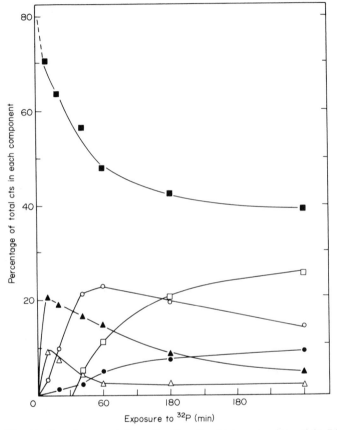

FIG. 4. The kinetics of accumulation of radioactive high molecular weight RNAs in the nuclear fraction from carrot discs. Carrot discs were incubated with [^{32}P]orthophosphate (as described in Fig. 3), and the nuclear fraction prepared. Total nucleic acid was extracted, separated by gel electrophoresis, and the fractionations used for the calculation of the accumulation of the various RNAs. —△—△— 2.8×10^6 daltons; —▲—▲— 2.2×10^6 daltons; —○—○— 1.4×10^6 daltons; —●—●— 0.7×10^6 daltons; —□—□— 1.3×10^6 daltons; —■—■— polydisperse AMP-rich RNA (Leaver and Key, 1970).

present after 20 min and increased in relative amount up to 1 h. By 4 h the radioactive profile closely followed the stable rRNAs, and there was only a trace of the large precursor molecules. The similarity in the kinetics of accumulation of the 2.8 and 2.2×10^6 molecules (Fig. 4) suggested that both may be initial transcription products, rather than show a precursor-product relationship analogous to the 4.1 to 3.1×10^6 conversion in HeLa cells. A rather similar situation has recently been reported in mung beans (*Phaseolus aureus*) (Grierson and Loening, 1972). The leaf of the seedling contained precursor rRNAs of 2·9, and 2·5 × 10⁶ daltons, whereas the root contained molecules of 2·7 and 2·5 × 10⁶ daltons. This situation is, however, complicated by the possible presence of precursors to chloroplast rRNA in the leaf preparations. In this

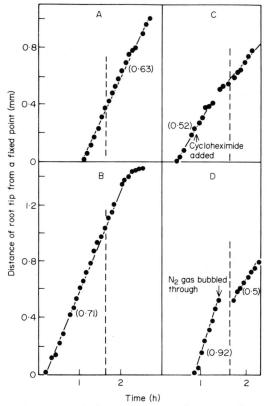

FIG. 5. Growth rates of pea roots. The linear growth rate of roots was determined by measuring, with a travelling microscope, the position of the root tip relative to a fixed point at approximately 10 min intervals. (a) Control, 60 μCi/ml [³²P]orthophosphate, (b) 160 μCi/ml [³²P]orthophosphate, (c) 60 μCi/ml [³²P]orthophosphate plus cycloheximide at 0·2 μg/ml, (d) incubation gassed with N_2 for 10 min, followed by the addition of 60 μCi/ml [³²P]orthophosphate. The arrows indicate the time of addition of cycloheximide or N_2 gassing, and [³²P]orthophosphate was added at the time indicated by the broken vertical line (Jackson and Ingle, 1972).

tissue the 2·5 × 10⁶ molecule was thought to arise from cleavage of either the 2·9 or 2·7 × 10⁶ initial precursor, this conclusion being based on the presence of a 0·45 × 10⁶ component, present in leaf preparations but not from roots, which could be the other product from cleavage of the 2·9 × 10⁶ precursor to the 2·5 × 10⁶ component. There are unfortunately no kinetic data to support this conclusion. The hypocotyl of the seedling, however, contained molecules characteristic of both leaf and root, 2·9, 2·7 and 2·5 × 10⁶ daltons, although it is not certain whether all these components exist within a single cell, or whether this simply reflects different cell types within the tissue. The situation in plants is therefore rather complicated by the presence of several large precursor molecules, the relationships between which are not very clear.

One final point must be made concerning the precursor rRNA molecules. Intact pea roots, grown under conditions of minimal disturbance, contained relatively little of the large precursor. (Jackson and Ingle, 1972). In certain experiments much more 2·5 × 10⁶ precursor rRNA was present, and this appeared to be correlated with disturbance of growth. Linear growth of the root was therefore measured continuously throughout the course of an experiment in which growth was disturbed by a high level of [³²P]orthophosphate, a low concentration of cycloheximide and N₂ gassing (Fig. 5). In the control sample, which contained 60 μCi/ml[³²P]orthophosphate, growth continued undisturbed during the course of the experiment. In the presence of 160 μCi/ml[³²P]orthophosphate growth continued for 30 min and then ceased. The cycloheximide and N₂ gassing both reduced the growth rate by about 50%. The distribution of RNA molecules synthesized during these treatments varied greatly from the control (Fig. 6). The most obvious difference was that in each of the disturbed samples the 2·5 × 10⁶ precursor rRNA represented a much larger percentage (13%) of the newly synthesized RNA than in the control (3·5%). The large, sharp precursor peaks usually seen in such experiments

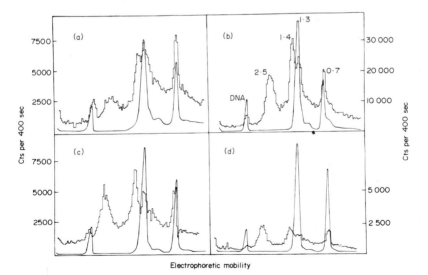

FIG. 6. Gel fractionations of total nucleic acid prepared from pea roots after various growth inhibitory treatments. Total nucleic acid was prepared 1 h after the addition of [³²P]ortho-phosphate to the samples described in Fig. 5, and fractionated by gel electrophoresis (Jackson and Ingle, 1972).

appear, therefore, to be the result of tissue disturbance and, or, disturbance caused by the very high levels of radioisotope used. These peaks probably have limited significance in the study of rRNA synthesis during normal growth.

III. THE rRNA GENE

A. THE NUMBER OF COPIES OF THE GENE

Because rRNAs are direct gene products, and because they are stable molecules with discrete sizes which can be prepared and purified easily, it is possible to go one stage back from the initial transcription product and look at the DNA sequences, or genes, from which the rRNAs are transcribed. This may be achieved by hybridization, whereby radioactively labelled rRNA is incubated with single stranded (denatured) DNA under conditions favourable for the formation of duplex structures, or hybrids, between the rRNA and the complementary sequence present in the DNA. It is possible to measure the percentage of the DNA which is complementary to rRNA by incubating the DNA with an excess of rRNA until all the complementary sites are saturated. From this value, and the amount of DNA present in the nucleus (or cell), the number of copies of the rRNA gene per nucleus (or cell) may be calculated. This varies from a single copy in *Mycoplasma*, through several copies in bacteria to hundreds of copies in animal and thousands in plant nuclei (Table II). The DNA containing these genes may be characterized to a certain extent, for example it can be resolved from the bulk of the DNA by equilibrium centrifugation in CsCl. (Fig. 7). The separation of the genes from the mainband was essentially independent of size of the DNA, complete separation being obtained with 20×10^6 dalton DNA, the largest which could be prepared from the plant tissue. This indicated that the genes were in clusters of at least 4 to 5. There is, in fact, good genetic evidence that all the rRNA genes are clustered within the nucleolar organizer DNA. The homozygous anucleolate mutant of the midwife toad (*Xenopus laevis*) contains essentially no rRNA genes, whereas the heterozygous toad contains half the number of genes of the wild type (Wallace and Birnstiel, 1966). Deletion of 800 genes by a single mutation can be explained only if the genes are clustered together at a single locus, in this case at the nucleolar organizer region (NOR) of the chromosome. Similarly, individuals containing 4, 3, 2 or 1 NOR per cell were produced in the fruit fly (*Drosophila melanogaster*) by genetic manipulation, and it was observed that the number of rRNA genes was strictly proportional to the number of NOR (Ritossa and Spiegelman, 1965). Recently a mutant of maize (*Zea mays*), having duplication of the NOR, has been shown to have twice as many rRNA genes as normal inbred lines (Phillips *et al.*, 1971).

Since the general area of rRNA genes has recently been reviewed by Birnstiel *et al.* (1971), I would like to concentrate on one particular aspect, that of utilization of these genes. The fact that different plants contain very different numbers of copies of the gene poses the question whether all the genes are involved in rRNA synthesis. Some light is thrown on this problem from a simple calculation of the amount, and rate, of rRNA synthesized in the pea root cell. This cell has a division cycle of about 10 h and contains 30×10^{-12} g of rRNA. Making the assumption that rRNA synthesis occurs throughout the cell cycle, 30×10^{-12} g rRNA are made per 10 h. This amount of rRNA may be

TABLE II

Number of copies of rRNA genes

Species	No. copies per cell
Bacterium *Mycoplasma* sp.	1
Bacterium *Escherichia coli*	4–7
Bacterium *Bacillus megaterium*	35–45
Yeast (*Saccharomyces carlsbergensis*)	240
Fungus (*Neurospora crassa*)	200
Fruit fly (*Drosophila melanogaster*)	260
Midwife Toad (*Xenopus laevis*)	1000–2500
Chicken	370
Rabbit	700–1300
Rat	480–1100
HeLa cells	1250

Species	No. copies/telophase nucleus
Artichoke (*Helianthus tuberosus*)	1580
Swiss chard (*Beta vulgaris* var. cicla)	2300
Maize (*Zea mays*)	6200
Pea (*Pisum sativum*)	7800
Cucumber (*Cucumis sativum*)	8800
Wheat (*Triticum aestivum*)	12 700
Onion (*Allium cepa*)	13 300

Data taken from Birnstiel *et al.* (1971) and Ingle and Sinclair (1972).

converted to 9×10^6 molecules of rRNA by using the relationship that a g mol.wt of any compound (2×10^6 g for rRNA) contains 6×10^{23} molecules (Avogadro's Number). The cell therefore makes, on average, 15 000 molecules of rRNA per min. However, a finite time is required for the transcription of each precursor molecule, which, in the case of HeLa cells, has been estimated as 2·5 min (Greenberg and Penman, 1966). If the transcription time in the pea cell is taken as 2 min then the cell must, on average, be in the process of transcribing 30 000 rRNA molecules at any one time. Since the cell has only 8000 copies of the gene (Table II), this suggests that several molecules are being transcribed from one gene at any one time. There is some support for this from the elegant pictures of amphibian oöcyte genes by Miller and Beatty (1969) (Fig. 8). Each matrix segment probably represents the DNA sequence coding for the precursor rRNA molecule, since the length, 2 to 3μ, would code for

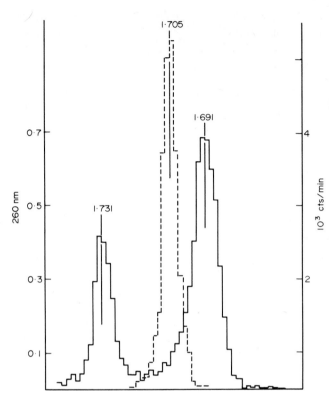

FIG. 7. The fractionation and hybridization of cytoplasmic rRNA genes from onion DNA. 100 μg of DNA from onion bulbs was fractionated by preparative CsCl density-gradient centrifugation. The DNA solution was adjusted to 1·720 mg.mm⁻³ and centrifuged at 40 000 rev/min for 68 h at 25°C in an MSE 50 rotor. *Micrococcus lysodiekticus* DNA, (1·731 mg.mm⁻³) was included as a marker. Five drop fractions were collected, and the 260 nm O.D. determined after the addition of 0·4 ml of 0·1 × SSC. The DNA in each fraction was alkaline denatured and fixed to a millipore filter. The filters from the gradient were hybridized for 2 h in 6 × SSC, 70°C with a 2:1 mixture of [³²P] labelled 1·3 × 10⁶ rRNAs (36000 cts/min/μg RNA) prepared from the bulbs of onion plants cultured for 10 days on nutrient solution containing [³²P]orthophosphate. The filters were then washed once with 6 × SSC, 3 times with 2 × SSC, incubated with 10 μg/ml ribonuclease at 25° for 15 min, dried and counted. The peak of hybridization was at 1·705 mg.mm⁻³ compared to the onion mainband at 1·691 mg.mm⁻³.

2 to 3×10^6 daltons of RNA, [precursor rRNA of *Xenopus* is $2·5 \times 10^6$ daltons, Loening *et al.* (1969)]. The matrix-free segments are interpreted as the spacer, or non-transcribed DNA associated with the rRNA genes (Birnstiel *et al.*, 1971). The matrix itself represents about 100 copies of the precursor rRNA in process of transcription from a single gene. The synthesis of 30000 rRNA molecules on 8000 copies of the gene could easily be accommodated if the amphibian oöcyte situation is extrapolated to the pea cell, in fact the cell may not use all its genes. There must be some hesitation about this extrapola-

tion, however, since this visualization of genes in action is from a preparation of the extrachromosomal nucleoli from the oöcyte, which represents a rather atypical situation. The oöcyte requires a high rate of rRNA synthesis, and it achieves this by specific amplification of the rRNA genes. For example, the normal somatic *Xenopus* cell, containing 6×10^{-12} g DNA, has 1500 copies of the rRNA gene, accounting for 0·1 % of the DNA. During oöcyte development the DNA per nucleus increases to 42×10^{-12} g which, correcting for the 4 C somatic content $(12 \times 10^{-12}$ g$)$ gives 30×10^{-12} g of amplified DNA. The

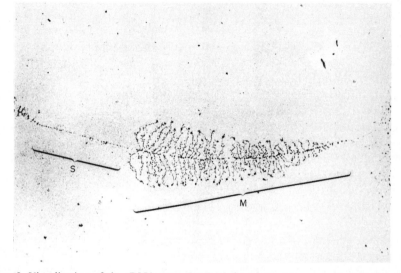

FIG. 8. Visualization of the rRNA gene. Portion of an extrachromosomal nucleolar core isolated from an amphibian (*Triturus viridescens*) oöcyte showing matrix units (M) separated by matrix-free segments (s) of the core axis (Miller and Beatty, 1969).

8×10^6 copies of the rRNA gene now represent 70% of the DNA, and are present in the nucleus as 1500 extrachromosomal nucleoli. (Perkowska *et al.*, 1968). The amphibian oöcyte therefore represents a rather special situation!

B. THE POSSIBILITY OF GENE AMPLIFICATION

Is there any evidence for such rRNA gene amplification during the development of higher plants? This problem can be approached at either the biochemical or cytological level. We have approached this in a gross way by preparing DNA from plant tissues during different phases of development, and determining the number of copies of the rRNA gene by hybridization with rRNA (Ingle and Sinclair, 1972). Seed germination represents a dramatic change from the dormant embryo to the rapidly growing seedling, and the rapid rate of RNA synthesis during germination may require amplification of the rRNA genes. The percentage of DNA which hybridized was, however, similar whether the DNA was prepared from the dormant embryo or from the embryo after 48 hours of germination, by which time the number of cells in the embryo had doubled (Table III). Moreover, no differences were found between DNA prepared from cells newly formed during germination (root tip and shoot) and those cells initially present in the seed (remainder of embryo). These results contrast with those of Chen and Osborne (1970) who reported a 30% deletion of genes during germination of wheat (*Triticum sativum*), and suggested that this may be the return to normal after amplification during seed development. We saw no significant differences during either the development or germination of wheat (Table IV). It should be noted that the amplification in the *Xenopus* oöcyte was in the order of 5000-fold, in *Colymbetes* (water beetle) it was 7-fold (Gall *et al.*, 1969) and in *Acheta* (house cricket) 15-fold (Lima-de-Faria *et al.*, 1969), so differences more of this order of magnitude should be considered, rather than changes of 30%.

There is therefore no evidence for gross amplification during these phases of plant development. It has, however, been reported that amplification occurs under certain conditions of physiological stress (Guille *et al.*, 1968, Quertier *et al.*, 1968). These reports are based on CsCl gradient analyses of DNA prepared from melon (*Cucumis melo*) seeds after storage in the cold for 20 days, or from tomato (*Lycopersicon esculentum*) seedlings germinated in the dark for long periods. After such treatment analyses showed an additional band of DNA with a buoyant density of $1\cdot720$ mg.mm^{-3}, which has been interpreted as a massive amplification of the rRNA genes in analogy with *Xenopus*, where the rRNA genes have a density of $1\cdot723$ mg.mm^{-3}. Since we could obtain no evidence for amplification during normal development it seemed a little peculiar that it should occur during these stress treatments, so some of these experiments were repeated. (Pearson and Ingle, 1972.) We found very little change during the etiolated growth of seedlings, but we did get the appearance of the $1\cdot720$ mg.mm^{-3} DNA during cold storage of melon seeds. Although the rRNA genes in *Xenopus* are at $1\cdot723$ mg.mm^{-3}, rRNA genes in all the plants we have studied have been between $1\cdot706$ and $1\cdot711$ mg.mm^{-3}, so DNA preparations from fresh and stored seeds were analysed for rRNA genes by CsCl fractionation (Fig. 9). The peak of hybridization was similar from both tissues, at $1\cdot711$ to $1\cdot712$ mg.mm^{-3}, and was obviously not associated with the stress-satellite at $1\cdot719$ mg.mm^{-3}, so that whatever the nature of the additional DNA it was not rRNA genes. During these experiments with stored seeds it was observed that the stress-satellite did not always appear, and when it did it represented a large net increase in the DNA per seed, rather than arising from interconversion of existing DNA. It was also noticed that when the stress-satellite appeared the seeds were highly contaminated with bacteria. When stored seeds were separated into the seed coat and cotyledons, the stress-satellite was found only in the seed coat (Fig. 10). DNA prepared from the washings from intact stored seeds contained only the stress-satellite component, indicating that this DNA was in fact arising on the outside of the seed. When

TABLE III

rRNA genes during germination of the maize embryo

Source of DNA	% DNA hybridized (6 × SSC, 70°C, 2·5 h)				
	Experiment 1	Experiment 2			
	1·3 × 10⁶ rRNA	1·3 × 10⁶ rRNA		0·70 × 10⁶ rRNA	
	5 μg/ml	5 μg/ml	2 μg/ml	5 μg/ml	2 μg/ml
0 h embryo	0·16	0·18	0·18	0·15	—
48 h complete embryo	0·15	—	—	—	—
48 h root tip	0·19	0·17	—	0·13	—
48 h shoot	—	0·16	0·18	0·13	0·12
48 h remainder of embryo	—	0·18	0·18	0·15	0·13

DNA was prepared from embryos dissected from dry seeds, and from embryos removed after 48-h germination. The 48-h embryo was also divided into the 10-mm root tip, the shoot and the remainder of the embryo. Fifteen seeds were germinated under sterile conditions in the presence of 1 mCi/ml of [^{32}P]orthophosphate for 60 h and RNA was prepared from the 15-mm root tips. The specific activities of the RNA used in experiments 1 and 2 were 140000 and 19000 cts/min/μg respectively. The values are calculated from the means of duplicate filters (Ingle and Sinclair, 1972).

TABLE IV

rRNA genes during development and germination of the wheat embryo

Source of DNA	% DNA hybridized (3 μg/ml, 6 × SSC, 70°C 2 h)	
	1·3 × 10⁶ rRNA	0·70 × 10⁶ rRNA
Ovule, 5 d after pollination	0·101	0·066
Embryo, 10 d after pollination	0·086	0·067
Embryo, 15 d after pollination	0·096	0·069
Embryo, from dry seed	0·087	0·068
Embryo, after 3 d germination	0·089	0·068

DNA was prepared from complete ovules 5 d after pollination, from the embryos 10 and 15 d after pollination, from the embryo of the dry seed and after 72 h germination. RNA was prepared from pea roots, which had grown in the presence of 100 μCi/ml [^{32}P]orthophosphate for 24 h. The specific activity of the RNA used was 58000 cts/min/μg, and the values are calculated from means of triplicate filters (Ingle and Sinclair, 1972).

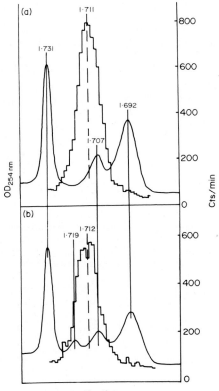

Fig. 9. Density of rRNA genes in melon DNA. DNA prepared from melon (*Cucumis melo*) seeds was fractionated on a preparative CsCl gradient in the M.S.E. 50 rotor at 38000 rev/min for 66 h at 25°C. The gradient was pumped from the tube and the ε 254 nm was continuously recorded. The DNA in each fraction was denatured, fixed to a millipore filter and hybridized with [^{32}P]rRNA. (a) DNA prepared from fresh seeds, showing a peak of hybridization at 1·711 mg.mm^{-3} relative to the nuclear mainband (1·692 mg.mm^{-3}) and satellite (1·707 mg.mm^{-3}) and *M. lysodiekticus* marker (1·731 mg.mm^{-3}) DNAs. (b) DNA prepared from seeds after 14 d cold-storage, showing the additional stress-satellite (1·719 mg.mm^{-3}) (Pearson and Ingle, 1972).

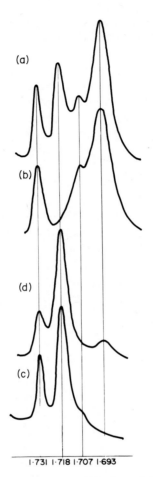

1·731 1·718 1·707 1·693

FIG. 10. Localization of the stress-satellite DNA within the seed. Seeds of melon, cold stored for 34 days, were separated into cotyledon and seed coat tissue. DNA was prepared from these tissues, and from the complete seed, and samples (2 to 3 μg) were adjusted to a density of 1·720 mg.mm^{-3} with CsCl and centrifuged at 44000 rev/min for 20 h at 25° in the Model E Analytical Ultracentrifuge; *M. lysodiekticus* DNA (1·731 mg.mm^{-3}) was included as marker. (a) DNA from complete seed, showing the nuclear mainband (1·693 mg.mm^{-3}) and satellite (1·707 mg.mm^{-3}), stress-satellite (1·718 mg.mm^{-3}) and marker (1·731 mg.mm^{-3}), (b) DNA from cotyledon tissue, showing only the nuclear mainband, satellite and marker; (c) DNA from the seed coat tissue showing the stress-satellite with only traces of mainband; (d) DNA prepared by washing the intact seeds in the detergent extraction medium, showing only the stress-satellite DNA (Pearson and Ingle, 1972).

the bacteria from the seed were cultured, the DNA prepared from the bacteria was identical to the stress-satellite DNA. The stress-satellite arising during the cold storage of seeds was certainly bacterial in origin, and therefore not evidence for rRNA gene amplification.

Certain observations at the cytological level may possibly be interpreted as gene amplification. During the development of the meiocytes in the lily (*Lilium henryi*), the ribosome population of the cell decreased dramatically (Table V) (Dickinson and Heslop-Harrison, 1970). The ribosomes reappeared between diakinesis and metaphase I. At about this time numerous nucleoloids, which are nucleolus-like bodies, appeared in the nucleus and then moved into the cytoplasm. From analogy with the amphibian oöcyte it is tempting to associate these nucleoloids with gene amplification. However, there are certain difficulties. The nucleoloids appeared a little too late, after the bulk of rRNA

TABLE V

Changes in total RNA, cytoplasmic ribosome numbers, nucleoli and nucleoloids during meiosis in *Lilium henryi* (Dickinson and Heslop-Harrison, 1970)

Meiotic stage	RNA per cell ($\mu\mu$g)	Ribosome count/unit vol	Nucleoli	Nucleoloids
Leptotene	60	15·4	1–4	0
Zygotene	60	9·8	1–2	0
Pachytene	31	2·9	1–2 Major, "cap" type, +0–3 supernumeraries	0
Diplotene	29	1·0	1–2 Major spherical + a few small supernumeraries	0
Diakinesis	29	1·67	1 (Rarely 2) major + a few supernumeraries; losing basiphilia	0
Metaphase I	57	9·2	None visible	0
Anaphase I	ND	9·8	Several among separating chromosome groups	15–30
Dyad	ND	ND	Up to 20 per nucleus; wide size range	15–30
Metaphase II	114·6	ND	None visible	Few visible
Telophase II	ND	ND	1–2 Major, up to 10 smaller supernumeraries	10–20 per cell
Young spore	ND	ND	1–2 Major, declining numbers of supernumeraries	6–10 per spore; numbers declining

synthesis, and their production was not restricted to the nucleolar organizing region of the chromosomes, indicating either that rRNA genes are located at positions other than the NOR or that the nucleoloids are not specifically related to NORs. Heslop-Harrison has concluded that the nucleoloids do not contain DNA, and do not represent rRNA gene amplification. A somewhat similar observation has been made during the development of polytene chromosomes in the suspensor cells of beans (*Phaseolus coccineus*) (Nagl, 1970), where the degree of endopolyploidy can be as high as $4096n$. During the early stages of this endoreduplication numerous micronucleoli appeared in the nucleus, and later moved out into the cytoplasm. This resembles, at least superficially, the situation in *Lilium* meiocyte development. The micronucleoli contained DNA on the basis of [^{3}H]actinomycin D staining, but their description as a spherical mass of ribonucleoproteins covered by a layer of DNA contrasts with normal nucleoli structure, where the DNA is contained within the cenrtal core (Avanzi *et al.*, 1970). Again the formation of micronucleoli was not limited to the NORs, and Feulgen spectrophotometry did not indicate any DNA replication other than increase in ploidy (Nagl, 1970). The appearance of the micronucleoli has, however, been interpreted in terms of amplification of rRNA genes (Avanzi *et al.*, 1970).

C. UTILIZATION OF THE GENES

The possibility of rRNA gene amplification in plants certainly cannot be ruled out, in fact essentially no studies have been made on the premeiotic ovule, the cell analogous to the oöcyte. However, at present there is very little convincing evidence for this phenomenon, and we are therefore left with the situation that different plants contain very different dosages of this gene. The calculation made for the pea cell suggests that perhaps only some of the rRNA genes are used. It is obvious from other systems that the full complement of rRNA genes is not always necessary for normal development and RNA synthesis. The heterozygote of the anucleolate mutant of *Xenopus*, containing only half the dosage of genes, is normal with respect to development and ribosome content (Wallace and Birnstiel, 1966) and different wild-type stocks of *Drosophila* contain different dosages of the rRNA gene (Ritossa and Scala, 1969). One approach to the problem of rRNA gene utilization in plants is to use the diploid-tetraploid series. Does the tetraploid, which presumably contains twice the dosage of genes, make twice as much rRNA as the diploid? The presence of twice the amount of rRNA in the tetraploid would suggest little regulation at the transcription level, whereas less than a doubling in rRNA could indicate regulation at this level.

There is relatively little information on this topic in the literature, and what there is does not form a unified picture (Table VI). In *Tradescantia ohioensis* the amount of RNA per tetraploid root cell was only about 20% higher than in the diploid, although the DNA was of course doubled (Sunderland and

TABLE VI

Relationship between DNA and RNA in polyploid cells

		10⁻¹¹ g/cell		
		DNA	RNA	RNA/DNA
Tradescantia ohioensis	2x	10·77	22·3	2·07
	4x	21·00	28·1	1·34

		10⁻¹¹ g/cell		
		DNAP	RNAP	RNA/DNA
Barley				
Hordeum vulgare. Asplund	2x	4·18	37·03	8·9
Asplund	4x	8·22	66·73	8·1
Jo	2x	4·97	31·29	6·3
Jo	4x	10·63	60·13	5·6
Timothy grass				
Phleum pratense	6x	0·89 (0·148)	5·06	5·7
	8x	1·14 (0·143)	8·19	7·2
	10x	1·52 (0·152)	13·44	8·8

DNA and RNA content was determined from 2-mm root tip segments of *Tradescantia* (Sunderland and McLeish, 1961) and from leaf material of *Hordeum* and *Phleum* (Skult, 1965).

McLeish, 1961). In barley (*Hordeum vulgare*), tetraploids of two varieties contained approximately twice as much RNA as their respective diploids, and in a polyploid series of timothy grass (*Phleum pratense*), the RNA content of the higher ploidy lines was greater than expected for the series, with consequent increase in the RNA/DNA ratio (Skult, 1965). Unfortunately in the *Hordeum* and *Phleum* studies RNA was prepared from leaf material which could contain 30 to 50% of chloroplast rRNA (Ingle et al., 1970), and little is known of the relationship between nuclear ploidy and chloroplast development. A some-what similar investigation has been made of RNA synthesis in the hexaploid wheat (*Triticum aestivum*) and its ancestral diploids (Jain et al., 1968). RNA synthesis, determined from grain counts of autoradiographs after incorporation of [³H]uridine by roots, was greater, on a NOR basis, in *T. monococcum* than in *Aegilops speltoides* (2 NOR pairs) or *A. squarrosa* (Table VII). The synthetic potentials of those parent genomes appeared to be essentially additive when incorporated into the hexaploid *T. aestivum*. A similar interpretation is possible from the values of total nucleolar mass (Pegington and Rees, 1970) which again was greatest in *T. monococcum*, and addition of the nucleolar mass from the three diploids was 80% that of the hexaploid (Table VII). The use of ditelosomic lines of the hexaploid (in which one specific NOR is re-

TABLE VII

RNA synthesis in hexaploid wheat and ancestral diploids

Species	Ploidy	Pairs NOR per complement	Grain count/cell (mean 5 roots)	Nucleolar mass $\times 10^{-11}$ g
Triticum monococcum	2x (A)	1 NOR	84·9	4·05
Aegilops speltoides	2x (B)	2 NOR	77·3	3·05
Aegilops squarrosa	2x (D)	1 NOR	53·6	3·28
			Total 215·8	10·38
Triticum aestivum	6x	4 NOR	197·6	7·91
Triticum aestivum (Chinese Spring)		4 NOR	35·2	
Ditelsome 1A^L		3 NOR	15·9	
Ditelsome 1B^L		3 NOR	34·7	
Ditelsome 6B^L		3 NOR	32·8	
Ditelsome 5D^L		3 NOR	39·1	

Roots of seedlings were incubated for periods from 0·5 to 4 h in [^{3}H]uridine. Root tips were then washed, fixed, softened in 5 % pectinase and squashed in 45 % acetic acid. Unincorporated label was removed with 2 % perchloric acid, and autoradiographs prepared (Jain *et al.*, 1968). The nucleolar mass was determined from isolated 4C nuclei (Pegington and Rees, 1970).

moved in each line) suggested a marked dominance effect on RNA synthesis by the NOR donated from *T. monococcum* which, as already noted, is approximately twice as active in RNA synthesis as the other NORs (Jain *et al.*, 1968). The interpretation of the autoradiographic studies of Jain *et al.* (1968) is, however, complicated since the grain counts reflect the total RNA synthesized during a short (1 h) incubation, and under these conditions probably only half of the incorporation was into rRNA (Ingle and Key, 1965).

We have approached this problem of gene utilization using cultivars of Hyacinth, which exists as diploids, triploids and tetraploid as well as many aneuploids, which may or may not involve the NOR chromosome (Timmis *et al.*, 1972) (Fig. 11). The number of copies of the rRNA gene in the euploids, 2x, 3x and 4x, was in the ratio of 2:3:4, suggesting a constant number of genes per NOR (8700) in these euploids (Table VIII). These cultivars are therefore suitable material for studying the rRNA synthesis and accumulation resulting from these different gene dosages. The aneuploid results are, however, more difficult to interpret. The loss of 1 NOR from the triploid (3x − 1) or from the tetraploid (4x − 2) certainly resulted in a decrease in rRNA genes per nucleus, but the number of genes was even lower than the expected diploid or triploid value. Moreover the additional NOR in 2x + 1 did not increase the number of

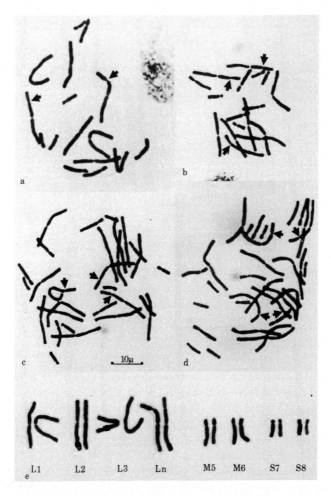

FIG. 11. Chromosome complements of Hyacinth cultivars. Feulgen stained metaphase squashes of (a) diploid, (2x) 16 chromosomes, 2 NOR; (b) aneuploid, (2x + 1) 17 chromosomes, 3 NOR; (c) aneuploid, (4x − 3) 29 chromosomes, 3 NOR; (d) tetraploid, (4x) 32 chromosomes, 4 NOR; (e) karyotype of diploid. The figure is shown by courtesy of Dr. J. N. Timmis.

TABLE VIII

The number of copies of rRNA genes in Hyacinth cultivars

Euploids	rRNA genes/ telophase nucleus	NORs/nucleus	rRNA genes/NOR
2x	17 800	2	8 900
3x	27 200	3	9 070
4x	32 900	4	8 220
		Mean	8 730
Aneuploids			
$3x - 1 (2x + 7)$	14 200	2	7 100
$2x + 1$	17 400	3	5 800
$3x + 3$	23 700	3	7 900
$4x - 2 (3x + 6)$	20 800	3	6 930
$4x - 3 (3x + 5)$	26 100	4	6 520
		Mean	6 850

Total DNA was prepared from these cultivars, purified, and fixed to filters. The DNA was hybridized with [^{32}P] labelled $1 \cdot 3 \times 10^6$ rRNA (246000 cts/min/μg RNA) at 5 μg rRNA/ml in $6 \times$ SSC at $70°$C for 2 h. The number of gene copies was calculated from the percentage of DNA hybridized to rRNA and the DNA content of the telophase nucleus, determined by comparative Feulgen microdensitometry (Timmis *et al.*, 1972).

genes from the diploid (1800) to the triploid (27000) state. The results suggested however, that the aneuploids should not be compared directly with the euploids but rather should be evaluated against other aneuploids, since the gene number per NOR for all the aneuploids was constant, and significantly lower than the redundancy of the euploids (e.g. 6850 for aneuploids compared to 8730 for the euploids). Addition of 1 NOR ($2x + 1$) increased the number of genes when compared with an aneuploid containing 2 NORs ($3x - 1$ or $2x + 7$) from 14 200 to 17400. Similarly the number of genes in $4x - 3$ ($3x + 5$, 4 NOR) was 26000 compared to 20700 in $4x - 2$ ($3x + 6$, 3 NOR).

The number of copies of rRNA genes per NOR therefore appears to be reduced in aneuploids relative to the euploids. Unfortunately the parentage of these cultivars is not recorded, so this cannot be eliminated as a possible cause for the variation between euploids and aneuploids. This variation in the gene redundancy per NOR may be analogous to the situation described for *Drosophila* where the number of genes per NOR varied with the nuclear environment. The NOR of the X chromosome contained 250 copies when present with another X or Y chromosome, but when present by itself, or present with a chromosome containing a NOR deletion, it contained approximately 400 genes (Tartof, 1971). Similarly the "bobbed" NOR (a mutation of the NOR in

which the number of rRNA genes is reduced) has been shown to rapidly accumulate rRNA genes when maintained within a "bobbed" nucleus, although
this number may be reduced if introduced into a wild type nucleus, or further
increased if present as a single X chromosome (Henderson and Ritossa, 1970).
These changes presumably occur as the result of disproportionate replication
of the rRNA genes at some point in the life cycle. The reduction of genes in the
Hyacinth aneuploids may be a result of the different nuclear (chromosomal)
environment of the NOR. Such disproportionate gene replication must be
associated with a mechanism capable of sensing and reacting to the nuclear
environment. The Hyacinth results therefore suggest some flexibility in the
redundancy of rRNA genes in a plant, although no dramatic examples of
amplification have yet been observed.

The regulation of synthesis of rRNA must therefore be considered in terms
of gene replication as well as at the normal transcriptional level.

REFERENCES

Avanzi, S., Cionini, P. G. and d'Amato, F. (1970). *Carylogia* 23, 605–638.
Birnstiel, M. L., Chipchase, H. and Spiers, J. (1971). *Progr. Nucleic Acid Res. molec. Biol.* 11, 351–389.
Burdon, R. H. (1971). *Progr. Nucleic Acid Res. molec. Biol.* 11, 33–79.
Chen, D. and Osborne, D. J. (1970). *Nature, Lond.* 225, 336–340.
Dickinson, H. G. and Heslop-Harrison, J. (1970). *Protoplasma* 69, 187–200.
Gall, J. G., MacGregor, H. C. and Kidston, M. E. (1969). *Chromosoma* 26, 169–187.
Greenberg, H. and Penman, S. (1966). *J. molec. Biol.* 21, 527–535.
Grierson, D. and Loening, U. E. (1972). *Nature, New Biol.* 235, 80–82.
Guille, E., Quertier, F. and Huguet, T. (1968). *C.r. hebd. Séanc. Acad. Sci., Paris* D. 266, 836–838.
Henderson, A. and Ritossa, F. M. (1970). *Genetics* 66, 463–473.
Ingle, J. and Key, J. L. (1965). *Pl. Physiol., Lancaster* 40, 1212–1219.
Ingle, J., Possingham, J. V., Wells, R., Leaver, C. J. and Loening, U. E. (1970). *Symp. Soc. exp. Biol.* 24, 303–325.
Ingle, J. and Sinclair, J. (1972). *Nature, Lond.* 235, 30–32.
Jackson, M. and Ingle, J. (1972). *Pl. Physiol., Lancaster* (In press).
Jain, H. K., Singh, M. P. and Raut, R. N. (1968). *In* "3rd Int. Symp. Wheat Gen." (K. W. Finlay and K. W. Shepherd, eds), pp. 99–104.
Jeanteur, P., Amaldi, F. and Attardi, G. (1968). *J. molec. Biol.* 33, 757–775.
Jeanteur, P. and Attardi, G. (1969). *J. molec. Biol.* 45, 305–324.
Leaver, C. J. and Key, J. L. (1970). *J. molec. Biol.* 49, 671–680.
Lima-de-Faria, A., Birnstiel, M. L. and Jaworska, H. (1969). *Genetics Suppl.* 61, 145–159.
Loening, U. E., Jones, K. W. and Birnstiel, M. L. (1969). *J. molec. Biol.* 45, 353–366.
Maden, B. E. H. (1971). *Progr. Biophys. molec. Biol.* 22, 129–174.
Miller, O. L. and Beatty, B. R. (1969). *Science, N.Y.* 164, 955–957.
Nagl, W. (1970). *J. Cell. Sci.* 6, 87–107.
Pearson, G. G. and Ingle, J. (1972). *Cell Differentiation* 1, 43–52.
Pegington, C. and Rees, H. (1970). *Heredity* 25, 195–205.
Perkowska, E., Birnstiel, M. L. and MacGregor, H. C. (1968). *Nature, Lond.* 217, 649–650.

Phillips, R. L., Kleese, R. A. and Wang, S. S. (1971). *Chromosoma* **36**, 79–88.

Quertier, F., Guille, E. and Vedel, F. (1968). *C.r. hebd. Séanc. Acad. Sci., Paris* D. **266**, 735–738.

Ritossa, F. M. and Scala, G. (1969). *Genetics Suppl.* **61**, 305–317.

Ritossa, F. M. and Spiegelman, S. (1965). *Proc. natn. Acad. Sci. U.S.A.* **53**, 737–745.

Rogers, M. E., Loening, U. E. and Fraser, R. S. S. (1970). *J. molec. Biol.* **49**, 681–692

Scherrer, K. and Darnell, J. E., Jr. (1962). *Biochem. biophys. Res. Commun.* **7**, 486–490.

Siev, M., Weinberg, R. A. and Penman, S. (1969). *J. Cell. Biol.* **41**, 510–520.

Skult, H. (1965). *Acta Acad. Abo.* (B) **25**, 1–14.

Sunderland, N. and McLeish, J. (1961). *Expt. Cell. Res.* **24**, 541–554.

Tartof, K. D. (1971). *Science, N.Y.* **171**, 294–297.

Timmis, J., Sinclair, J. and Ingle, J. (1972). *Cell Differentiation* (In press).

Wagner, E., Penman, S. and Ingram, V. (1967). *J. molec. Biol.* **29**, 371–388.

Wallace, H. and Birnstiel, M. L. (1966). *Biochim. Biophys. Acta* **114**, 296–310.

Warner, J. R. and Soeiro, R. (1967). *Proc. natn. Acad. Sci. U.S.A.* **58**, 1984–1990.

Weinberg, R. A., Loening, U. E., Willems, M. and Penman, S. (1967). *Proc. natn. Acad. Sci. U.S.A.* **58**, 1088–1095.

CHAPTER 5

Plant Cell Cultures:
Their Potential for Metabolic Studies

H. E. STREET

Botanical Laboratories, School of Biological Sciences,
University of Leicester, Leicester, England

I. Introduction

Studies of metabolism involving whole plants are usually faced with lack of uniformity between individuals and always with difficulties of interpretation. Interpretation usually involves some consideration of the individual contributions to the overall metabolism of the different tissues and an even greater number of separate cell types. Quantitative knowledge regarding these cellular components of overall plant metabolism is at present very limited. Various workers have recognized that the whole plant system can be simplified and the complex interrelationships between organs eliminated by working with isolated organs, and advances in our knowledge of particular aspects of metabolism have followed from work with isolated roots, leaves, stem apices and immature flowers and fruits. The value of such studies has been increased where continuing development of the isolated organ has been achieved by techniques of aseptic culture (Street, 1969). Further simplification has been attempted by use of organ discs or slices though such discs usually contain a number of different tissues and from the time of their isolation are destined to lose, more or less rapidly, their metabolic competence. Injury effects may also significantly alter their metabolism, and such discs are difficult to protect from rapid contamination with microorganisms.

Plant Cell Cultures (Suspension Cultures) have the very obvious attractions of sterility, and the ability to grow in defined media and under standardized conditions of temperature, illumination and gaseous exchange. They represent a significant step towards working at the level of a population of uniform cells. The first of these criteria is achieved to a high degree. Cell cultures do not contain significant numbers of microorganisms because any contaminant capable of increasing in numbers under the cultural conditions is readily detected by microscopic scrutiny, development of medium opacity or anomalous growth and metabolism. Contaminants which do not grow or grow very slowly can be readily detected by appropriate sterility tests involving enriched media. Cells from an increasing number of species can now be cultured in defined media containing only a utilizable carbohydrate, a mixture of inorganic salts, and a small number of pure vitamins and plant growth-regulating hormones.

The question of how far such cell cultures can be regarded as populations of cells identical in genetic competence and physiological state calls for more detailed consideration. The present paper is concerned to examine this question and other limitations of the technique of cell culture as at present practised. In developing this critical appraisal there will at the same time be presented material illustrative of current achievements in the field of plant metabolism coming from work with cell cultures and an assessment of their potential for future work.

II. GROWTH AND DEGREE OF CELL SEPARATION

The most generally adopted technique of growing plant cell suspensions involves initially the transfer of a piece of friable cultured callus to a liquid medium caused to swirl in an appropriate Erlenmeyer flask on a reciprocal platform shaker (Street and Henshaw, 1966). Under these conditions the callus piece breaks up and cells and cell aggregates derived from it are dispersed through the medium and continue to grow as agitation and incubation proceed. After a time the cell density (cells per unit volume) stabilizes and from this point the suspension can be continuously propogated by transferring a measured volume to new culture medium and repeating the incubation. For instance the stock sycamore (*Acer pseudoplatanus*) cell suspension grown in our laboratory is subcultured every 21–24 days by being submitted to a 10-fold dilution. Before subculture the stock suspension will contain 2–3×10^6 cells ml^{-1} and 20–30% of its volume will consist of cells. Its growth following subculture, when monitored by determination of cell number per ml, shows a growth curve of the form shown in Fig. 1. The phase of exponential growth is short and, under the conditions just described, only between 3 and 4 cell generations are passed through before the stock sycamore culture reaches stationary phase. In sycamore cell suspension the mean generation time (g) during exponential growth falls into the range 40–70 h. Mean generation times during transient exponential growth in other cell suspensions are reported as 48 h for tobacco (*Nicotiana tabacum*) (Filner, 1965), 36 h for *Rosa* sp. cv.

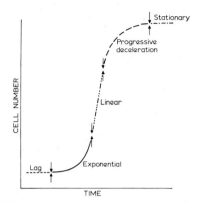

FIG. 1. Model curve relating cell number per unit volume of culture to time in a batch-grown plant cell suspension culture (after Wilson *et al.*, 1971).

Paul's Scarlet (Nash and Davies, 1972), 24 h for bean (*Phaseolus vulgaris* cv. Contender) (Liau and Boll, 1971) and 22 h for *Haplopappus gracilis* (Eriksson, 1967).

During the phase of exponential growth in such batch cultures the cells do not remain unchanged in composition. Figure 2 shows that although an exponential increase occurs during this period in cell number, in cell protein and in cell dry weight, the cells continuously decline in protein and dry weight

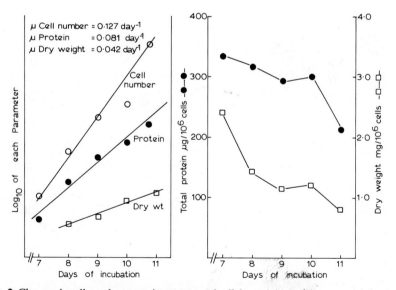

FIG. 2. Changes in cell number, protein content and cell dry weight during exponential growth in a batch culture of a sycamore cell suspension. μ = specific growth rate (previously unpublished data of P. J. King).

content. Further, these three parameters depart from exponential increase at different points in the growth cycle.

The average cell volume of batch-propagated sycamore cells reaches a minimum at the time of the transition from exponential to linear increase in cell number. At this same point the degree of cell aggregation is also maximal (Fig. 3). When the stationary phase is reached the proportion of free cells is at its highest value; at the end of the phase of exponential growth it is at its

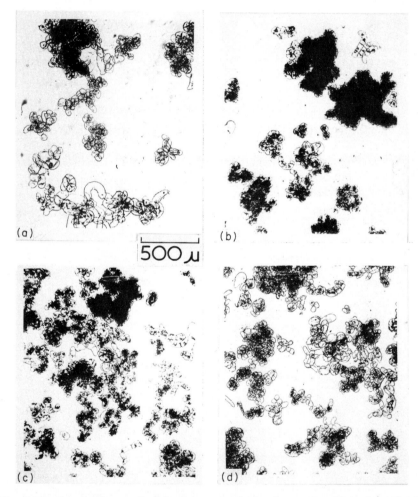

FIG. 3. The changing degree of cellular aggregation during the progress of growth in a batch culture of a sycamore cell suspension. (a) = early in lag phase; (b) = during exponential phase; (c) = during phase of declaration of growth; (d) = at beginning of stationary phase (previously unpublished data of K. J. Mansfield).

lowest value. At no stage, however, is the culture free from cell aggregates. This is true for all cell suspensions at present in culture; the sycamore suspension described is one showing, by comparison with many other suspension cultures, good cell separation at stationary phase and limited aggregation during the phase of rapid increase in cell number. Clearly however there are, during the progress of growth of a batch culture, not only changes in division rate and in the extent of cell expansion but also in cell association (Henshaw et al., 1966). Cells in aggregates are surely in quite a different environment to the free floating single cells. Cytochemical examination of the chains of cells formed during the early stages of the growth of a tobacco cell suspension has shown that individual cells in such simple aggregates may differ greatly in enzyme activities (de Jong et al., 1967). Thus there is not, at any stage, a uniform cell population although variation (in cell size and shape, in degree of aggregation and also in enzyme activities) is at a minimum in the mature stationary phase culture.

Can cell aggregation be eliminated? Aggregation can certainly be increased and decreased by changes in culture medium composition and speed of agitation during incubation (Rajashekar et al., 1971). However, the aggregation shown in Fig. 3 is that of a culture growing under cultural conditions favouring cell separation. Can a proportion of the total cell population be selected which does not aggregate or aggregates less? When such a selection is attempted by subculturing only from the upper part of the culture, previously allowed to sediment for 30–60 sec, some improvement in dispersion results (though this improvement in our experience is quite limited). This subject of whether the cell populations of suspension cultures are mixed populations containing genetic variants which can be isolated by appropriate selection (including variants with a genetic lesion in the middle lamella-forming system?) will be discussed later in relation to the assessment of the feasibility of obtaining a stable truly free cell culture. Meanwhile a very highly dispersed actively-growing suspension, at least of sycamore cells, can be obtained by incorporating into the culture medium low concentrations of cell-wall degrading enzymes (a cellulase and a pectinase) and sorbitol to increase the osmotic potential (and hence prevent cell bursting) (Fig. 4).

As will be discussed later, such highly dispersed suspensions are being developed to assist the isolation of single-cell clones. They are, however, described here to emphasize that such suspensions (at least over 2 or 3 culture periods) show a similar pattern of growth to that described for the untreated suspension. Aggregation is not essential for growth and cell division and there is no convincing evidence that isolated cells cannot divide as rapidly as cells in aggregates. Such a dispersed culture therefore approaches closer to the goal of a viable uniform cell population; its cells are more uniform in morphology, are in more intimate contact with the controlled environment and are presumably more uniform in physiological activity.

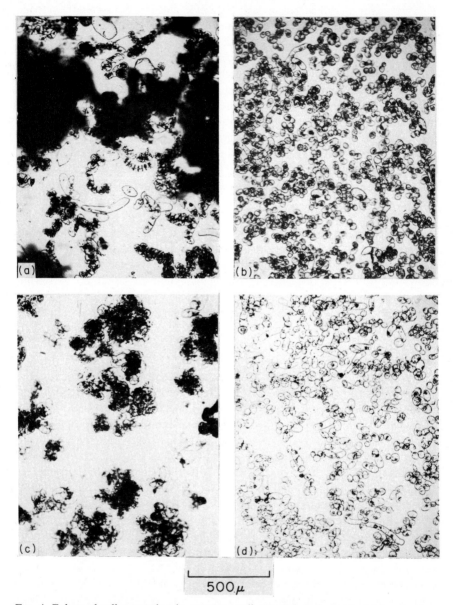

500 μ

FIG. 4. Enhanced cell separation in sycamore cell suspensions by appropriate enzymes. Photographs of control cultures (a and c) and cultures growing in presence of macerozyme (0·05%), cellulase (0·05%) and sorbitol (8%) (b and d) harvested after 13 days' (a and b) and after 21 days' (c and d) incubation (previously unpublished data of K. J. Mansfield).

III. METABOLIC STUDIES WITH BATCH-PROPAGATED CELL SUSPENSIONS

During the last five years a number of studies have been made of changes in cellular fine structure, in content of cellular constituents (such a protein, RNA, carbohydrates and particular enzymes) and in physiological activities which occur during the progression undergone by suspension cultures from initiation to stationary phase. These studies have involved minimal media and media supplemented with particular growth-regulating hormones. Such studies reveal that the metabolic activity of the cells is in continuous change; differing physiological states follow one another in succession. The contents of particular structural and enzymic components rise to peaks and then decay; physiological processes show similar large change in activity during the growth cycle. These patterns of metabolic change are interlocked with the pattern of growth; the physiological activity of the cells can be predicted from the growth stage of the culture. This situation might perhaps have been predicted, it can now be demonstrated. The nature of the changes observed can be illustrated by studies on sycamore cell suspensions, the data are expressed, where appropriate, on a per cell basis; contents of nucleic acids (Fig. 5a), protein (Fig. 5b), and nucleotides (Fig. 5c), rates of respiration (Fig. 5b) and ethylene production

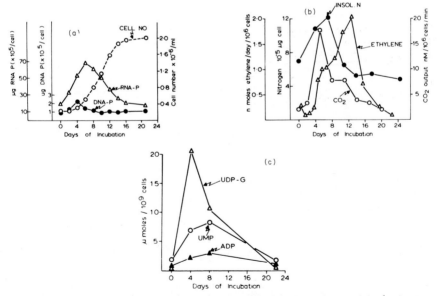

FIG. 5. Changes in growth and certain aspects of metabolism of batch propagated sycamore cell suspensions. (a) Increase in cell number, RNA-P and DNA-P per cell (data of Short *et al.*, 1969). (b) Insoluble N, CO_2 output and ethylene release per cell (data for insoluble-N from Givan and Collin (1967) and for ethylene from Mackenzie and Street (1970). (c) Contents of uridine diphosphate glucose (UDP-G), uridine monophosphate (UMP) and adenosine diphosphate (ADP) per cell (data from Brown and Short, 1969).

(Fig. 5b), activities of respiratory enzymes in the Embden-Meyerhof-Parnas EMP and pentose phosphate pathways (Fig. 6a, b, c), of acid and neutral invertase (Fig. 7) of various phosphatases (Fig. 8a) and peptidases (Fig. 8b), and of mitochondrial dehydrogenases (Fig. 9) are illustrated. Similar changes in the activity of a number of dehydrogenases during the growth cycle of tobacco cultures have been reported by de Jong *et al.* (1967).

To illustrate very briefly the kind of problems in metabolic regulation posed by these studies two examples may be taken. First the very sharp peak of ethylene production (Fig. 5b) occurs late in the cell-division phase (substantially after the peaks in protein and RNA content) (Mackenzie and Street, 1970). At this stage the culture is highly aggregated but immediately following this stage the aggregates begin to break up and the cells enlarge. Is ethylene production related to aggregation, does it trigger off the break-up of the aggregates? Following the peak of ethylene production its rate of production declines very steeply. What is the nature of the strong repression mechanisms operating at this point in the growth cycle? Secondly, the studies in the activities of the EMP and pentose phosphate pathways (Fowler, 1971, Fig. 6) suggest that phosphofructokinase, glucose 6-phosphate dehydrogenase and transketolase may be involved in metabolic regulation and this interpretation of the enzyme data is supported by labelling patterns of the release of $[^{14}C]CO_2$ from $[1\text{-}^{14}C]$- and $[6\text{-}^{14}C]$glucose. During the early stage of growth the preponderance of

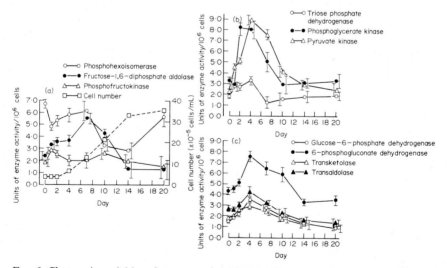

FIG. 6. Changes in activities of enzymes of the EMP pathway and the pentose phosphate pathway in extracts from sycamore cells harvested at intervals during the progress of growth of batch-propagated suspension cultures. A unit of enzyme activity is defined as the utilization of 1 nmole of substrate/min at 25°C. The lengths of the vertical lines represent twice the standard errors of the mean values plotted with the mean as mid point. The activities of phosphohexoisomerase and phosphoglycerate kinase presented are one-half of the measured activities of these enzymes (data from Fowler, 1971).

label from [1-^{14}C]glucose in the released CO_2 over that from [6-^{14}C]glucose indicates that both pathways are making appreciable contributions to carbohydrate oxidation, whereas in the later stages of growth the data indicates that the carbohydrate oxidation is mainly via the EMP pathway. Wilson's studies

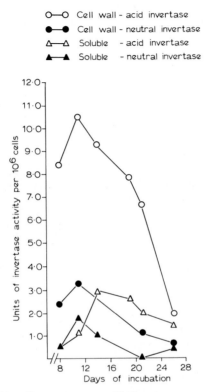

FIG. 7. Change in activities of invertases extractable from sycamore cells harvested at intervals during the progress of growth of batch-propagated suspension cultures. One unit of invertase activity is that amount catalysing the transformation of 1 μmole sucrose in 2 h at 25°C (data from Copping and Street, 1972).

(1971) with mitochondria, isolated from cells at different stages in the growth cycle, also suggest that in the early stages of growth the tricarboxylic acid cycle is geared to the production of substrates for biosynthesis; the high activity of the pentose-phosphate pathway at this stage may be providing the NADPH necessary for this synthesis. Later electron transport becomes geared to the provision of large amounts of ATP; at this time the activity of the pentose pathway decreases relative to the EMP pathway. A final point from this work; very similar changes in activity occur in the four pentose-pathway enzymes studied—do these enzymes form a genetically regulated "constant proportion group" as envisaged by Pette et al. (1962)?

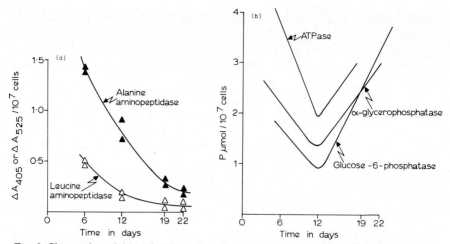

FIG. 8. Changes in activities of various phosphatases (a) and peptidases (b) extractable from sycamore cells harvested at intervals during the progress of growth in batch-propagated suspension cultures. Activity of phosphatases expressed as μmole P released by extract equivalent to 10^7 cells under the conditions of assay. Substrate for leucine aminopeptidase was leucyl-p-nitroanilide and for alanine aminopeptidase was alanyl-β-naphthylamide HCl. For leucine aminopeptidase extinction measured at 405 nm, for alanine aminopeptidase at 525 nm (data from Simola and Sopanen, 1970).

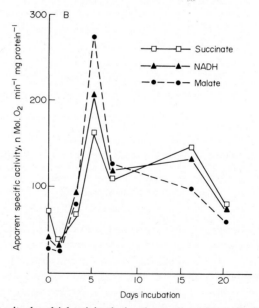

FIG. 9. Changes in mitochondrial activity during the progress of growth of batch-propagated sycamore cell suspensions. The period of high relative activity of the cyanide-sensitive respiratory pathway coincides with the peaks of specific activity of the three dehydrogenases. The apparent specific activities (nmoles $O_2 \cdot \text{min}^{-1}$, mg mitochondrial protein^{-1}) were calculated from the State 3 rates of O_2 uptake in the presence of the respective substrates (data from Wilson, 1971).

Nash and Davies (1972) have undertaken a similar general survey of the changes in concentration of major cell constituents and in physiological activities which occur during the growth cycle of batch-propagated Paul's Scarlet Rose suspensions with the declared objective of gaining an insight into the potential value of their particular cell culture for the study of metabolic processes. Similar marked changes in cell composition and physiological activity geared to the growth cycle were reported. Of particular interest is their observation that the accumulation of phenolic materials follows a pattern distinct from all other parameters in that it shows maximum accumulation late in the growth cycle and reaches a peak as the cells enter the stationary phase. This raises the important question, further discussed later in this paper, of whether special cultural environments will be needed to elicit the activity of particular metabolic sequences, especially those concerned with the synthesis of secondary metabolites.

The transient nature, in batch cultures, of each growth phase and of each associated physiological state limits the scope of studies on regulatory mechanisms in metabolism. However over sufficiently short periods of time the cultures do provide large numbers of cells with a uniformity of metabolic activity. Exploitation of this can be greatly enhanced by using, instead of the small (culture volume 50–100 ml) flask cultures, a larger stirred culture (4·5 l capacity) receiving controlled aeration and capable of being continuously monitored for certain physiological activities and frequently sampled at a sample size large enough to measure growth parameters and undertake analytical and biochemical work (Wilson et al., 1971).

Such cultures have already been used to follow continuously O_2 uptake and CO_2 evolution and to study growth limitation by O_2 availability and depletion of glucose (Street et al., 1971). It has also been possible, using such cultures, to construct carbon balance sheets and work out economic coefficients. By use of such equipment it is also possible to monitor more precisely the important changes which occur when the enlarged, highly vacuolated, metabolically quiescent stationary phase cells are transformed during lag phase into cells capable of exponential growth (Sutton-Jones and Street, 1968; Davey and Street, 1971) and to carry out metabolic studies during the short period of exponential growth.

The use of batch suspension culture for the study of particular aspects of metabolism can be illustrated from work on RNA synthesis on sycamore cells using [2-^{14}C]uridine as a specific precursor (Cox et al., 1973). Studies of uptake during the phase of active cell division showed immediate incorporation of label into RNA and almost immediate reduction in rate following depletion of precursor from the medium (Fig. 10). The radioactivity of the acid-soluble pool (whose major component is 5'-UMP) increases throughout the period of uridine uptake and then declines very slowly following depletion of the precursor from the medium (or following a "chase" with a high concentration of unlabelled uridine). When uridine is fed at a higher concentration

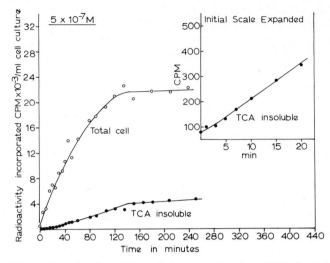

FIG. 10. Initial uptake of radioactivity and incorporation into RNA from 5×10^{-7} M [2-^{14}C]uridine (62 mCi/mM) by sycamore suspension cultures on day 6 of incubation (cell density *ca* 10^6 cells ml^{-1}, dry wt *ca* 4·3 mg.ml^{-1}). The inset shows on an expanded scale, incorporation into RNA during the first 20 min after addition of uridine (data of Cox, Turnock and Street, 1973).

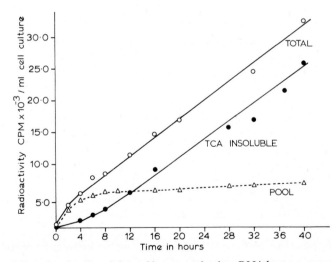

FIG. 11. Total uptake of radioactivity and incorporation into RNA by a sycamore suspension culture which received 5×10^{-4} M [2-^{14}C]uridine (0·62 mCi/mM) on day 6 of incubation. The size of the acid-soluble pool is also shown (data of Cox *et al.*, 1973).

$(5 \times 10^{-4}$ M) the duration of incorporation can be extended to 48 h. Under these conditions the radioactivity of the acid soluble pool reaches a level after 8 h which remained constant through the remaining 40 h of the experiment (Fig. 11). These, and other results not now presented, can be explained on the hypothesis that [^{14}C]uridine taken up from the medium passes through a small pool with a high rate of turnover to be incorporated predominantly into RNA while radioactivity builds up more slowly in the major pool of pyrimidine nucleotides. When [^{14}C]uridine is not entering this small pool from the medium a slower rate of [^{14}C]uridine incorporation occurs from the larger pool of nucleotides (Fig. 12).

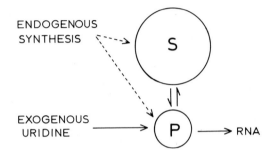

FIG. 12. Proposed pathway of [^{14}C] incorporation from uridine into RNA.

These studies have also demonstrated that both the uracil and cytosine residues of RNA are labelled by [2-^{14}C]uridine (at high concentrations of external uridine these residues are almost exclusively derived from the externally fed precursor), and that nucleic acids are the only macromolecules which receive radioactivity from [2-^{14}C]uridine. When [^{14}C]uridine incorporation is studied at intervals throughout the growth cycle it is found that a high rate of RNA turnover occurs even when, as occurs late in the growth cycle, the cells contain a fixed low level of RNA. Balance sheets accounting for the fate of the [^{14}C] of [2-^{14}C]uridine show that a considerable proportion (a very high proportion at the higher levels of uridine feeding) of this carbon is released as [^{14}C]CO_2 (Table I). The fact that these studies of the kinetics of [^{14}C] incorporation into RNA from [2-^{14}C]uridine by cultured sycamore cells very briefly outlined here, are amongst the most definitive hitherto undertaken with higher plant material illustrates the amenability of plant cell cultures for this kind of study.

This work has now been extended to study the processing of cytoplasmic ribosomal RNA and has led to the detection of high molecular weight ribosomal RNA precursors (estimated molecular weights-daltons-3.8×10^6 for the 45 s molecule, 2.3×10^6 for the 38 s, 1.4×10^6 for the 28 s and 0.9×10^6 for the 20 s) (Cox and Turnock, unpublished data).

TABLE I

Balance sheet tracing the fate of [2-^{14}C]uridine lost from the medium when supplied to actively dividing sycamore cell cultures (data of Cox et al., 1973)

Concentration and specific activity of uridine supplied to the culture	Duration of incubation period	Total loss of radioactivity from the medium, cts/min per culture	Total radioactivity present in the cells after incubation, cts/min per culture	Total [^{14}C]CO$_2$ radioactivity collected, cts/min per culture	[^{14}C]CO$_2$ evolved as a percentage of total loss of radioactivity from the medium	Radioactivity recovered from [^{14}C]CO$_2$ and from cells as a percentage of total loss from medium
10^{-6} M (62 mCi mmol^{-1})	2·0 h	107 828	70 171	23 725	25·3%	89·3%
10^{-5} M (0·5 mCi mmol^{-1})	2·0 h	4855	1876	2314	55·2%	93·7%
10^{-4} M (0·5 mCi mmol^{-1})	8·0 h	23 225	7443	20 002	72·9%	104·1%
10^{-3} M (0·5 mCi mmol^{-1})	8·0 h	103 535	28 201	67 569	76·1%	92·5%

The studies discussed above have involved cells growing in a basal or minimal medium. However, a number of workers have used batch suspension cultures to study the response of cultured cells to plant growth regulating hormones or particular metabolites either with the object of gaining an insight into the mechanism of action of plant hormones or in attempts to open up some special aspect of metabolism normally associated with specialized function (differentiation). Two examples will illustrate this approach. Thus Simola and Sopanen (1971), using suspension cultures of a strain of deadly nightshade (*Atropa belladonna* cf. *lutea*) cells initiated in our laboratory, have studied the levels of activity of a number of enzymes at various stages in the growth cycle of cells cultured without added auxin and with addition of α-naphthaleneacetic acid (NAA) and with addition of α-naphthoxyacetic acid. NAA was shown to enhance strongly the activity of most of the enzymes studied (particularly of aldolase (Fig. 13a) and aminopeptidase (Fig. 13b) and glutamate-oxaloacetate transaminase during the phase of rapid cell division and of ribonuclease activity (Fig. 13c) in stationary phase cells).

An example more directly related to the problem of initiating differentiation are studies on the influence of sucrose concentration and additions of a synthetic auxin (2,4-D) and a synthetic cytokinin (kinetin) on growth and lignin production in cultured sycamore cells (Carceller *et al.*, 1971). Cell wall material and lignin as a percentage of cell dry weight increase as the concentration of sucrose was raised from 2 to 15% in basic medium. Lignin content was increased further when the kinetin(K) level was raised (from 0·25 to 10 mg/l) either in the absence of 2,4-D or in the presence of an enhanced level of this auxin (10 mg/l instead of 1·0 mg/l) (Table II). Lignin synthesis and the synthesis of extracellular polysaccharide (Street *et al.*, 1968) proceed most actively during the phase of rapid cell division and the hormone treatments promotive of highest lignin formation also increase cellular aggregation. Studies of this kind will be further commented upon in Section V of this chapter.

The cultures so far discussed have been initiated at high cell densities (in the case of the sycamore suspension at an initial density of *ca* 2×10^5 cells/ml) and have in consequence only shown a transient period of exponential growth. It is, however, possible to initiate successful cultures in the minimal media from a density of *ca* 2×10^4 cells/ml and in a modified synthetic media (Stuart and Street, 1971) from a density of *ca* 5×10^3 cells/ml. When cultures are initiated at these low densities from mature, stationary phase cells (28 day old cultures) the lag phase is lengthened and the culture shows evidence of undergoing a sequence of synchronous cell divisions (Fig. 14). Such cultures would appear to offer unique material for the study of the metabolic events of the cell cycle. We have, however, yet to characterize adequately their synchrony. Preliminary observations, as illustrated in Fig. 15 by the activities of two enzymes, are suggestive of a synchrony of metabolism linked to the cell cycle as monitored by cell counting.

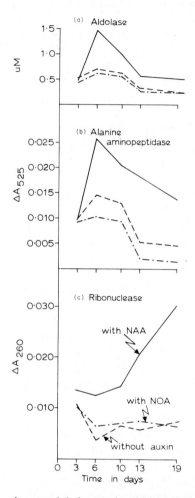

FIG. 13. Influence of the auxins α-naphthyleneacetic acid (NAA) (2 mg/l) and α-naphthoxy-acetic acid (NOA) (4 mg/l) on the activities of various enzymes during the progress of growth in a batch-propagated suspension culture of *Atropa belladonna* cv. *lutea* Doll. Key to (a) and (b) as for (c). Activities related to mg dry weight of cells. Enzyme assays as described by Simola and Sopanen (1970) (data of Simola and Sopanen, 1971).

TABLE II

Growth and composition of sycamore cells grown in media containing 15% sucrose and various levels of 2,4-D and kinetin (data from Carceller *et al.*, 1971)

2,4-D mg/l	Kinetin mg/l	Cell no. 10^6 ml^{-1}	Cell dry weight (mg) per 10^6 cells	Extractive-free weight[a] (mg) per 10^6 cells	Lignin by 2,6-dichloroquinone[b] as % extractive free weight
1·0	0·25	1·9	7·3	2·9	24·6
0	10·0	1·7	4·2	2·0	50·8
10·0	10·0	1·9	3·4	1·8	62·9

[a] Cell residue after extracting with water in a boiling water bath, followed by ethanol/benzene (1:2) extraction of the dry residue (as Thornber and Northcote, 1961).
[b] Stafford, 1960.

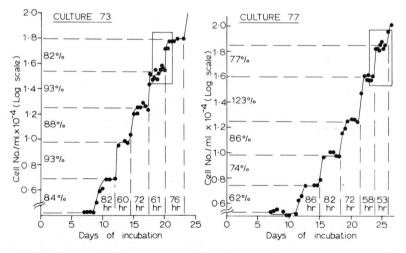

FIG. 14. Synchrony of cell division observed in sycamore cell suspension cultures initiated from low initial cell density with 28 d stationary phase cells. % increase in cell number at each division shown against log scale of cell number ml^{-1}. Duration of each cell cycle shown against scale of days of incubation. Note extended lag phase before the first division (previously unpublished data of P. J. King).

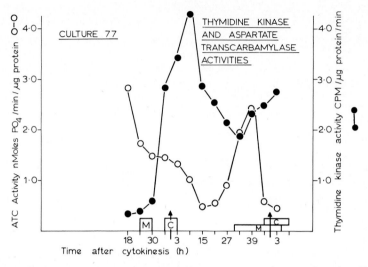

FIG. 15. Changes in the activities of thymidine kinase and aspartate transcarbamylase (ATC) during the cell cycle enclosed in the rectangle for Culture 77 in Fig. 14. C = duration of cytokinesis from the cell count data. M = duration of mitosis from mitotic index data. Thymidine kinase assayed by method of Hirraga, Igarashi and Yana (1967) and aspartate transcarbamylase by method of Shepardson and Pardee (1960) (previously unpublished data of M. W. Fowler).

IV. CONTINUOUS CULTURE SYSTEMS AND THE ACHIEVEMENT OF STEADY STATES OF GROWTH AND METABOLISM

A steady state (a state of balanced growth), where the culture has not only a constant specific growth rate (μ) (relative growth rate) but the cells retain a constant average composition and metabolic activity, does not occur in suspension cultures of sycamore cells initiated at their normal density (despite their short-lived period of exponential growth—see Fig. 2). With cultures initiated at much lower densities the duration of any such steady state, if it can be shown to occur, will be strictly limited in duration because it is occurring simultaneously with progressive nutrient depletion. In pursuit of stable steady states, the 4·5 l batch culture vessel, previously mentioned, can readily be made into the culture vessel of an open continuous system (Wilson *et al.*, 1971). Such a system has enabled us to test to what extent the growth kinetics of a sycamore cell suspension conform to the theory developed by Monod and by Novick and Sziland for the chemostat propagation of microorganisms (Málek and Fencl, 1966). We have found in such a system, where growth is not limited by O_2 availability, temperature or pH, that a fixed flow rate of new medium (and balancing harvest of culture) results in a stable population of dividing cells with a constant biomass consisting of cells of constant composition and constant general metabolic state. We have arrived at a dynamic equilibrium between the dilution rate [D] (Volume of new medium day^{-1}/

culture volume) and the specific growth rate (μ) of the population. Data from one such steady state experiment, achieved with a sycamore cell suspension in chemostat culture, are shown in Fig. 16. This steady state was studied over 312 h. The flow rate was 0·194 culture volumes per d^{-1} and hence the mean generation time (g) of the culture was ca 85 h (μ is numerically equal to the dilution rate and related to g by the equation $\mu = \log_e 2/g = 0·69/g$). A further important characteristic of a steady state culture demonstrated here is the establishment of steady state equilibrium concentrations of nutrients. Steady state levels of enzyme activity can also be demonstrated in such steady state cultures. Data showing the changing levels of nitrate reductase activity recorded at intervals during the growth of a batch culture and the constant levels of activity of this enzyme in a chemostat are shown in Fig. 17. A rapid transient change in enzyme activity occasioned by a change in dilution rate in the chemostat is also illustrated. The absence of detectable nitrate in the medium during the steady state and the fact that following change in dilution rate there is a rise in the cellular nitrate level before nitrate can be detected in the medium both indicate the high affinity of the cells for nitrate. During the batch culture the levels of assayed enzyme activity correspond well with the calculated rates of nitrate assimilation by the cells. In the continuous culture the enzyme activity appears to be significantly in excess (about 3-fold excess) of that required for assimilation of the nitrate.

Steady state cultures with *different* specific growth rates can be obtained by changing the dilution rate (Fig. 18) (each point represents the mean of the data collected at intervals through a steady state for each parameter). This strongly suggests that the growth rate of the cells is being regulated by the steady state level of a single limiting nutrient present in the inflowing medium and we are currently investigating its identity in our basal medium. The steady state population density declines with increased dilution rate; a smaller proportion of cells is supported growing a faster rate. This suggests that the cells have a relatively low affinity for the limiting nutrient.

The trends shown in Fig. 18 indicate that there is an upper limit to the dilution rate compatible with a stable population density of cells even though at higher dilution rates availability of any limiting nutrient is high and the cell population low. The decline in population density with increase in dilution rate is not linear but becomes progressively steeper as this critical dilution rate is approached. The data in Fig. 19 are from a culture being diluted at a rate of 0·274 d^{-1}. No stable population is produced. Instead the cells are undergoing progressive wash-out at a rate of 0·038 d^{-1}. From this wash-out rate it can be calculated that the critical dilution rate $= 0·236$ d^{-1}. This corresponds to a minimum mean generation time of 70 h. This minimum mean generation time agrees with growth rates recorded in open continuous culture of the turbidostat type previously reported (Wilson *et al.*, 1971) and with growth rates during exponential periods of growth in batch cultures in the same basal medium. Growth rate of cells in this condition is independent of the availabilit of

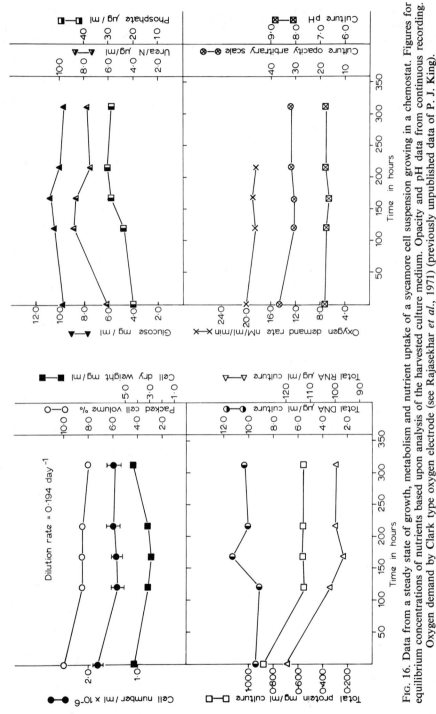

Fig. 16. Data from a steady state of growth, metabolism and nutrient uptake of a sycamore cell suspension growing in a chemostat. Figures for equilibrium concentrations of nutrients based upon analysis of the harvested culture medium. Opacity and pH data from continuous recording. Oxygen demand by Clark type oxygen electrode (see Rajasekhar et al., 1971) (previously unpublished data of P. J. King).

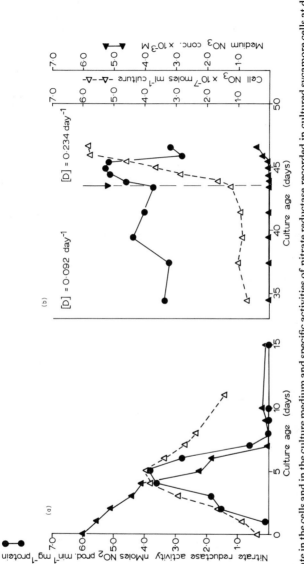

FIG. 17. Levels of nitrate in the cells and in the culture medium and specific activities of nitrate reductase recorded in cultured sycamore cells at different stages of growth in batch culture (a) and in a steady state (b). Nitrate reductase was assayed by monitoring nitrite formation under anaerobic conditions in the presence of NADH and benzylviologen (previously unpublished data of M. Young).

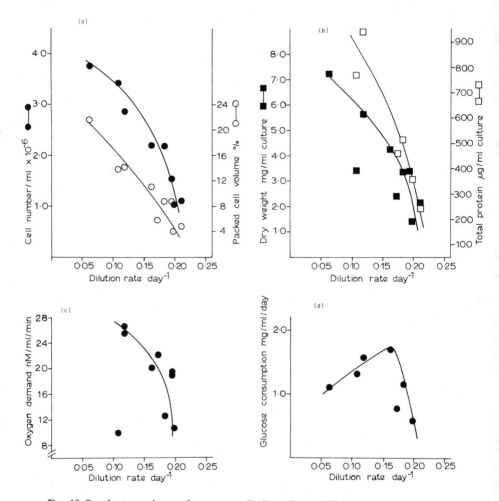

FIG. 18. Steady state cultures of sycamore cells. Data from 9 different steady states covering a range of dilution rates and established in 5 different chemostat cultures. Each point represents the mean of a parameter measured at intervals during each steady state. The standard errors of these means have been calculated but are not shown in the Figure; these standard errors did not exceed 5–10% of the mean values. Note occasional points outside of the normal trend occurring with change in dilution rate (previously unpublished data of P. J. King).

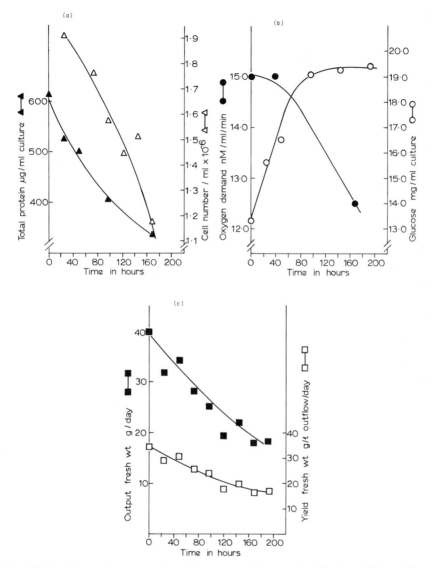

FIG. 19. Data from a chemostat culture of sycamore cells undergoing "wash-out" by application of a dilution rate (0.274 d^{-1}) in excess of the critical dilution rate. The wash-out rate is 0.038 d^{-1} indicating a value for maximum specific growth rate (μ_{max}) of *ca* 0.236 d^{-1} (corresponding to a mean generation time of 70 h) (previously unpublished data of P. J. King).

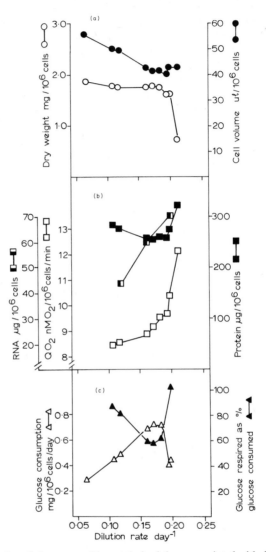

FIG. 20. Differences in cellular composition and physiology associated with different steady states of chemostat cultured sycamore cells. Values calculated from the trend lines in Fig. 17 (previously unpublished data of P. J. King).

nutrients supplied by the culture medium and is presumably regulated by some intrinsic metabolic parameter. Thus the chemostat system allows the stabilization and investigation of the growth-limiting conditions which produce growth rates within the growth rate range from the end of the exponential phase through to the beginning of stationary phase in batch cultures (see Fig. 1).

If, by reference to the trend lines in Fig. 18, growth parameters are calculated on a per cell basis, the data indicates that the cells in the different steady states differ significantly in their composition and physiology (Fig. 20). There is a decline in dry weight per cell as growth rate increases and this is associated with a decline in cell volume at the high growth rates (Fig.20a). Q_{O_2} and RNA values rise progressively with growth rate (Fig. 20b); protein content per cell increases at the highest dilution rate tested. Glucose consumption reaches a peak at an intermediate growth rate (Fig. 20c) and at this peak of glucose consumption the % of the glucose respired is at a minimum (Fig. 20c).

If in fact cell populations of distinctly differing metabolism can be selected simply by altering the potentiometer on a metering pump, it may be that the "special cultural environments needed to elicit the activity of particular metabolic sequences" can be achieved in a chemostat system. This idea is illustrated by the changes in protein yield in a chemostat culture of sycamore cells shown in Fig. 21. If, for example, total protein was the desired product, then cultures established at intermediate growth rates would be economically most productive. However, if a specific quality of protein was required this might call for some other dilution rate.

With a particular cell line and composition of culture medium the biomass of a steady state chemostat culture should be reproducible on different occasions and predictable from the dilution rate. We have not found this to be entirely so (Table III). Two of the steady states quoted in this table illustrate

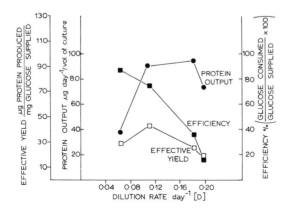

FIG. 21. Yield data from four steady states of a sycamore cell suspension culture established in a chemostat (previously unpublished data of P. J. King).

TABLE III

Data for steady states obtained at similar dilution rates in cultures established at different times (previously unpublished data of P. J. King)[a]

Culture no.	54	78	90
D (days^{-1})	0·182	0·189	0·194
g.(h)	90	87	85
Properties of culture			
P.C.V. %	8·7 ± 0·53 (10)	13·0 ± 0·35 (10)	8·7 ± 0·34 (5)
Cell no. × 10^{-6}.ml^{-1}	2·19 ± 0·06 (9)	3·62 ± 0·05 (11)	1·53 ± 0·07 (5)
Dry wt mg.ml^{-1}	3·33 + 0·16 (10)	6·20 ± 0·15 (11)	3·38 ± 0·27 (5)
Total protein μg.ml^{-1}	513 ± 34·0 (8)	1136 ± 16·0 (10)	556 × 1·9 (5)
Total DNA μg.ml^{-1}	—	19·3 ± 0·84 (10)	10·04 ± 0·35 (5)
Total RNA μg.ml^{-1}	—	169·1 ± 1·9 (10)	103·5 ± 3·7 (5)
O$_2$ demand nM.ml^{-1} min^{-1}	12·5 × 1·0 (7)	35·3 ± 2·4 (8)	19·0 ± 1·4 (5)
Medium constituent levels			
Glucose mg.ml^{-1}	13·7 ± 0·39 (9)	7·2 ± 0·22 (11)	10·1 ± 0·20 (5)
Nitrate/N μg.ml^{-1}	—	0·5 ± 0·1 (10)	8·0 ± 0·68 (3)
Urea/N μg.ml^{-1}	—	3·7 ± 0·39 (11)	7·8 ± 0·43 (5)
Phosphate μg.ml^{-1}	—	2·0 ± 0·28 (11)	26·8 ± 2·0 (5)
Yield of protein (moles protein N per mole nutrient)			
Glucose	0·146	0·161	0·101
Nitrate/N	—	1·63	0·86
Urea/N	—	1·77	0·91
Phosphate/PO$_4$	—	13·9	9·5

[a] Standard error of means and in brackets number of samples examined during the steady state shown against each value. D = dilution rate in culture volumes. g = mean generation time.

the normal degree of reproducibility (Culture 54 and Culture 90). The variation between these is such as might be expected from failure to reproduce exactly the conditions of culture. By contrast with these steady states is shown a third state obtained on a further occasion at a similar dilution rate (Culture 78). This latter culture has a higher biomass and consumes more nutrients and produces a higher yield of protein per unit of each nutrient monitored. The cells of this culture are more efficient in converting supplied nutrients into cellular material. The recognition that such cultures are encountered naturally leads on to the final sections of this Chapter which are concerned with gene expression and genetic variants in cultured plant cells.

V. GENE EXPRESSION IN PLANT CELL CULTURES

This subject was reviewed in some detail by Krikorian and Steward (1969). These workers attempted to answer two important questions: is it possible to reproduce at will the particular biochemistry of any particular organ, tissue or cell? Can one induce potentially totipotent cells to express the biochemistry that they normally achieve in a given morphological setting, without the necessity of reproducing that setting intact by the growth of the whole organism? Their broad conclusion was that it is very difficult to cause cultured cells to recapitulate, in isolation, the metabolism they exhibit in the intact plant. Only when we understand more completely the whole plant system will it become possible to subject such isolated cells to the complex stimuli which they experience when they are part of a differentiated organism.

The cells of suspension cultures do not correspond exactly in morphology to any of the types of tissue cell of the plant body. They could be described under the general heading of parenchyma cells and when growing exponentially at close to their maximum growth rate they clearly resemble cells recently initiated from apical meristems and which "*in situ*" normally undergo further divisions as they differentiate into cells of the ground tissues of plant axes. They cannot yet be shunted into a normal pathway of differentiation. The nearest approach to the appearance in suspensions of a differentiated tissue cell type is seen when certain suspensions are induced to turn green when cultured in moderately high light intensity and come to contain many chlorenchyma-like cells containing chloroplasts. However, in our experience (Davey *et al.*, 1971) the callus masses of the same culture strain always contain higher levels of chloroplast pigments than suspensions cultured under similar conditions of illumination and even very green callus masses contain less pigment and less chloroplasts per cell than leaf cells of the species and are only capable of relatively low photosynthetic activity. Callus cultures capable of sustained growth in the absence of a supply of sugar have not yet been obtained (Hildebrandt *et al.*, 1963; Wilmar *et al.*, 1964; Vasil and Hildebrandt, 1965). It has been found that high levels of carbohydrate and particularly of sucrose can markedly inhibit chloroplast differentiation (Davey *et al.*, 1971; Edelman and Hanson, 1971a, b). Chloroplast development is also sensitive to the kind and level of auxin (Bergmann, 1967; Davey *et al.*, 1971). There however remains some unidentified aspect of the culture environment preventing the uniform development of functional chloroplasts and the development of the required number of such chloroplasts per cell to render the cells autotrophic. The situation therefore conforms to the conclusion of Krikorian and Steward although this particular aspect of differentiation proceeds in culture to such a promising extent as to suggest that only perhaps a single factor might have to be changed for chloroplast differentiation to be fully achieved in culture.

The observation that callus greens better than the derived suspension and that green suspensions are always rather highly aggregated suggests that cell association may be important in achieving the necessary environment for chloroplast development. Cell association is clearly important for other aspects of differentiation; callus cultures often show vascularization and other specialized cells and the tracheid-like cells observed in some cell suspensions arise in cell aggregates. Cell aggregates, whenever they are built up of more than a few cells, always show some differentiation into peripheral and central cells (Thomas *et al.*, 1972), the peripheral cells often being smaller and dividing more frequently than the central cells. Central cells can show more obvious morphological differentiation, e.g. into tracheid-like cells or sieve-tube like structures. Within aggregates there can arise organized growing points capable of developing roots and shoots. Superficial cells of aggregates can give rise to embryoids by a sequence closely paralleling the development of the zygotic embryo. It is very important to bear this in mind in considering claims for the synthesis of particular secondary plant products by callus and suspension cultures. Callus cultures of common rue (*Ruta graveolens*) synthesize volatile oil; this is the only case known of volatile oil synthesis by a cultured tissue (Corduan and Reinhard, 1972). However this callus is a highly organized structure; the oil occurs in schizogenous passages just below the surface of the callus and these canals closely resemble those in the intact plant. Unorganized cultures of *Scopolia parviflora* (Tabata *et al.*, 1972) and of *Atropa belladonna* (Raj Bhandary *et al.*, 1969; Thomas and Street, 1970) do not produce detectable amounts of the main tropane alkaloids of the species. However, as soon as root primordia are initiated in the aggregates of the suspension cultures, these alkaloids are synthesized. Where growth regulating hormones apparently invoke or suppress secondary product synthesis it is important to examine how far their effects are mediated by changes in culture organization. Thus in a recent paper (Tabata *et al.*, 1971) it is reported that kinetin in the absence of auxin promotes nicotine synthesis in tobacco callus. However it is clear that nicotine synthesis in such cultures is almost exactly paralleled by an increase in the number of shoot buds initiated per culture. Similarly inhibition of nicotine synthesis in these cultures by 2,4-D was associated with suppression of bud formation. These are typical of many culture systems where secondary product formation depends upon organization creating the necessary microenvironment for the appropriate biochemical differentiation.

A recent review (Puhan and Martin, 1971) emphasizes that a wide range of secondary plant products have now been detected in plant cultures (mainly callus cultures) but at levels far below these occurring in the whole organism. It would clearly be of great interest to examine, in such systems, whether the low rate of synthesis of the particular metabolites per unit of biomass is because their synthesis is occurring in only a few cells activated by being at a particular site within the callus or in the floating aggregates of the suspension cultures or whether it results from a very slow rate of synthesis (or active degradation) in the whole cell population.

Despite the general importance of organization in relation to secondary product formation there are a few reports of the production of significant quantities of particular secondary metabolites by cultures in which apparently no obviously specialized cells are formed and in which neither organ initiation nor embryogenesis are occurring. Examples of such cultures are those of

Panax ginseng producing panaxatriol (Furuya *et al.*, 1970), those of various species of *Dioscorea* (Mehta and Staba, 1970) and of *Solanum* (Vágujfalvi *et al.*, 1971) producing diosgenin and of cultures of various species producing anthocyanins (Straus, 1960; Blakely and Steward, 1964; Reinert *et al.*, 1964; Ibrahim *et al.*, 1971). There are also several reports of the active synthesis in culture of secondary products not formed in detectable amounts in the whole organism. Thus Butcher and Connolly (1971) detected three sesquiterpenes (paniculides) in callus of *Andrographis paniculata* although the three typical diterpenes (andrographides) characteristic of the whole plants were absent from the cultured tissue. Other workers have shown the capacity of plant cell culture to effect particular biotransformations. Thus suspension cultures of tobacco and of *Sophora angustifolia* convert progesterone to 5-α-pregnanolone and 5-α-pregnanolone palmitate and convert pregnenolone to pregnenolone palmitate and 5-α-pregnanolone palmitate (Furuya *et al.*, 1971). Along somewhat similar lines P. How and A. R. Macrae (unpublished work at Unilever Research Laboratories) have shown that rose suspension cultures not only fail to produce the monoterpenes characteristic of rose flowers but rapidly metabolize these terpenes, for instance rapidly converting geraniol and geranyl pyrophosphate to geranial, neral, nerol and citronellal.

It should perhaps be emphasized that hitherto the whole emphasis in plant cell culture research has been towards achieving rapid culture growth in defined media and that such conditions may be antagonistic at least to certain pathways of cellular differentiation. Earlier, attention was drawn to the observation of Nash and Davies (1972) that their cultures accumulated phenolic materials very late in the growth cycle when cell division was ceasing. Ogutuga and Northcote (1970) reached a similar conclusion regarding caffeine production by their cultures of tea (*Camellia sinensis*). Similarly, in studies on the synthesis of chlorophyll in suspension of *Atropa belladonna* cells it was shown that the chlorophyll content fell during the stage of initiation of culture growth and later continued to rise linearly after growth of the culture, as measured by cell density and cell dry weight, had ceased (Fig. 22). Many papers reporting the production of secondary products by callus cultures have involved chemical analysis after periods in culture clearly in excess of the predictable duration of active growth. These considerations suggest that more work should now be undertaken to study the metabolic activities which can be invoked in "maintained" cultures, cultures in stationary phase prevented from senescence and cell death by appropriately controlled nutrient supply. Certain observations have also pointed to the possibility of evoking a particular aspect of culture growth by applying sequential treatments (Kent and Steward, 1965). This has been particularly demonstrated in relation to the induction of organogenesis in asparagus (*Asparagus officinalis*) cultures (Steward *et al.*, 1967) but may equally be applicable to the induction of biosynthetic sequences. Fosket and Torrey (1969) have also stressed the importance of the cultural treatment during the cell division phase for subsequent cellular differentiation. These considerations suggest that various specialized aspects of metabolism might be activated by the approach of first establishing conditions promotive of growth but involving various applications of growth-regulating substances, followed by conditions capable of "maintaining" the stationary culture and also combined with the application of different combinations of growth regulators. The very empirical (hit and miss) nature of this approach presents a rather daunting prospect to the investigator and leads us on to consider whether the suppression or activation of particular metabolic pathways in cultured cells can be approached in any other way.

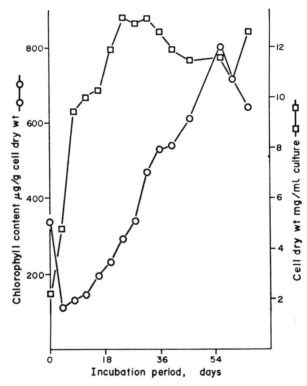

FIG. 22. Relationship between growth (expressed in terms of cell dry wt) and chlorophyll content for a cell line of *Atropa belladonna* cv. *lutea* Doll. grown in light in batch culture (data from Davey, Fowler and Street, 1971).

VI. THE ISOLATION OF GENETIC VARIANTS OF CULTURED CELLS

By obtaining suspensions of free cells either by an appropriate filtration or enzyme treatment (already described) of cell suspension cultures and by plating these out on agar medium in petri dishes (Bergmann, 1960) it is possible to obtain cell colonies of single cell origin. A good percentage yield (plating efficiency) of such clones from low density platings can now be achieved by using media previously "conditioned" by having previously supported the growth of cells (Stuart and Street, 1969) or synthetic media elaborated on the basis of chemical analysis of conditioned media (Stuart and Street, 1971). Such single cell clones, when isolated from established cell structures, have been shown to differ from one another indicating the presence of variants of various kinds in the stock culture cell population. These differences may be expressed in differences in growth rate on minimal media, in friability, and cell morphology, and in various biochemical properties. As examples of the biochemical differences observed between such clones may be cited their capacity to form tracheid-like cells in culture (Earle and Torrey, 1965), their ability to utilize different carbohydrates (Sievert and Hildebrandt, 1965), their anthocyanin content (Nickell and Tulecke, 1959; Blakely and Steward, 1961; Eriksson, 1967), their capacity for chloroplast differentiation (Davey et al., 1971), and their resistance to acriflavin (Blakely and Steward, 1964) and to various antimetabolites (unpublished data). Only a beginning has yet been made to expose the full range of such biochemical variants developing in continuously propagated plant cell cultures.

A new and even more promising approach to the isolation of variants has been opened up by the demonstration that haploid cell lines can be obtained from pollen cultures of various species (Sunderland, 1971). Enhanced mutagenesis in haploid cell suspensions by appropriate mutagen treatment, followed by mutant selection and diploidization with colchicine, should make available a range of homozygous mutant cell lines.

If the opening up of particular secondary biochemical pathways would follow inactivation of particular repressor molecules (say by accumulation of particular primary metabolites), or the inactivation of the regulator genes responsible for synthesis of the repressors, then appropriate mutant cells might be expected to express these pathways without having to be subjected to the very special microenvironments needed for their activation in normal cells. Further experience from work with microorganisms would suggest that once a pathway is opened up it can be intensified both by modification of cultural conditions and by further selection of mutants. The availability of mutants of this kind in which the biochemical sequences affected are those of pathways unique to plant cells would not only greatly advance our knowledge of plant metabolism and its regulation but could well make plant cells into industrial "organisms" of considerable importance. In the present state of our knowledge, this approach may therefore be more rewarding than trying to reproduce in culture apparatus and in an appropriately timed sequence the complex of interacting factors which operate to control differentiation in developing plant organs.

ACKNOWLEDGEMENTS

Grateful acknowledgement is made to present and former research colleagues for permission to quote data from their published work. I am particularly indebted to P. J. King, K. J. Mansfield, M. W. Fowler, M. Young, P. How and A. R. Macrae for permission to present their previously unpublished data or refer to their unpublished findings and to Mrs V. J. Lambert for expert technical assistance. Part of the work reported here was undertaken with the aid of an S.R.C. Research Grant to one of us (H. E. S.). One of us (P. J. K.) was able to participate in the work through a C. A. P. S. award of the S.R.C. sponsored by Unilever Ltd. Text figures have been kindly prepared by Miss S. Pearcey and Mr G. G. Asquith.

REFERENCES

Bergmann, L. (1960). *J. gen. Physiol.* **43**, 841–851.

Bergmann, L. (1967). *Planta* **74**, 243–249.

Blakely, L. M. and Steward, F. C. (1961). *Am. J. Bot.* **48**, 351–358.

Blakely, L. M. and Steward, F. C. (1964). *Am. J. Bot.* **51**, 809–819.

Brown, E. G. and Short, K. C. (1969). *Phytochemistry* **8**, 1365–1372.

Butcher, D. N. and Connolly, J. D. (1971). *J. exp. Bot.* **22**, 314–322.

Carceller, M., Davey, M. R., Fowler, M. W. and Street, H. E. (1971). *Photoplasma* **73**, 367–386.

Copping, L. G. and Street, H. E. (1972). *Physiologia Pl.* **26**, 346–354.

Corduan, G. and Reinhard, E. (1972). *Phytochemistry* **11**, 917–822.

Cox, B. J., Turnock, G. and Street, H. E. (1973). *J. exp. Bot.* **24**.

Davey, M. R., Fowler, M. W. and Street, H. E. (1971). *Phytochemistry* **10**, 2559–2575.

Davey, M. R. and Street, H. E. (1971). *J. exp. Bot.* **22**, 90–95.

De Jong, D. W., Jansen, E. F. and Olson, A. C. (1967). *Expl Cell Res.* **47**, 139–156.

Earle, E. D. and Torrey, J. G. (1965). *Pl. Physiol., Lancaster* **40**, 520–528.

Edelman, J. and Hanson, A. D. (1971a). *Planta* **98**, 150–156.

Edelman, J. and Hanson, A. D. (1971b). *Planta* **101**, 122–132.

Eriksson, T. (1967). *Physiologia Pl.* **20**, 507–518.

Filner, P. (1965). *Expl Cell Res.* **39**, 33–39.

Fosket, D. E. and Torrey, J. G. (1969). *Pl. Physiol., Lancaster* **44**, 871–880.

Fowler, M. W. (1971). *J. exp. Bot.* **22**, 715–724.

Furuya, T., Hirotani, M. and Kawaguchi, K. (1971). *Phytochemistry* **10**, 1013–1017.

Furuya, T., Kojima, H., Syono, K. and Ishii, T. (1970). *Chem. Pharm. Bull.* **18**, 2371–2372.

Givan, C. V. and Collin, H. A. (1967). *J. exp. Bot.* **18**, 321–331.

Henshaw, G. G., Jha, K. K., Mehta, A. R., Shakeshaft, D. J. and Street, H. E. (1966). *J. exp. Bot.* **17**, 362–377.

Hildebrandt, A. C., Wilmar, J. C., Johns, H. and Riker, A. J. (1963). *Am. J. Bot.* **50**, 248–254.

Hirraga, S., Igarashi, K., and Yana, T. (1967). *Biochim. biophys. Acta.* **145**, 41–51.

Ibrahim, R. K., Thakur, M. L. and Permanand, B. (1971). *J. nat. Products (Lloydia)* **34**, 175–182.

Kent, A. E. and Steward, F. C. (1965). *Am. J. Bot.* **52**, 619.

Krikorian, A. D. and Steward, F. C. (1969). *In* "Plant Physiology" (F. C. Steward, ed.), Vol. 5b. pp. 227–326, Academic Press, New York and London.

Liau, D. F. and Boll, W. G. (1971). *Can. J. Bot.* **49**, 1131–1139.

Mackenzie, I. A. and Street, H. E. (1970). *J. exp. Bot.* **21**, 824–834.

Málek, I. and Fencl, Z. (Eds) (1966). "Theoretical and Methodological Basis of Continuous Culture of Microorganisms." Publ. House Czechoslovak. Acad. Sci., Prague.

Mehta, A. R. and Staba, E. J. (1970). *J. Pharm. Sci.* **59**, 864–865.

Nash, D. T. and Davies, M. E. (1972). *J. exp. Bot.* **23**, 75–91.

Nickell, L. G. and Tulecke, W. (1959). *Bot. Gaz.* **120**, 245–250.

Ogutuga, D. B. A. and Northcote, D. H. (1970). *J. exp. Bot.* **21**, 258–273.
Pette, D., Luh, W. and Bucher, T. L. (1962). *Biochem. biophys. Res. Commun.* **7**, 414–419.
Puhan, Z. and Martin, S. M. (1971). *Progr. Industrial Microbiol.* **9**, 13–39.
Raj Bhandary, S. B., Collin, H. A., Thomas, E. and Street, H. E. (1969). *Ann. Bot.* **33**, 647–656.
Rajashekhar, E. W., Edwards, M., Wilson, S. B. and Street, H. E. (1971). *J. exp. Bot.* **22**, 107–117.
Reinert, J., Clauss, H. and Ardenne, R. V. (1964). *Naturwissenschaften* **51**, 87.
Shepardson, M. and Pardee, A. B. (1960). *J. biol. Chem.* **235**, 32–33.
Short, K. C., Brown, E. G. and Street, H. E. (1969). *J. exp. Bot.* **20**, 579–590.
Sievert, R. C. and Hildebrandt, A. C. (1965). *Am. J. Bot.* **52**, 742–750.
Simola, L. K. and Sopanen, T. (1970). *Physiologia Pl.* **23**, 1212–1222.
Simola, L. K. and Sopanen, T. (1971). *Physiologia Pl.* **28**, 8–15.
Stafford, H. A. (1960). *Pl. Physiol., Lancaster* **35**, 108–114.
Steward, F. C., Kent, A. E. and Mapes, M. O. (1967). *Ann. N.Y. Acad. Sci.* **144**, 326–334.
Straus, J. (1960). *Pl. Physiol., Lancaster.* **35**, 645–650.
Street, H. E. (1969). *In* "Plant Physiology" (F. C. Steward, ed.), Vol. 5b, pp. 3–324. Academic Press, New York and London.
Street, H. E., Collin, H. A., Short, K. and Simpkins, I. (1968). *In* "Biochemistry and Physiology of Plant Growth Substances" (F. Wightman and G. Setterfield, eds) pp. 489–504. Runge Press Ltd., Ottawa.
Street, H. E. and Henshaw, G. G. (1966). *In* "Cells and Tissues in Culture" (E. N. Wilmer, ed.), Vol. 3, pp. 459–532. Academic Press, London and New York.
Street, H. E., King, P. J. and Mansfield, K. J. (1971). *In* "Les Cultures de Tissus de Plantes", pp. 17–40. C.N.R.S., Paris.
Stuart, R. and Street, H. E. (1969). *J. exp. Bot.* **20**, 556–571.
Stuart, R. and Street, H. E. (1971). *J. exp. Bot.* **22**, 96–106.
Sunderland, N. (1971). *Sci. Prog. Oxf.* **59**, 527–549.
Sutton-Jones, B. and Street, H. E. (1968). *J. exp. Bot.* **19**, 114–118.
Tabata, M., Yamamoto, H., Hiraoka, N. and Konoshima, M. (1972). *Phytochemistry* **11**, 949–955.
Tabata, M., Yamamoto, H., Hiraoka, N., Marumoto, Y. and Konoshima, M. (1971). *Phytochemistry* **10**, 723–729.
Thomas, E., Konar, R. N. and Street, H. E. (1972). *J. Cell Sci.* **11**, 95–109.
Thomas, E. and Street, H. E. (1970). *Ann. Bot.* **34**, 657–669.
Thornber, J. P. and Northcote, D. H. (1961). *Biochem. J.* **81**, 449–455.
Vágujfalvi, D., Maróti, M. and Tétényi, P. (1971). *Phytochemistry* **10**, 1389–1390.
Vasil, V. and Hildebrandt, A. C. (1965). *Science, N.Y.* **150**, 889.
Wilmar, J. C., Hildebrandt, A. C., and Riker, A. J. (1964). *Nature, Lond.* **202**, 1235.
Wilson, S. B. (1971). *J. exp. Bot.* **22**, 725–734.
Wilson, S. B., King, P. J. and Street, H. E. (1971). *J. exp. Bot.* **22**, 177–207.

CHAPTER 6

Ethylene and Protein Synthesis

DAPHNE J. OSBORNE

A.R.C. Unit of Developmental Botany, Cambridge, England

I. INTRODUCTION

Plants are remarkable amongst living things in that their rate of growth and their type of growth are subject to modifications by the 2-carbon olefinic gas ethylene. Further, their growth is modified in very obvious ways by concentrations in the air of as little as a few parts per million—or even less for many species (Crocker, 1932; Miller *et al.*, 1970).

We now know that this gas is a natural product of all parts of higher plants and I think in this year 1972 there remains no doubt in the minds of plant physiologists that ethylene is one of the naturally produced, endogenous regulators of plant growth and development. Ethylene is, therefore, a *plant hormone*. But since the ethylene production of one plant can affect the growth of its neighbours, the gas is, in many respects, analogous to the volatile regulatory products (pheromones) of insects, so we can also think of ethylene as a *plant pheromone*. One of the most sensitive plants to externally applied (Crocker, 1948), or endogenously produced (Goeschl *et al.*, 1967) ethylene is the dark grown pea seedling and typical growth responses were described in detail by Neljubow as long ago as 1901 (Neljubow, 1901). The present paper reviews some of our more recent information concerning the effects of ethylene on cell growth in shoots of the etiolated pea plant with particular reference to protein metabolism and synthesis and certain enzyme activities in the extending tissues. These results will be integrated with those from other workers in different plant systems. The paper also includes information concerning another type of tissue response in the etiolated pea that may look, at first sight, similar to that induced by ethylene but which we believe to be induced by auxin in a quite different way.

All our experiments have been conducted on the dark grown pea seedling (*Pisum sativum*, var. Alaska) grown in sand at 24°C and exposed only to green light during watering. Six days from sowing, when the third internode is 10 to 20 mm long, the boxes of plants were either (a) transferred to large glass tanks (20–40 l), sealed, and ethylene gas introduced to the appropriate concentration; (b) sprayed carefully with an aqueous solution of indole-3-acetic acid (IAA) 3×10^{-3} M so that the apex was completely wetted; or (c) left as untreated control plants. In some experiments, before the plants were treated, the apical bud was removed at the point of inflection of the hook.

When intact plants are exposed to 100 μl/l ethylene for 24 h extension growth is arrested, the shoot is tightly recurved and the region of expanding cells below the hook is visibly swollen (Fig. 1c). If treatment with ethylene is continued for 4 days a few mm of elongation growth occurs and the whole extending region below the hook shows a marked lateral expansion (Fig. 1d). The plants show increasing effects between 0·1–10 μl/l and respond maximally to concentrations between 10 and 500 μl/l. No toxicity effects are apparent at concentrations as high as 2000 μl/l and on removal from ethylene, normal growth is resumed.

Pea plants sprayed with 3×10^{-3} M IAA (Fig. 1b) look (after 24 h) rather similar to those treated with ethylene for 4 days (Fig. 1d). There are however some specific differences in the visible response. Although elongation growth is somewhat reduced, the hook of auxin treated plants is not recurved and the swelling of the tissue occurs sooner than in ethylene.

If the apical buds of pea plants are removed, their responses to auxin and ethylene are again quite dissimilar. When the cut stump is treated with 0·2% IAA in petroleum jelly it is seen (Fig. 2a) that considerable swelling occurs in the region below within 24 h and increases markedly up to 4 days (Fig. 2b). No swelling occurs if decapitated plants are kept in air or in ethylene. From these visible responses alone it is clear that the auxin induced, and ethylene induced swelling of the expanding cells of pea shoots is differently controlled.

1. Cell Shape and Size

If fully mature swollen tissue from auxin treated or ethylene treated plants is sectioned longitudinally and the transverse area and the length of cortical cells compared with that from the equivalent fully expanded control internodes

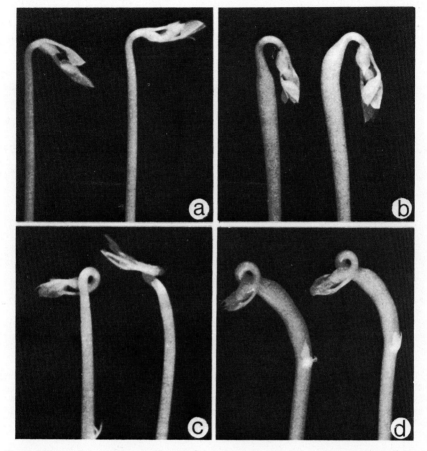

FIG. 1. Effects of an auxin spray (3×10^{-3}M) or ethylene (100 μl/l) on the growth of etiolated pea shoots. a, control in air (enclosed with pellets of KOH to absorb CO_2, and dishes of 0·25 M mercuric perchlorate in 0·25 M perchloric acid (MP) to absorb ethylene). b, 24 h after auxin treatment (+KOH and MP). c, 24 h after ethylene treatment (+KOH). d, 4 d after ethylene treatment (+KOH).

(Table I), further differences between auxin and ethylene treated cells become apparent. Whereas the volumes of ethylene treated cells do not differ significantly from controls, those in the auxin treatment are increased nearly three-fold, and although the volumes in ethylene and control cells are very similar, expansion in ethylene occurs predominantly in the lateral direction so that their shape is short and squat compared with the normally elongated controls (Ridge and Osborne, 1969).

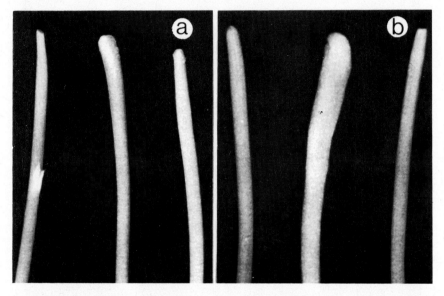

Fig. 2. Decapitated shoots of 6 day old pea plants treated as follows L → R. 2 μl petroleum jelly applied to cut stump; plants exposed to ethylene 100 μl/l (KOH pellets included in the containers—see legend to Fig. 1); 2 μl 0·2% IAA in petroleum jelly in air (KOH and MP included); 2 μl petroleum jelly, control (KOH and MP included). Response after (a) 24 h and (b) 4 d.

2. *Ultrastructural Characteristics of Cell Walls*

Electron micrographs of mature ethylene-swollen tissue or control tissue shows marked differences in cell wall structure (Osborne *et al.*, 1972). The longitudinal wall of an ethylene-treated cortical cell is nearly two and a half times as thick as that of the controls though the transverse walls appear somewhat less thickened (Fig. 3). In contrast, cells from auxin swollen tissue are only slightly thickened at maturity. In time course experiments of cell wall changes induced by auxin and ethylene, J. A. Sargent and D. J. Osborne (in preparation) have measured the thickness of the walls during treatment and found that within 24 h both longitudinal and transverse walls of auxin treated cells may even be thinner than the controls whereas those in ethylene show a 40–60% increase in thickness (Table II).

TABLE I

Dimensions of mature cortical cells from the marked apical region of plants after treatment for 8 d with IAA (10–15 μg in lanolin per plant) or for 5 d with ethylene, 500 μl/l. Control tissue is taken from the fully extended internode at 8 d. Dimensions are given in Projectina Units

	Cell dimension		
	IAA	Control	C_2H_4
Average area (T.S.)	594	108	316
Average length (L.S.)	36·5	57·6	23·0
Average volume	21087	6220	7273
	±2249	±599	±695

Data from I. Ridge and D. J. Osborne, 1969.

Clearly then a cell swelling caused by auxin is fundamentally different from that induced by ethylene and three major differences can be summarized:

(a) Rate of growth is reduced by ethylene compared with that of auxin treated plants and control plants;

(b) Final size of cortical cells is greater in auxin treatments than those in ethylene or the controls;

(c) Wall thickness of cortical cells is increased by ethylene compared with controls but thickness may be initially reduced by an auxin treatment.

II. Protein Metabolism in Ethylene

Following an ethylene treatment (Table III) the levels of total soluble and cell wall proteins per marked segment are lower than controls. The lowered protein levels (per marked segment or per mg wall dry weight) can be recognized within 24 h and are maintained for the whole of the time the plant remains in ethylene. The reduction in protein levels is borne out by the results from incorporation experiments with segments of tissue cut from control or ethylene intact peas. Incorporation into protein is reduced some 30–50% in plants treated with 500 μl/l ethylene, although ethylene does not affect the total uptake of the labelled precursors into the segments (Ridge and Osborne, 1970a).

Although total incorporation and total protein levels are reduced by ethylene, the content of one type of protein is considerably enhanced, namely that which is hydroxyproline-rich. The increase in these proteins together with the activity

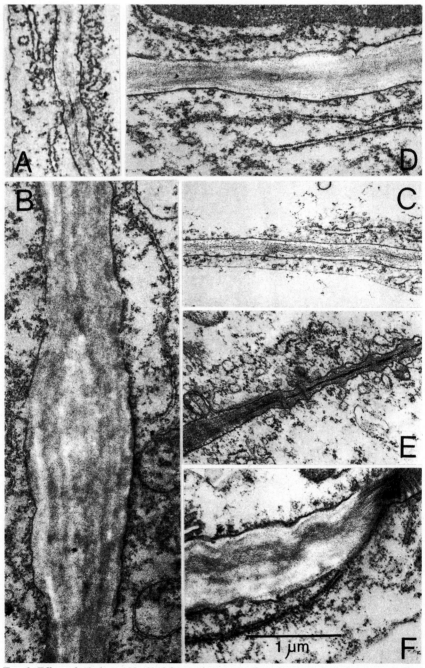

FIG. 3. Effect of ethylene 100 µl/ml for 4 d on walls of cortical cells from extending apical region of etiolated pea shoots. A. L.S. Longitudinal wall. Control. B. L.S. Longitudinal wall. Ethylene. C. L.S. Transverse wall. Control. D. L.S. Transverse wall. Ethylene. E. T.S. Control. F. T.S. Ethylene.

TABLE II

Relative thickness of cortical cell walls after treatment of intact plants with ethylene, 100 μl/l or IAA 3×10^{-3} M. Values are the means for 400–500 walls of the fourth cell from the epidermis in apical swollen tissue

	Treatment				
	IAA			C_2H_4	
Cell wall	Day 1	Day 6	Control	Day 1	Day 4
T.S.	97	133	100	140	207
L.S.	96	136	100	159	242

Data from J. A. Sargent, A. Atack and D. J. Osborne, 1972.

TABLE III

Effect of ethylene on levels of total soluble and cell wall protein in marked 5 mm segments of apical tissue following treatment of intact plants with ethylene. Methods of protein determination and preparation of cell walls as in Ridge and Osborne (1970a, b, 1971)

Ethylene concentration μl/l	Duration of treatment	Control	C_2H_4
		mg Total protein.segment^{-1}	
500	24 h	0·15	0·13
100	24 h	0·21	0·18
		mg Soluble protein.segment^{-1}	
500	24 h	0·085	0·071
50	24 h	0·070	0·064
		mg Protein.100 mg wall dry wt^{-1}	
500	0	8·7	—
	1 d	—	5·5
	3 d	—	6·5
	5 d	—	5·2

I. Ridge and D. J. Osborne, unpublished.

DAPHNE J. OSBORNE

TABLE IV

Peroxidase activity, hydroxyproline and protein levels in cytoplasm and cell walls of control peas and peas treated for 5 d with ethylene 500 μl/l

Fraction	Hydroxyproline (μg.100 μg^{-1} protein)		Peroxidase activity (mg.protein^{-1})		Protein (*mg.g^{-1} fresh weight) (†mg.100 mg^{-1} wall dry weight)	
	Control	C$_2$H$_4$	Control	C$_2$H$_4$	Control	C$_2$H$_4$
Cytoplasm	0·20	0·23	44·8	285·7	16·2*	6·0*
Cell walls	5·01	16·50	3·1	21·2	8·9†	5·8†

Data and methods from I. Ridge and D. J. Osborne, 1971.

of peroxidase enzymes are shown in Table IV. These levels remain unchanged for at least seven days after transfer of plants back to air (Ridge and Osborne, 1970b).

A. THE HYDROXYPROLINE-RICH PROTEINS OF THE CELL WALL

The fractions of the proteins of the cell wall which are hydroxyproline-rich and which are increased by ethylene appear to be covalently linked to some polysaccharide moieties for they are removed by enzymatic treatment (pronase or cellulase) and not by ionic or detergent extraction.

When washed cell walls are freed of ionically bound protein and extracted with cellulase, enzymatic activity is retained in the liberated wall protein. Ridge and Osborne (1971) showed that the separation pattern of these hydroxy-proline-rich proteins on DEAE-cellulose columns corresponded to the separation pattern of the wall bound peroxidase. It is therefore interesting in this regard to note the earlier reports by Shannon et al. (1966) of a hydroxyproline containing isoenzyme of peroxidase from horseradish (*Armoracia rusticana*) and of Shimizu and Morita (1966) for the presence of hydroxyproline in Japanese horseradish peroxidase and a more recent study by Liu and Lamport (1968) describing a hydroxyproline-rich peroxidase in horseradish cell walls which contains an hydroxyproline-*O*-arabinosidic linkage.

The role of hydroxyproline-rich proteins in plant cell walls has attracted considerable attention over recent years. The finding that the levels of these proteins and their attendant peroxidase activities are regulated by ethylene and, as will be seen later, by auxin, arouses further speculation as to the function of such proteins in the control of the growth of cells. It is instructive therefore to review some points in our knowledge of these proteins.

1. *Hydroxyproline-rich Proteins*

They were first demonstrated in cell walls of sycamore (*Acer pseudoplatanus*) callus by Lamport and Northcote (1960) and in tobacco (*Nicotiana tabacum*) callus by Dougall and Shimbayashi (1960). King and Bayley (1965) identified L-4-hydroxyproline in the cell walls of artichoke (*Helianthus tuberosus*) tubers, pea stems and oat (*Avena sativa*) coleoptiles. Cleland and Karlsnes (1967) noted the increase in hydroxyproline protein in pea stem cell walls once the tissue reached full expansion, though little hydroxyproline could be detected in the cytoplasm. From these early reports it appeared that the hydroxyproline-rich proteins were concentrated in the cell wall, probably as structural proteins, with relatively little present elsewhere in the cell.

There has been some dispute about the distribution of the iminoacid between cytoplasmic and wall proteins. Steward and Chang (1963) found hydroxy-proline-rich proteins with hypro/pro ratios varying from 0·53 to 1·28 in the soluble fractions of protein from carrot (*Daucus carota*) root cultures. More recently Puztai *et al.* (1971) found that although 70% of the hydroxyproline in green leaves of broad bean (*Vicia faba*) was associated with the cell wall, 30% was present in the cytoplasm and could be extracted with 6·7% TCA in the cold. The cytoplasmic polysaccharide associated fraction contained heterogenous glycoproteins with hypro/pro ratios varying from 0·3 to 6·0 together with small amounts of hydroxyproline-containing glycolipids. Srivastava (1970) has reported a 3% hydroxyproline content of the F_1 histone of tobacco callus nuclei and results of Thompson *et al.* (1971) show that four of the twelve proline residues in the cytochrome *c* of *Abutilon theophrasti* may be hydroxylated.

Although functional roles for special hydroxylated cytoplasmic proteins has been suggested (Srivastava, 1970), most of the hydroxyproline-rich peptides of the cytoplasm are, it would seem, in transit to the cell wall.

The work of Cleland (1968), of Chrispeels (1969, 1970), Sadava and Chrispeels (1971) and of Dashek (1970) has shown that radioactive proline is first incorporated into peptides within the cytoplasm of the cell. There follows a lag of some ten minutes before radioactive hydroxyproline appears in the cell wall protein. From a series of kinetic studies and "pulse-chase" experiments by these and other workers it is clear that glycosylation of proline occurs after the peptide leaves the ribosome, and appears to take place in association with membranous bodies that separate in density gradients with smooth bodies of the endoplasmic reticulum. The glycosylated peptide is subsequently released through the plasmalemma to the cell wall and there becomes an integral part of the protein-polysaccharide components of the wall structure.

Clearly, ethylene enhances some or all of these processes in the pea stem (Table IV), for not only are the hydroxyproline-rich protein and peroxidase activity of the cell walls increased, small increases are consistently found in the content of hydroxyproline within the cytoplasmic fractions (Table IV).

However, since the total content of protein per unit wall weight is less than that in control tissue of the same age (Tables III and IV) ethylene must stimulate the movement of relatively more polysaccharide than protein to the wall.

2. *Hydroxylation of Protein*

The proline hydroxylase of the cytoplasm of *Acer pseudoplatanus* cultures is present in the supernatant after pelleting of the ribosomes at $105000 \times g$ (Dashek, 1970) and this may be a general feature of plant hydroxylases. By supplying $[^{18}O]O_2$ to peas, it has been shown that molecular oxygen is a requirement for hydroxylation (Cleland, 1968). The enzyme is relatively unspecific for it will hydroxylate chick protocollagen in cell free systems in the presence of Fe^{2+}, ascorbate and α-ketoglutarate, and a chick enzyme will hydroxylate plant prolyl peptides (Sadava and Chrispeels, 1971). However, whereas the plant enzyme can utilize α-ketoglutarate, pyruvate or oxaloacetate, the animal enzyme preferentially uses α-ketoglutarate (Rhoads and Udenfriend, 1968). Further, the animal and plant peptide contents differ considerably for the resulting glycoproteins of animals are glycine-rich whilst those from plants are serine-rich.

The hydroxylation reactions appear to be limited to the cytoplasm, with no further increase of hydroxyproline content occurring once the protein is transferred to the cell wall. When washed cell wall preparations containing $[^{14}C]$proline proteins were incubated in the appropriate media containing Fe^{2+}, ascorbate and α-ketoglutarate the hydroxylation of protein *in situ* could not be enhanced either with or without the addition of soluble cytoplasmic fractions (Ridge and Osborne, 1971). Clearly then, the increase in hydroxyproline-rich protein in the cell walls of pea plants treated with ethylene results from an ethylene-induced change in cytoplasmic turnover of such proteins and/or of a hydroxylation and glycosylation of such proteins and their subsequent release to the cell wall through some "smooth Er"-type body. Blobel and Sabatini (1970) and Sabatini and Blobel (1970) have suggested that the products of membrane bound ribosomes might be transported within the lumen of the endoplasmic reticulum. It is possible therefore that ethylene could regulate hydroxylation of the proline within the endoplasmic reticulum by regulating the binding of ribosomes rather than through the modification of the hydroxylase reaction itself. We have preliminary information (J. A. Sargent and D. J. Osborne, unpublished) that ethylene does regulate changes in the orientation of the endoplasmic reticulum and distribution of associated ribosomes.

B. HYDROXYPROLINE-RICH GLYCOPROTEINS AND CELL GROWTH

Cleland and Karlsnes (1967) in their study of etiolated pea stems showed that the deposition of hydroxyproline-rich protein in walls of pea stems was associated with the reduction of growth rate at maturity. King and Bayley (1965) first reported that the hypro/pro ratio in cell walls increased with age in the excised tissue of artichoke tubers, noting that the ratio rose from 1·6 to 2·3 in 36 h.

Lamport in a series of papers and reviews (Lamport, 1969, 1970) has proposed that the hydroxyproline-rich glycoproteins cross-link the proteins of the cell wall through the hydroxyl group of hydroxyproline in a sugar-ether-linkage to the polysaccharides of the wall matrix. He indicates that the greater the number of cross linkages, the less extensible the wall should become. One molecule of hydroxyproline-rich protein containing in its sequence two hydroxyprolines could in this way cross link two polysaccharide chains (Lamport, 1967). The hydroxyproline-rich glycoproteins can form a large percentage of the protein of the wall, particularly in tissue cultures. In tomato (*Lycopersicon esculentum*) cell cultures the walls may be 50% protein and 50% carbohydrate and Lamport (1969) reports that 20–30% of the amino acid residues of the wall can be hydroxyproline, of which over 90% can be glycosylated.

Enlargement of the cell is eventually determined by the stretching of its wall, and Ridge (in preparation) has shown that both elastic and plastic extensibility of the cell walls of the etiolated pea is reduced after as little as 6 h of ethylene treatment. One may ask, therefore, whether tissue cultures produce high levels of ethylene to account for the relatively high content of hydroxyproline-rich proteins in their cell walls. Mackenzie and Street (1970) have reported that suspension cultures of *Acer pseudoplatanus* do produce ethylene and at the stationary phase of growth (i.e. full cell expansion) the content of ethylene within the culture flasks may rise as high as 10 μl/l.

It seems clear (Lamport and Miller, 1971) that the walls of higher plants contain peptides with contiguous polyhydroxyproline residues linked to arabinose and/or galactose constituents of a carbohydrate moiety. The exact nature of the binding with the cell wall matrix still remains in some doubt. Puztai and Watt (1969) suggest hydroxyproline association with wall pectic materials, Lamport (1967) proposes a covalent linkage with cellulosic or polysaccharide components of the wall matrix. In contrast, Heath and Northcote (1971) have shown that since only peptide bonds and no carbohydrate to carbohydrate bonds are broken by hydrazinolysis of cell walls, despite the fact that 84% of the hydroxyproline is thereby released as small glycopeptides, it is unlikely that the hydroxyproline rich glycopeptide can be linked to any major wall polysaccharide by covalent linkages.

1. Hydroxyproline Glycoproteins and Cell Wall Extensibility

If this class of glycoproteins is of importance in the regulation of cell growth then their distribution in cell walls should reflect the response of the apical stem tissue of the pea to (a) ethylene, when the cells grow more slowly, expand laterally, and have thicker walls, and to (b) auxin, when the cells grow larger, faster and initially have thin cell walls. Both the enzyme activity of the cell wall proteins and their hydroxyproline content should be modified by the concentrations and combinations of ethylene and auxin to which the tissue is exposed. The relatively lower protein to carbohydrate content of cell walls of ethylene treated plants is borne out by the lowered incorporation of [^{14}C]leucine into both cytoplasmic and wall fractions of segments of apical tissue of pea seedlings 24 h after the whole plant is exposed to ethylene. In contrast, auxin enhances the incorporation (Table V). No differences from controls can be detected in the total uptake of precursors by the hormone treatments. Comparable results are obtained for the incorporation of other amino acid precursors into cell wall protein when intact plants treated with ethylene or when decapitated plants are treated with auxin (Table VI). Increased incorporation of amino acids into wall protein in the presence of auxin has been noted also by Kuraishi et al. (1967).

For a regulation of wall extensibility, it might be expected that the ratio of hypro/pro should be modified by the hormone treatments. These ratios can be determined in washed cell walls preparations by hydrolysis of the protein, separation of the amino acids by paper chromatography or electrophoresis, and measurement of the radioactivity in the hydroxyproline and proline spots. Little difference in this ratio is found between control and ethylene treated material incubated with [^{14}C]proline. Since cell walls contain a variety of proteins and glycoproteins the estimation of hypro/pro ratios in total protein could well be too insensitive to reveal changes in any special fractions of ethylene treated tissues after only 24 h. However, the hypro/pro ratio is lower in the walls of auxin treated plants. This result might accord with a greater plasticity and expansion of the auxin treated tissue, and perhaps also with the observed initial thinness of the cell wall (Table II).

It has been shown (Table IV and Ridge and Osborne, 1970b; 1971) that changes in hydroxyproline content of cell walls are reflected in parallel changes in the activity of wall peroxidases. These in turn are a reflection of the activity of peroxidases in the cytoplasm. Estimations of cytoplasmic peroxidases in pea shoots treated with auxin or ethylene show that auxin does not cause a rise in peroxidase activity in the hook and bud region of the shoot and slightly reduces that in the elongating apical segment below (Table VII), whereas ethylene induces large increases in peroxidase activity which are dependent upon the concentration of ethylene supplied (see also Ridge and Osborne, 1970a, b). For the hydroxyproline-rich proteins to function as regulators of cell expansion their levels in cell walls should reflect the ethylene/auxin balance to which the tissue is exposed. So also, the peroxidase activities should parallel the hydroxyproline values. It is seen from Table VIII that little increase occurs in auxin treated tissues after 24 h, though large rises occur with ethylene, but when the auxin and ethylene treatments are combined the increases in peroxidase and hydroxyproline in the walls attributable

TABLE V

Incorporation of [^{14}C]leucine into the protein of cytoplasm and cell walls in 5 mm segments cut from apical tissue of plants exposed for 24 h to ethylene 10 μl/l or 24 h after spraying with IAA 3 × 10^{-3} M. Segments incubated in [^{14}C]leucine for 2 h, then rinsed, homogenized in 1 M NaCl plus 1% Triton X 100 and centrifuged at 2000 × g. Cytoplasmic protein estimated as TCA insoluble material in the supernatant fraction. The 2000 × g pellet was washed and dried to give the cell wall fraction. Radioactivity in both cytoplasmic and wall fractions determined as in Ridge and Osborne (1971)

Protein fraction	cpm.mg^{-1} TCA insoluble material		
	IAA	Control	C$_2$H$_4$
Cytoplasmic fraction	119	93	89
	cpm.mg^{-1} wall dry weight		
Cell wall	276	126	97

D. J. Osborne, unpublished.

TABLE VI

Incorporation of [^{14}C]amino acids into cell wall proteins of 5 mm segments cut from apical tissue of (a) intact plants exposed for 24 h to air or ethylene 100 μl/l and (b) decapitated plants in which the surface of the cut stump was treated for 30 h with plain petroleum jelly or petroleum jelly containing 0·5% IAA. Segments incubated in the [^{14}C]amino acid for 4 h. Washed cell walls prepared as described by Ridge and Osborne (1971), hydrolysed in 6N HCl at 110°C, the amino acids separated by two dimensional chromatography and the radioactive spots cut out and the radioactivity determined by standard scintillation counting procedures

	cts/min.mg^{-1} Wall dry wt			Hypro/pro
Treatment	Lysine	Leucine	Proline	Ratio
(a) Intact plants				
Control	215	140	164	0·20
C$_2$H$_4$	177	84	100	0·19
(b) Decapitated plants				
Control	74	22	17	0·64
IAA	211	111	98	0·53

Data from K. Rowan and D. J. Osborne, unpublished.

TABLE VII

Peroxidase activity of cytoplasmic extracts of sections taken from intact pea plants 20 h after exposure to different concentrations of ethylene or to a spray of IAA 3×10^{-3} M

	Peroxidase per g fresh wt	
Treatment	Hook + bud	Apical segment
*II. Control	35·0	4·9
IAA	35·0	3·0
II. Control	187·2	10·4
C_2H_4 (5 μl/l)	247·5	12·7
C_2H_4 (10 μl/l)	316·1	15·4
III. Control	68·4	3·0
C_2H_4 (500 μl/l)	137·8	5·6

* Different batch of peas from I and III.
Data from J. A. Sargent, A. Atack and D. J. Osborne, 1972.

TABLE VIII

Effect of ethylene 500 μl/l or IAA (0·5% in petroleum jelly) or a combined ethylene and IAA treatment on levels of hydroxyproline and peroxidase activity in cell walls of the apical segments cut from decapitated peas treated for 24 h

	Hydroxyproline, μg.		Peroxidase activity . 100 mg^{-1} wall dry wt
Treatment	100 μg^{-1} wall dry wt	100 μg^{-1} wall protein	
0 h	0·45	5·0	27·3
24 h IAA	0·49	5·5	29·1
C_2H_4	0·85	7·1	81·5
IAA + C_2H_4	0·44	6·4	29·0

Data from I. Ridge and D. J. Osborne, unpublished.

to ethylene are almost abolished. Parallel with these results, the auxin induced swelling of the apical part of these *decapitated* pea shoots is greatly reduced when ethylene is also supplied for the 24 h period of treatment.

III. Conclusions

Ethylene reduces the rate of expansion growth of cells and it is suggestive that the rate of expansion is controlled at least in part by the synthesis of hydroxyproline rich glycopeptides that are secreted with other polysaccharide material through the plasmalemma into the cell wall, thereby enhancing the thickness of the cell wall and also rendering it poorly extensible. In combination, auxin would appear to counteract the effect of ethylene in this respect, for although auxin enhances the synthesis of protein and the content in the cell walls, as well as causing some increase in wall thickness, it reduces the amount of hydroxyproline reaching the wall. Such effects may be instrumental in enhancing wall plasticity, the rate of expansion and the final cell size.

These results indicate that ethylene and auxin together afford a dual regulatory system exerted through a control of a specific part of the protein synthetic pathway, the products of which regulate the rate of expansion, and the potential for expansion, of the plant cell wall.

REFERENCES

Blobel, G. and Sabatini, D. D. (1970). *J. Cell Biol.* **45**, 130–145.
Chrispeels, M. J. (1969). *Pl. Physiol., Lancaster* **44**, 1187–1193.
Chrispeels, M. J. (1970). *Pl. Physiol., Lancaster* **45**, 223–227.
Cleland, R. (1968). *Pl. Physiol., Lancaster* **43**, 865–870.
Cleland, R. and Karlsnes, A. M. (1967). *Pl. Physiol., Lancaster* **42**, 669–671.
Crocker, W. (1932). *Proc. Am. Phil. Soc.* **71**, 295–298.
Crocker, W. (1948). "Growth of Plants." Reinhold, New York.
Dashek, W. V. (1970). *Pl. Physiol., Lancaster* **46**, 831–838.
Dougall, D. K. and Shimbayashi, K. (1960). *Pl. Physiol., Lancaster* **35**, 396–404.
Goeschl, J. D., Pratt, H. K. and Bonner, B. A. (1967). *Pl. Physiol., Lancaster* **42**, 1077–1080.
Heath, M. F. and Northcote, D. H. (1971). *Biochem. J.* **125**, 953–961.
King, N. J. and Bayley, S. T. (1965). *J. exp. Bot.* **16**, 294–303.
Kuraishi, S., Vematsu, S. and Yamaki, T. (1967). *Pl. Cell Physiol.* **8**, 527–528.
Lamport, D. T. A. (1967). *Nature, Lond.* **216**, 1322–1324.
Lamport, D. T. A. (1969). *Biochemistry* **8**, 1155–1163.
Lamport, D. T. A. (1970). *Ann. Rev. Pl. Physiol.* **21**, 235–270.
Lamport, D. T. A. and Miller, D. H. (1971). *Pl. Physiol., Lancaster* **48**, 454–456.
Lamport, D. T. A. and Northcote, D. H. (1960). *Nature, Lond.* **188**, 665–666.
Liu, E. and Lamport, D. T. A. (1968). *Pl. Physiol., Lancaster* **43**, 5–16.
Mackenzie, I. A. and Street, H. E. (1970). *J. exp. Bot.* **21**, 824–834.
Miller, P. M., Sweet, H. C. and Miller, J. H. (1970). *Am. J. Bot.* **57**, 212–217.
Neljubow, D. (1901). *Bot. Zbl. Beinhefte* **10**, 128–139.
Osborne, D. J., Ridge, I. and Sargent, J. A. (1972). "Plant Growth Substances" 1970. (D. J. Carr, ed.), pp. 534–542. Springer-Verlag, Berlin.

Puztai, A. and Watt, W. B. (1969). *Europ. J. Biochem.* **10**, 523–532.

Puztai, A., Begbie, R. and Duncan, I. (1971). *J. Sci. Fd. Agric.* **22**, 514–519.

Rhoads, R. E. and Udenfriend, S. (1968). *Proc. natn Acad. Sci. U.S.A.* **60**, 1473–1478.

Ridge, I. and Osborne, D. J. (1969). *Nature, Lond.* **223**, 318–319.

Ridge, I. and Osborne, D. J. (1970a). *J. exp. Bot.* **21**, 720–734.

Ridge, I. and Osborne, D. J. (1970b). *J. exp. Bot.* **21**, 843–856.

Ridge, I. and Osborne, D. J. (1971). *Nature, Lond.* **229**, 205–208.

Sabatini, D. D. and Blobel, G. (1970). *J. Cell Biol.* **45**, 146–151.

Sadava, D. and Chrispeels, M. J. (1971). *Biochim. biophys. Acta* **227**, 278–287.

Sargent, J. A., Atack, A. and Osborne, D. J. (1972). *Planta.* (In press).

Shannon, L. M., Kay, E. and Lew, J. Y. (1966). *J. biol. Chem.* **241**, 2166–2172.

Shimizu, K. and Morita, Y. (1966). *Agric. biol. Chem.* **30**, 149–154

Srivastava, S. (1970). *Biochem. biophys. Res. Commun.* **41**, 1357–1361.

Steward, F. C. and Chang, L. O. (1963). *J. exp. Bot.* **14**, 379–386.

Thompson, E. W., Notton, B. A., Richardson, M. and Boulter, D. (1971). *Biochem. J.* **124**, 787–791.

CHAPTER 7

Biosynthesis of Gibberellins

CHARLES A. WEST

Division of Biochemistry, Department of Chemistry,
University of California, Los Angeles, California, U.S.A.

I. Introduction

The role of the gibberellins as endogenous regulators of development in higher plants has come to be recognized on the basis of research results from many laboratories. Thirty-five different, but closely related, members of this family of plant hormones have been isolated (March, 1972) from higher plants themselves and from cultures of the phytopathogenic fungus *Fusarium moniliforme* (*Gibberella fujikuroi*) and their structures determined. Doubtless there are other members of this family yet to be characterized. The general features of the biosynthetic pathways involved in the elaboration of this group of hormones is also understood in, at least, broad outline. The development of this field and the documentation of ideas about the biosynthesis of the gibberellins have been reviewed recently by Cross (1968), Hanson (1968), Lang (1970) and MacMillan (1971).

II. Biosynthetic Pathway

Figure 1 summarizes the present understanding of intermediates and interconversions involved in gibberellin biosynthesis. This is a composite view derived from the work of a number of investigators. Much of this pattern has been developed from studies with *F. moniliforme,* although work with higher plants has contributed in some instances. A number of points should be kept in mind with reference to this Figure. It appears that similar pathways operate in the fungal and higher plant systems through the steps shown in Figure 1A. However, based on the qualitative analytical data available to date, it seems likely that any one organism or tissue may synthesize only a selected few of the large number of possible gibberellins. Thus, there may be considerable divergence of pathways and a number of possible alternative schemes for the interconversions of gibberellins in different organisms or even in different tissues within one organism. Even less information is available on which to base speculations about the patterns of interconversions of the gibberellins. Therefore, Figs 1B and 1C attempt to interrelate only those gibberellins found

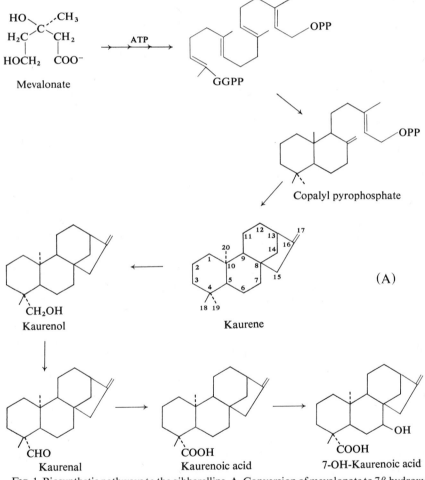

(A)

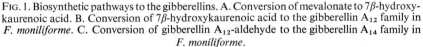

FIG. 1. Biosynthetic pathways to the gibberellins. A. Conversion of mevalonate to 7β-hydroxy-kaurenoic acid. B. Conversion of 7β-hydroxykaurenoic acid to the gibberellin A$_{12}$ family in *F. moniliforme*. C. Conversion of gibberellin A$_{12}$-aldehyde to the gibberellin A$_{14}$ family in *F. moniliforme*.

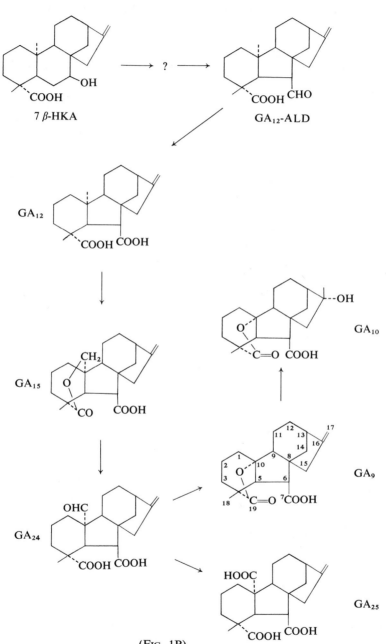

(Fig. 1B)

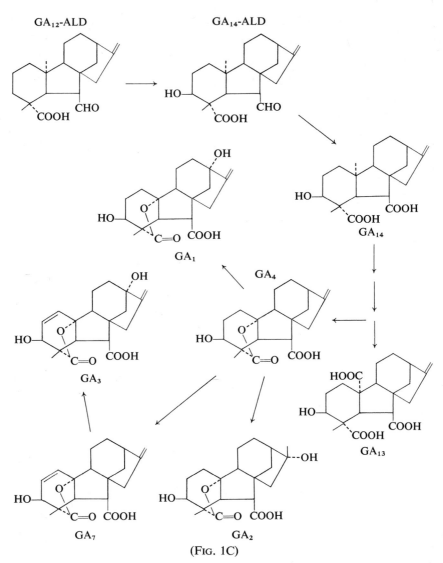

(Fig. 1C)

in *F. moniliforme* cultures with which most of the work on this phase of metabolism has been done. The sequences of gibberellin interconversions in higher
plants are likely to differ significantly from this pattern.

The discussion which follows in this section is intended only to bring out
some of the background which led to the formulation of the pathways in Fig. 1,
and is not a comprehensive survey of the literature in this field. The characteristics of labeling of gibberellin A_3 formed from [1-^{14}C]acetate and [2-^{14}C]meva-
lonate in *F. moniliforme* cultures supported the proposal that the gibberellins
were biogenetically related to diterpenes (Birch *et al.*, 1958). The role of the

diterpene hydrocarbon kaurene* as an intermediate was first indicated by its identification among the neutral metabolites produced with the gibberellins by *F. moniliforme* cultures. This was confirmed by the demonstration of a direct conversion of [17-^{14}C]kaurene to [17-^{14}C]gibberellin A$_3$ in fungal cultures (Cross *et al.*, 1964). The acyclic precursor of kaurene was shown to be geranylgeranylpyrophosphate (Fig. 1A) (Upper and West, 1967), thus confirming the prediction of Ruzicka (1953) that this alcohol or some derivative of it was the precursor of cyclic diterpenes. Subsequent work established copalylpyrophosphate as an intermediate in this conversion (Hanson and White, 1969a; Shechter and West, 1969). The intermediates in the formation of geranylgeranylpyrophosphate from mevalonate appear to be identical with those which participate in farnesylpyrophosphate (hence sterol and triterpene) synthesis; however, at least in some cases, the specificity of prenyl transferase determines the C$_{20}$ instead of the C$_{15}$ prenyl pyrophosphate as the major product of chain elongation (Oster and West, 1968).

The further metabolism of kaurene is via a successive series of oxidation steps (see Fig. 1A) whereby the C-19-methyl group is transformed into a carboxylic acid group (Dennis and West, 1967; Coolbaugh and Moore, 1971; Graebe *et al.*, 1972; Briggs, 1972) and the resulting kaurenoic acid is hydroxylated in the 7β-position (Hanson and White, 1969b; Lew and West, 1971). The role of these compounds as intermediates in gibberellin biosynthesis is suggested by their activity in gibberellin bioassay systems and their conversion to gibberellins in *F. moniliforme* cultures. The natural occurrence of kaurene, kaurenol, kaurenal and kaurenoic acid together, in isolable quantities, in the resin of *Espeletia grandiflora* has recently been reported (Piozzi *et al.*, 1971).

The mechanism of contraction of the B ring of the *ent*-kaurene skeleton to the *ent*-gibberellane skeleton has been the subject of considerable interest and speculation. Ring contraction is believed to occur at or near the stage shown in Fig. 1B because of the demonstrated existence of gibberellin A$_{12}$ with its methyl group at C-10 still present and no oxidized positions in the molecule other than those already oxidized in 7β-hydroxykaurenoic acid. Cross *et al.* (1968) proposed gibberellin A$_{12}$-aldehyde (*ent*-gibberellan-7-al-19-oic acid) as the product of ring contraction. This was based on a comparative study of the efficiency of conversion of several intermediates labeled with [^{14}C], including the synthetic aldehyde, to labeled gibberellic acid and other products in *F. moniliforme* cultures. Hanson and White (1969b) provided additional evidence for the role of the aldehyde and for its natural occurrence in the fungus. Further

---

* Kaurene refers to *ent*-kaur-16-ene (also referred to as (−) kaur-16-ene). Kaurenol, kaurenal and kaurenoic acid refer to *ent*-kaur-16-en-19-ol, *ent*-kaur-16-en-19-al and *ent*-kaur-16-en-19-oic acid, respectively. 7-β-hydroxykaurenoic acid refers to *ent*-kaur-16-en-7αol-19-oic acid; note that the designation of configuration in the *ent*-kaurene system is opposite to that previously used and thus is a source of possible confusion. In this paper the β-designation for this hydroxyl group will be used as in the earlier literature. Copalol is a trivial name for *ent*-labda-8(17), 13-dien-15-ol. The trivial designation of the gibberellins and the *ent*-gibberellane system as described by MacMillan (1971) will be employed.

consideration of the nature of the ring contraction step will be presented later.

Hedden and MacMillan (1971) cite work which indicates that gibberellin A_{12} aldehyde is converted efficiently to gibberellin A_{14} without prior conversion to gibberellin A_{12}. Presumably gibberellin A_{12}-aldehyde is first hydroxylated in the 3β position to form gibberellin A_{14}-aldehyde which then acts as the immediate precursor of gibberellin A_{14}. Gibberellin A_{14}-aldehyde has been synthesized so that this proposal can be tested directly (Hedden and MacMillan, 1971).

Thus, there appear to be two general families of gibberellins produced by *F. moniliforme* cultures—one of them derived directly from gibberellin A_{12}-aldehyde without hydroxylation in the ring system and the other, all members of which possess a hydroxyl group in the 3β position, derived from gibberellin A_{14}-aldehyde. The interconversions pictured among the gibberellins within these two families are in part speculative and based on the structural relationships of the different compounds, although support for some of the steps has come from experiments in which a specifically labeled precursor has been fed to *F. moniliforme* cultures and one or more metabolites have been examined for radioactivity. Such experiments by Cross *et al.* (1968) showed that gibberellin A_{14} is precursor of both gibberellin A_3 and gibberellin A_{13} as pictured. However, gibberellin A_{13} appeared not to be a precursor of gibberellin A_3, a finding which has been taken to indicate that the elimination of the carbon substituent at C-10 to form C_{19}-gibberellins occurs at some oxidation stage prior to its conversion to the carboxylic acid. The C_{20}-gibberellins with hydroxymethyl and formyl groups at C-10, which presumably occur as intermediates between gibberellin A_{14} and A_{13}, have not been isolated and identified to date.

The kinetics of labeling of gibberellins A_1, A_3, A_4 and A_7 after feeding [17-^{14}C]kaurenol to fungal cultures indicated that A_4 and A_7 served as precursors to A_1 and A_3 and this was confirmed by experiments in which a mixture of [^{14}C]-labeled A_4 and A_7 were isolated and re-fed to the fungus (Verbiscar *et al.*, 1967). Pitel *et al.* (1971) have shown the relationship of these gibberellins to be that shown in Fig. 1C in which A_4 gives rise to both A_7 and A_1, while A_7 serves as the direct precursor to A_3. A_1 is only very poorly converted to A_3; A_3 appears to be a relatively stable end-product in the fungus.

The relationships pictured in Fig. 1B for the gibberellin A_{12} family have not been tested very thoroughly. The formation of gibberellin A_{10} from gibberellin A_9 has been shown and the failure of gibberellin A_9 to serve as a precursor of gibberellin A_3 has been noted (Cross *et al.*, 1964). One finding which is not accounted for by the scheme in Fig. 1B is the conversion, albeit in low yield, of gibberellin A_{12} to gibberellins A_{13} and A_3 (Cross *et al.*, 1968).

III. Other Aspects of Gibberellin Biosynthesis

Studies of the natural occurrence and physiological properties of the gibberellins have led to a general picture of these as a family of chemical messengers which serve as part of the mechanism by which the plant responds to its environment and its own intrinsic genetic information—in short they are one of several families of hormones acting in higher plants. Knowledge of how they are biosynthesized will contribute to our understanding of how they play this role. This requires more than knowledge of the sequence of transformations in a mechanistic sense. It is essential that we also know the characteristics of the enzymes catalysing these transformations. This latter information should provide a basis for understanding other interesting aspects of the functions of gibberellins as plant hormones.

A central question concerns the factors which regulate the levels and kinds of gibberellins produced in plant tissues. It is logical to assume that if the gibberellins are to serve as growth regulators, their own synthesis must be under strict qualitative and quantitative control. Such questions as the development of the enzymatic machinery for gibberellin biosynthesis, the distribution of these enzymes in different plant parts and within the cell, control of the concentrations of the enzymes, the regulation of the activity of the enzymes in response to environmental influences, and the relationship of the intermediates of this pathway to the formation of other classes of isoprenoid compounds such as sterols and triterpenes, carotenes, monoterpenes, sesquiterpenes and other diterpenes known also to be produced in higher plants must be considered. The key to understanding some aspects of gibberellin function will surely come from answers to these questions.

Unfortunately, the amount of information in this area is still very limited. We are attempting to approach questions of the control of gibberellin synthesis through studies of the characteristics of some of the enzymes involved in their biosynthesis. Presented below are the results of some of our still incomplete efforts in this direction along with some related developments made by other investigators.

A. ENZYMIC CYCLIZATION OF GERANYLGERANYLPYROPHOSPHATE TO KAURENE

Kaurene synthetase is of interest as a potential site of regulation for the gibberellin synthesis pathway. The geranylgeranylpyrophosphate (GGPP) pool may serve the synthesis of carotenes and other diterpenes as well as kaurene, and thus GGPP may be a branch point metabolite. Kaurene synthetase has also been shown to be inhibited by some synthetic plant growth retardants (Dennis *et al.*, 1965).

Kaurene synthetase has been partially purified from *F. moniliforme* mycelia and a number of its properties studied (Fall and West, 1971). The catalytic activities for the cyclization of either GGPP (activity A) and copalylpyrophosphate (activity B) as substrates copurified with a constant ratio of the two activities throughout. The apparent molecular weight of both activities was shown to be 460000 ± 30000. None of several approaches to separation of the two activities was successful. However, differences in their properties suggested that separate catalytic sites were involved in activities A and B. It was concluded that the two activities must be associated in some sort of enzyme complex.

A similar investigation of the properties of kaurene synthetase from a higher plant source, the endosperm of immature wild cucumber (*Echinocystis macrocarpa*) seed, has been undertaken by R. Frost. Enzyme activities were assayed essentially by the method described by Fall and West (1971). Activity A was determined from the amounts of radioactive copalol plus kaurene formed from radioactive GGPP after incubation under a prescribed set of conditions and subsequent treatment with alkaline phosphatase to release copalol from copalylpyrophosphate while further cyclization activity was inhibited. Activity B was determined from the amount of radioactive kaurene formed from radioactive copalylpyrophosphate. The radioactive lipids were measured after extraction from the incubation mixture and resolved from one another by thin layer chromatography.

Purification of kaurene synthetase has not been as successful for the *E. macrocarpa* enzyme as it was for the *F. moniliforme* enzyme. Many types of purification procedures led to large losses in activity and an unstable enzyme, and therefore could not be employed. A procedure starting with reconstituted lyophilized endosperm has been developed which employs chromatography on QAE-Sephadex and then on hydroxylapatite columns. This leads to a relatively small purification (5- to 20-fold) and a low recovery of the activities. It should also be noted that, unlike the case with kaurene synthetase from *F. moniliforme*, activities A and B do not purify with a constant or reproducible ratio of specific activities. In spite of these limitations, the purified enzyme has the important advantage over the crude preparation that most of the phosphatase activity, which interferes with kinetic studies, has been removed.

Some general properties of activities A and B of the purified kaurene synthetase of *E. macrocarpa* are presented in Table I in comparison with those of the *F. moniliforme* enzyme. Although some differences in properties, have been detected such as the susceptibility to inhibition of activity A by deoxychelate, the pattern of apparent inhibition by substrates and the details of divalent metal ion activation, the similarities are more striking. The comparison is of some interest in connection with speculations about the evolutionary origin of kaurene synthetase in this phytopathogenic fungus since it would appear to be one of very few, if not the only fungus, capable of kaurene synthesis.

As indicated in Table I, no evidence has been obtained for feed-back inhibition of these kaurene synthetases by gibberellins.

A series of experiments were performed with *F. moniliforme* kaurene synthetase (Fall and West, 1971) which indicated at least some degree of metabolic channeling of the copalyl moiety, generated as the product of activity A, directly to the catalytic site for activity B without loss to the surrounding

TABLE I

Comparison of properties of purified kaurene synthetase preparations from *E. macrocarpa* and *F. moniliforme*

	Activity A		Activity B	
	E. macrocarpa	*F. moniliforme*	*E. macrocarpa*	*F. moniliforme*
K_m (app) for substrate	1·8 μM	0·8 μM	0·6 μM	1·0 μM
Inhibition by substrate	+	–	–	+
pH optimum	7·3	7·5	6·9	6·9
Activation by metal ions	$Mg^{2+} > Mn^{2+} > Co^{2+}$ (stimulation)	$Mg^{2+} > Ni^{2+} = Co^{2+}$ (stimulation)	$Mg^{2+} > Mn^{2+} > Co^{2+}$ (required)	$Mg^{2+} > Co^{2+} > Mn^{2+} = Ni^{2+}$ (required)
Stimulation by dithiothreitol				
−EDTA	+	+	++	++
+10^{-4} M EDTA	None	None	None	+
% Inhibition by 10^{-3} M sulfhydryl reagents plus 10^{-4} M EDTA				
p-Hydroxymercuribenzoate	68	94	100	99
N-Ethylmaleimide	43	12	54	98
$HgCl_2$	75	93	100	96
$CuSO_4$	88	95	100	98
Inhibition by gibberellins	None	None	None	None
% Inhibition by 5×10^{-4} M deoxycholate	100	3	76	43

medium. In these experiments, the ratio of [^{14}C] to [^{3}H] in the kaurene produced was determined at a specified time after [^{14}C]GGPP and [^{3}H]-copalylpyrophosphate were added simultaneously as substrates for kaurene synthetase. Three to four times more [^{14}C] than [^{3}H] is found in the kaurene than is predicted by a model in which the copalyl moiety generated as the product of activity A completely equilibrates with the exogenous pool of copalylpyrophosphate before reacting to form kaurene.

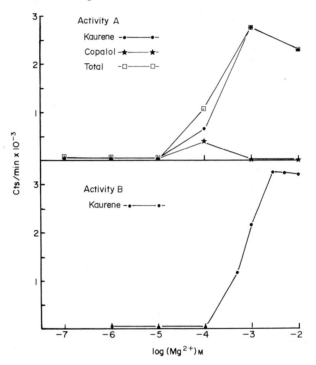

FIG. 2. Products formed from geranylgeranylpyrophosphate (activity A) or copalylpyrophosphate (activity B) by purified kaurene synthetase from *E. macrocarpa* as a function of Mg^{2+} concentration. Activity A was determined from incubation of the following for 5 min at 30°C in 0·5 ml total volume at the concentrations indicated: Tes, pH 7·3 (50 mM), dithiothreiol (0·1 mM), [^{14}C]geranylgeranylpyrophosphate (1·0 μM), MgCl$_2$ as indicated, and enzyme protein (5 μg). At the end of this time *p*-hydroxymercuribenzoate was added to 2 mM to stop further synthetase action and 100 μg *E. coli* alkaline phosphatase was added and incubation continued for 90 min at 30°C to hydrolyse phosphate esters. Activity B was determined with the same incubation mixture except that 1 mM [^{14}C]copalylpyrophosphate was substituted as substrate and 50 mM TES, pH 6·9, was employed as buffer. There was no phosphatase treatment in this case. Radioactive products were extracted, separated by TLC and assayed as described by Fall and West (1971).

Comparable experiments have not yet been performed with the *E. macrocarpa* kaurene synthetase. However, a study of the nature of the products formed when GGPP is supplied as substrate for the purified *E. macrocarpa* enzyme as a function of Mg^{2+} concentration (Fig. 2a) is of interest in this same connection. All the GGPP which has reacted has been converted to kaurene with no detectable copalylpyrophosphate at concentrations of 10^{-3} M or higher Mg^{2+}. Comparable amounts of kaurene and copalylpyrophosphate are seen at 10^{-4} M Mg^{2+}. Figure 2b presents for comparison the profile for activity B as a function of Mg^{2+} concentration. It is interesting to note that activity B is not detectable at 10^{-4} M Mg^{2+} and the range of Mg^{2+} concentrations which promote production of kaurene as the sole product of activity A is the same as the range in which activity B is stimulated. These results indicate that a substantial pool of copalylpyrophosphate may never accumulate under physiological conditions.

Although it is not so clearly established for the *E. macrocarpa* enzyme, it seems likely that both kaurene synthetases exist as complexes of activities A and B (perhaps other proteins may also be present). Such complexes may facilitate transfer of intermediates from one catalytic site to the next in a pathway without permitting the wasteful accumulation of large pools of these intermediates. The existence of such complexes in other areas of biosynthetic metabolism has been increasingly recognized in recent years. It is possible that others of the "soluble cytoplasmic" enzymes involved in the conversion of mevalonate to kaurene are complexed *in vivo*, although there is no evidence for this to date. We propose to examine this possibility.

Lang (1970) has reviewed the background and evidence which indicates that certain plant growth retardants may act in both *F. moniliforme* and higher plants to inhibit gibberellin biosynthesis. Dennis *et al.* (1965) pointed to kaurene synthetase as a site of inhibition of amo 1618, phosfon D and other closely related substances from studies using crude preparations from *E. macrocarpa* endosperm. Shechter and West (1969) subsequently found activity A to be the major site of inhibition of kaurene synthetase by these substances. A more quantitative evaluation of the interaction of these growth retardants with purified *F. moniliforme* kaurene synthetase has recently been completed (Fall and West, 1971). Inhibition at this site may account for the observed effects of retardants on plant development in some instances, although it should be emphasized that not all retardants act here, and the observed physiological effects of retardants which do have an effect on kaurene synthetase, and thus on gibberellin biosynthesis, cannot always be readily explained on this basis (see for example Reid and Crozier, 1970).

A more quantitative indication of the inhibitory interaction of some retardants and other inhibitory substances with the purified kaurene synthetase of *E. macrocarpa* endosperm is given in Table II for activity A and Table III for activity B. Closely related analogues of substances shown in these Tables were tested, but these showed similar (although generally less) inhibition and are not included. The general pattern of inhibition seen was quite similar to those observed earlier with the *F. moniliforme* enzyme. SKF 525-A was more inhibitory at low concentrations with the *E. macrocarpa* enzyme than had been seen with *F. moniliforme* kaurene synthetase whereas amo 1618 was less

TABLE II

Effect of growth retardants and related inhibitors on kaurene synthetase activity A
from *E. macrocarpa*

Inhibitor	% Inhibition at M						
	5×10^{-8}	5×10^{-7}	5×10^{-6}	5×10^{-5}	5×10^{-4}	5×10^{-3}	5×10^{-2}
SKF 525A[a]	17	56	100	100	100		
Phosfon D[b]	7	20	94	100	100		
Q-64[c]	0	17	86	100	100		
Amo-1618[d]	0	16	65	100	100		
Delcosine			11	39	78		
Nicotine			0	11	65		
CCC[e]					4	51	
Acetylcholine						36	78

[a] Dimethylaminoethyl-2,2-diphenylpentanoate.
[b] Tributyl-2,4-dichlorobenzylphosphonium chloride.
[c] 2-(N,N-dimethyl-N-octylammonium bromide)-p-methan-1-ol.
[d] 2'-Isopropyl-4'-(trimethylammonium chloride)-5'-methylphenyl piperidine-1-carboxylate.
[e] 2-Chloroethyltrimethyl ammonium chloride.

active as an inhibitor with the *E. macrocarpa* enzyme. Q-64 gave no significant inhibition of activity B, while significant inhibition by this substance was observed with *F. moniliforme* activity B. There was, in fact, relatively little inhibition of *E. macrocarpa* activity B by any of these substances.

Delcosine is a tetracyclic diterpenoid alkaloid with some structural features in common with *ent*-kaurene. It was tested here because of a report that delcosine and a related diterpenoid alkaloid have growth inhibitory properties in decapitated pea (*Pisum sativum*) seedlings. (Waller and Burström, 1969). Some inhibition of activity A was seen at the relatively high concentration of 5×10^{-4} M. It is difficult to assess whether this observation has any physio-

TABLE III

Effect of growth retardants and related inhibitors on kaurene
synthetase activity B from *E. macrocarpa*

Inhibitor	% Inhibition at M		
	5×10^{-6}	5×10^{-5}	5×10^{-4}
SKF 525 A	8	28	63
Phosfon D	2	8	34
Q-64	6	0	0
Amo-1618	11	12	
Delcosine	0	0	0
Nicotine	0	0	0
CCC	0	0	0
Acetylcholine	12	6	9

logical significance, but it is interesting to consider whether some natural *ent*-kaurene structural analogue might have a regulatory role. Nicotine was another alkaloid tested because of a report that it inhibits the formation of cyclic carotenes in mycobacteria (Hower and Batra, 1970). Since the cyclization process in carotenes and the first stages of GGPP cyclization to copalyl-pyrophosphate are mechanistically analogous, it was interesting to see what effect nicotine might have on kaurene synthetase. Also it should be noted that among the earliest plant growth retardants reported were quaternary nicotinium salts (Mitchell *et al.*, 1949). Only at the relatively high concentration of 5×10^{-4} M was any significant inhibition of activity A by nicotine seen. It seems doubtful that this inhibition has any physiological significance.

The rationale for testing acetylcholine chloride as a potential inhibitor was as follows. Certain choline analogues do act as growth retardants. Newhall (1969) reported that synthetic quaternary ammonium derivatives based on limonene showed good correlation in their activities as plant growth retardants and *pseudo*choline esterase inhibitors. More recently Jaffe (1971) has suggested that amo 1618 may function by inhibiting cholinesterase, thus raising endogenous levels of acetylcholine which act more directly as the growth retardant. In our present tests we found significant inhibition by acetylcholine of kaurene synthetase activity A only at the high concentration of 5×10^{-3} M or greater. Thus, it seems unlikely that acetylcholine would be acting to regulate kaurene synthetase *in vivo* in this system. It had about the same activity as CCC and some other choline analogues which are also not very effective against the purified kaurene synthetase. Of course, it is entirely possible that other systems might respond differently or that acetylcholine might have some other site of action as an inhibitor of the pathway of gibberellin synthesis.

The synthetic growth retardants tested with the *F. moniliforme* kaurene synthetase all showed non-competitive inhibition with respect to GGPP or copalylpyrophosphate. Thus, it is likely that they are acting at some regulatory site in kaurene synthetase separate from the catalytic site. This finding leads to the speculation that these synthetic substances are acting at a binding site whose normal ligand is a natural regulatory agent which controls kaurene synthetase activity. There was an indication that delcosine, nicotine and acetylcholine might have the potential to regulate kaurene synthetase, and therefore these natural substances were tested; however, their activity as inhibitors was such that they presumably do not serve this role. Nonetheless, there is interest in continuing the search for natural growth-retardants which function at this site.

B. OXIDATIVE REACTIONS IN GIBBERELLIN BIOSYNTHESIS

Most, if not all, of the steps leading from kaurene to end-product gibberellins are oxidative in nature as contrasted with the completely non-oxidative formation of kaurene from intermediates of the central metabolic pathways. The characteristics which have been examined to date suggest that these oxidative steps have many common features and, indeed, may share electron transport components and sites of oxygen binding. This provides some interesting possibilities for regulation of the overall pathway.

1. *Oxidation of Kaurene to 7β-Hydroxykaurenoic Acid*

The characteristics of the enzymes which catalyse the sequence of oxidative steps leading from kaurene to 7β-hydroxykaurenoic acid (Fig. 1A) have been studied primarily in the endosperm of *E. macrocarpa* seed (Murphy and West, 1969). All four steps are catalysed by preparations of microsomes and have the same general properties including requirements for a reduced pyridine nucleotide and oxygen, which indicate that they are mixed-function oxidases. The participation of a cytochrome P-450 species in the reactions is strongly indicated by the characteristics of carbon monoxide inhibition which shows maximal photoreversal at 450 nm and also the interaction with other inhibitors. Although it was not surprising to find mixed-function oxidases for the two hydroxylation steps, it was novel to find that the alcohol and aldehyde oxidation steps have these same characteristics.

Recently the mechanism of oxidation of the 4α-methyl group to a carboxyl group prior to its elimination by decarboxylation in the sterol synthesis pathway has been clarified (Miller *et al.*, 1971). Each of the three steps of oxidation, including the ones involving the alcohol and aldehyde groups, has the characteristics of a microsomal mixed-function oxidase and thus is analogous to the oxidation of the 4α-methyl group of kaurene. As is probably not surprising, there are many similarities in the biosynthetic pathways leading to the steroid hormones in animals and the gibberellin hormones in plants.

Some further characteristics of the electron transfer components associated with the microsomal mixed-function oxidation of kaurene in *E. macrocarpa* have been examined by Dr. E. Hasson in our laboratory. Added flavin adenine dinucleotide (FAD) at optimal concentrations of 10^{-5} M stimulated these activities. It had been demonstrated earlier that either NADPH or, less efficiently, NADH could satisfy the requirement for a reduced pyridine nucleotide. However, the best rates of oxidation were consistently achieved by supplying a mixture of NADH, NADP$^+$ and ATP, with the presence of all three components necessary. It has now been demonstrated directly that these microsomes catalyse a pyridine nucleotide transhydrogenation reaction in which NADPH is formed at the expense of NADH in a reaction dependent on the addition of ATP (an energy-dependent transhydrogenase). By the use of 2,4-dinitrophenol as a selective inhibitor, it has been possible to show that this transhydrogenase can supply NADPH which is utilized in the mixed-function oxidation of kaurene. It has further been shown that the relatively low levels of both NADH and NADPH present in the transhydrogenase system are far better in supporting mixed-function oxidation than either reduced pyridine nucleotide alone at a comparable concentration. This may well reflect more nearly the actual physiological situation. These latter findings are very similar to those reported for pyridine nucleotide requirements in mammalian liver microsomal mixed-function oxidases by Cohen and Estabrook (1971).

2. Oxidation of Kaurenoic Acid to Gibberellins

(a) *Characteristics of a cell-free enzyme system from* F. moniliforme. Studies of a cell-free extract from *F. moniliforme* mycelium capable of catalysing substantial oxidation of kaurenoic acid were initiated by R. R. Fall and have been carried forward by D. Nakata and W. Schook. The cell-free extract is prepared by crushing frozen mycelia from a 4 to 5 day old culture of *F. moniliforme* (strain ACC 917 M419) in a Sager's press at 15000 to 20000 lbs/sq in applied pressure. The crushed preparation is resuspended in 0·1 M Tricine buffer (pH 8) containing 0·29 M sucrose and 10 mM 2-mercaptoethanol. The suspension is centrifuged first at $10000 \times g$ for 20 min and the resulting supernatant fraction (S_{10}) is further centrifuged at $150000 \times g$ for 60 min. The pellet (P_{150}) is resuspended in the same buffer solution, at a volume equal to one-half of the original volume of S_{10} from which it was derived. The P_{150} suspension and supernatant fraction (S_{150}) are quick-frozen and stored separately at $-20°C$.

The results of a typical thin layer chromatographic analysis of an incubation mixture with kaurenoic acid as the substrate are presented in Fig. 3. The relative amounts of the different fractions produced varies to some extent depending on the conditions employed. The A region at the origin in this basic system can be further resolved into two major components (A-1 and A-2) plus a small amount of origin radioactivity by rechromatography on silica gel in an acidic solvent system [isopropyl ether-glacial acetic acid (95:5, v/v)].

Certain of the radioactive components in this incubation mixture have been identified, some conclusively, and others tentatively. The products of a large number of incubations were pooled to provide sufficient material for these tests. Details of the characterizations will be published elsewhere. Fraction D appears to be unreacted kaurenoic acid. Fraction C is chromatographically identical to 7β-hydroxykaurenoic acid previously characterized in *E. macrocarpa* incubations (Lew and West, 1971) and has been identified from its mass spectral properties after resolution of a mixture of extracted components containing [^{14}C] fraction C by gas chromatography. It has also been shown that 7β-hydroxykaurenoic acid prepared in *E. macrocarpa* and Fraction C give rise to the same spectrum of products when incubated in the *F. moniliforme* cell-free system. Fraction A-1 has been identified as gibberellin A_{14} from gas chromatography-mass spectrometry in comparison with an authentic reference sample. Fraction A-2 has been tentatively identified as gibberellin A_{12} from its chromatographic behavior and the fact that it coincides with a major product of synthetic gibberellin A_{12}-aldehyde metabolism (see later). Fraction

B-1 contains a major component which co-chromatographs with authentic 6β,7β-dihydroxykaurenoic acid and also yields a product on periodate oxidation which co-chromatographs with the product formed from the authentic diol. Fraction B-1 also contains four or five other minor radioactive components which can be resolved from the major component in suitable solvent systems. Fraction B-2 is similarly complex, but has a major component with the chromatographic behavior of gibberellin A$_{12}$-aldehyde and yields the same

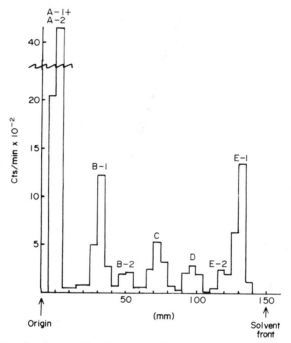

FIG. 3. Radioactive fractions produced from [17-^{14}C]kaurenoic acid by an *F. moniliforme* P$_{150}$ preparation. The incubation mixture included at the final concentrations indicated: [17-^{14}C]kaurenoic acid (2 μM; 16000 cts/min), NADPH (0·5 mM), Tricine buffer, pH 8 (50 mM), sucrose (140 mM), 2-mercaptoethanol (5 mM), 0·25 ml of P$_{150}$ suspension and 0·25 ml of boiled S$_{150}$ in a total volume of 1·0 ml. Incubation was under air at 30°C with shaking in a water bath for 30 min. Reaction was stopped by the addition of acetone and the lipid products were extracted from acidified solution into benzene-acetone. The extract was subjected to chromatography on a silica gel thin layer developed with benzene-ethanol-7·2N ammonia (65:34:1, v/v). The cts/min associated with 0·5 cm regions of the chromatogram were determined by liquid scintillation spectrometry.

products as authentic gibberellin A$_{12}$-aldehyde after chemical oxidation or reincubation with P$_{150}$ preparations (see below). Approximately 50% of the radioactivity of fraction E-1 co-crystallizes to constant specific radioactivity with added 7β-hydroxykaurenolide; at least a part of the remaining radioactivity is due to a contaminant produced non-enzymically from the substrate. Fraction E-2 has not been identified to date. Thus, the major components are believed to be those pictured in Fig. 4. It should be emphasized that some of the

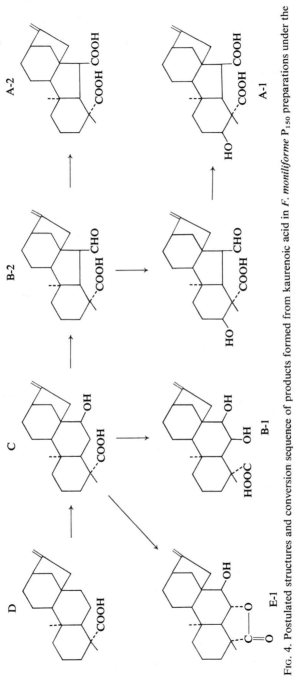

Fig. 4. Postulated structures and conversion sequence of products formed from kaurenoic acid in *F. moniliforme* P_{150} preparations under the conditions described in Fig. 3.

fractions are complex and other products, some produced enzymically and others probably non-enzymically, are also present.

The production of these fractions from kaurenoic acid has been studied as a function of time through two hours under the conditions described in the legend to Fig. 3. The results are generally consistent with the pattern of interconversions postulated in Fig. 4, although a decrease noted in the level of B-1 after about 30 min is not accounted for by this scheme. The results of reincubations of isolated fractions obtained from initial kaurenoic acid incubations are also consistent with the pathways pictured in Fig. 4. Fraction C gives rise to all products except Fraction D, with A-1 as the major product. Fraction B-2 is converted to A-1 as the major product along with smaller amounts of A-2. Small amounts of B-1 are converted to E-2, but most of it remains unconverted. Fractions A-1, A-2 and E-1 appear to be end-products in the isolated system since they are not converted to other detectable products.

Also consistent with the scheme in Fig. 4 are the results of incubations of [17-^{14}C]gibberellin A_{12}-aldehyde prepared from 7β-hydroxykaurenolide by a modification of the method of Cross et al. (1968). This substrate is very rapidly converted to a mixture of A-1 and A-2 when incubated with the fungal enzyme system under the same conditions described in Fig. 3. In this behavior, gibberellin A_{12}-aldehyde resembles the major component of Fraction B-2; this is a major argument for assuming the presence of gibberellin A_{12}-aldehyde in Fraction B-2.

Graebe et al. (1972) have recently reported the formation of kaurenoic acid, 7β-hydroxykaurenoic acid, gibberellin A_{12}-aldehyde and small amounts of gibberellin A_{12} from mevalonate in a cell-free system prepared from immature pumpkin (*Cucurbita pepo*) seed endosperm. They were uncertain whether the gibberellin A_{12} was formed enzymically or non-enzymically from the aldehyde. Nakata has recently obtained chromatographic evidence (unpublished results) for 7β-hydroxykaurenoic acid, $6\beta,7\beta$-dihydroxykaurenoic acid, gibberellin A_{12}-aldehyde and gibberellin A_{12}, as products of kaurenoic acid metabolism in lyophilized *E. macrocarpa* endosperm preparations. In this case it could be demonstrated that gibberellin A_{12} was a product of an enzyme-catalysed oxidation of gibberellin A_{12}-aldehyde. The interconversions of these compounds in this cell-free system are identical with those portrayed in Fig. 4 for the *F. moniliforme* system; however, no evidence for gibberellin A_{14} or 7β-hydroxykaurenolide formation was found in these higher plant systems.

(b) *The ring contraction problem*. The labeling studies of Birch et al. (1958) showed that the carboxyl carbon attached to ring B of gibberellic acid was derived from the 2-position of mevalonate. It was therefore proposed that the six-membered B-ring of the diterpenoid precursor must undergo ring contraction to form the 5-membered B ring of the *ent*-gibberellanes with the appended carbon originating from the 7-position of cyclic diterpene. There is considerable interest in the mechanistic course of this unusual reaction.

As reviewed above, 7β-hydroxykaurenoic acid is the most oxidized of the kaurene derivatives and gibberellin A_{12}-aldehyde is the least oxidized of the *ent*-gibberellane derivatives which have been strongly implicated as intermediates in gibberellin biosynthesis. The evidence from studies in cell-free systems is also consistent with the view the 7β-hydroxykaurenoic acid is an immediate precursor of gibberellin A_{12}-aldehyde. This conversion would

ROUTE A

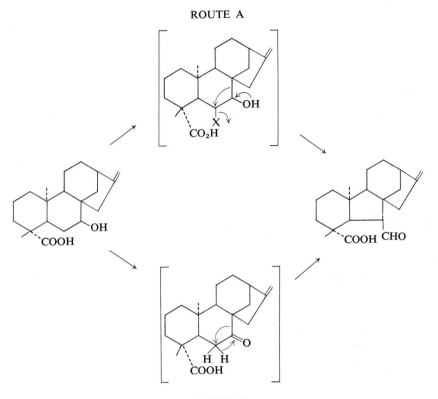

ROUTE B

FIG. 5. Possible mechanisms for contraction of ring B.

require an oxidative ring-contraction involving the transfer of two electrons. Two general ways in which this might be formulated are illustrated in Fig. 5.

Reaction course A in Fig. 5 pictures a preliminary oxidation of the 6β-position followed by a rearrangement in which the newly introduced 6β-substituent (X) is eliminated by a nucleophilic displacement. The configuration of the 6β-carbonyl substituent in the *ent*-gibberellane product makes it most likely that the elimination of X^- would occur from the 6β position in this rearrangement. Cross and Stewart (1970) tested the hypothesis that X might be a hydroxyl group by feeding synthetic [17-^{14}C]kaur-6β,7β-diol-19-oic acid to cultures of *F. moniliforme*. They failed to find significant conversion of this substance to labeled gibberellin A_3 even though it was converted in reasonable yield to the metabolite fujenal in which the B-ring has been opened by cleavage between the 6 and 7 positions. Hanson and Hawker (1971) similarly failed to find any conversion of this diol to gibberellin A_3 and Jefferies *et al.* (1970), did not see significant conversion of the $6\beta,7\beta$-diol of kaurenol to gibberellin A_3 in fungal cultures. Our own results in cell-free extracts of *F. moniliforme* indicated that, although a fraction (B-1) which is believed to contain largely the $6\beta,7\beta$-diol is formed from kaurenoic acid, B-1 is not further converted to gibberellin A_{12}-aldehyde, gibberellin A_{12} or gibberellin A_{14}. Thus, there is no evidence to support reaction course A when X is a hydroxyl group. Cross and Stewart suggested that X might be some derivative of a hydroxyl such as a pyrophosphate ester. The $6\beta,7\beta$-epoxide has also been suggested as a possible substrate for contraction. However, these proposals have not been tested.

Reaction course B in Fig. 5 envisions a preliminary oxidation of the 7β-hydroxyl group to a 7-keto function followed by a rearrangement in which an intramolecular transfer of a hydride from the 6β to the 7 position takes place. Hanson and his associates have provided evidence in support of some version of this route. First it was inferred from double-labeling experiments involving $[2\text{-}^{14}C, (5R)\text{-}5\text{-}^3H_1]$ mevalonate and $[2\text{-}^{14}C, 1\text{-}^3H_2]$geranylpyrophosphate as precursors of several diterpenoid metabolites in *F. moniliforme* that the 6β-H is lost and the 6α-H is retained in the overall synthesis of gibberellin A_3 (Evans *et al.*, 1970). More recently it has been shown that both hydrogens at C-6 are retained in gibberellin A_{12}-aldehyde, although one of these is lost on the further chemical oxidation of this product to gibberellin A_{12} (Hanson and Hawker, 1971).

Although Hanson and Hawker do not suggest free 7-ketokaurenoic acid as an intermediate in the ring contraction, this would seem an obvious possibility from the above results. However, 7-ketokaurenoic acid was prepared and found not to have any significant gibberellin-like biological activity when applied to the *dwarf-5* mutant of maize (*Zea mays*) (Lew and West, 1971). Recently, we have prepared $[17\text{-}^{14}C]$7-ketokaurenoic acid by the oxidation of $[17\text{-}^{14}C]7\beta$-hydroxykaurenoic acid (fraction C) with chromic oxide. This substrate was then incubated with the cell-free preparation of *F. moniliforme* capable of catalysing kaurenoic acid oxidation. Thin layer chromatography of the reaction products in a basic system is shown in Fig. 6A. The distribution of products formed from $[17\text{-}^{14}C]$gibberellin A_{12}-aldehyde under similar conditions are shown in Fig. 6B for comparison. Although most of the 7-ketokaurenoic acid is converted to other products, the pattern differs from that seen for the products from gibberellin A_{12}-aldehyde. Furthermore, when the radioactive products from 7-ketokaurenoic acid at the origin are eluted and rechromatographed in an acidic system (Fig. 7A), they do not appear to contain significant amounts of gibberellin A_{14} (A-1) or gibberellin A_{12} (A-2) as does the analogous fraction formed from gibberellin A_{12}-aldehyde (Fig. 7B).

These data do not favor the idea that exogenous 7-ketokaurenoic acid can serve as a precursor of gibberellin A_{12}-aldehyde. However, in view of the results of Hanson and Hawker, some version of reaction course B seems more likely than reaction course A. It is possible that oxidation and ring contraction both occur while the substrate is bound to the enzyme without loss of an intermediate to the surrounding medium. Thus, even though oxidation of the 7β-hydroxy group may be a requisite for contraction, added 7-ketokaurenoic acid may not

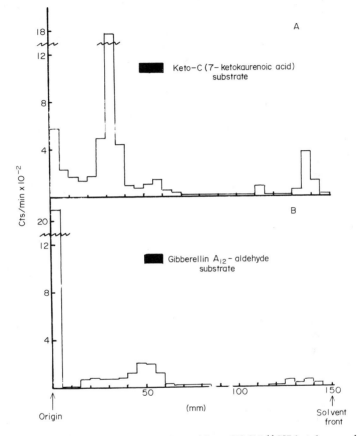

FIG. 6. Distribution of radioactive products formed from (A) [17-^{14}C]7-ketokaurenoic acid and (B) [17-^{14}C]gibberellin A_{12}-aldehyde as substrates by *F. moniliforme* P_{150} preparations. The conditions of incubation other than substrate and analysis of products were the same as described in the legend of Fig. 3. The black bar indicates the position of unreacted substrate.

penetrate to the reaction site and serve as a substrate for contraction. This would seem best to accommodate the data obtained up to the present.

(c) *Characteristics of the enzymes involved in kaurenoic acid oxidation.* Requirements for the oxidations of kaurenoic acid, 7β-hydroxykaurenoic acid (Fraction C) and gibberellin A_{12}-aldehyde as substrates for the P_{150} preparation at pH 8 as the source of enzyme were determined under the general conditions described in the legend to Fig. 3. The addition of NADPH is absolutely required in all cases. The reactions were severely inhibited when conducted in an nitrogen atmosphere instead of air, a finding which suggests that O_2 is required. Carbon monoxide inhibited the reactions in the presence of O_2. The addition of 10 mM cyanide not only failed to inhibit, but actually stimulated the rate of conversion of kaurenoic acid to gibberellin A_{12} and gibberellin A_{14}.

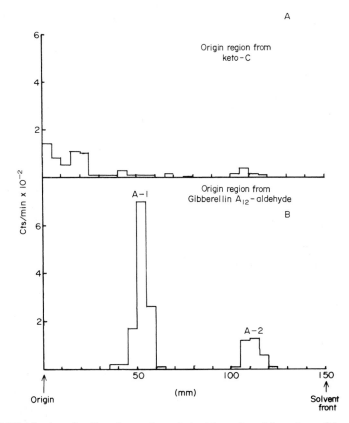

FIG. 7. Distribution of radioactive products derived from the origin regions of the chromatograms pictured in Fig. 6 after rechromatography on silica gel thin layers developed with isopropyl ether-glacial acetic acid (95:5, v/v). A is the origin material formed from 7-keto-kaurenoic acid and B is the origin material formed from gibberellin A_{12}-aldehyde. Fractions A-1 and A-2 have been identified as gibberellins A_{14} and A_{12}, respectively.

The combination of these properties suggests that the reactions are mixed-function oxidases of the general type shown to be involved in kaurene oxidation in *E. macrocarpa* endosperm. However, no indication of the presence of a cytochrome P-450 in the fungal P_{150} fractions has as yet been seen. The addition of a boiled $150000 \times g$ supernatant fraction (S_{150}) to the P_{150} preparation stimulates the overall conversion of kaurenoic acid over that seen with the pellet alone. The addition of flavin adenine dinucleotide at 10^{-5} M to the P_{150} preparation gives a similar stimulation, although it is not quite as effective in this regard as the boiled S_{150}.

A study of the effect of pH in the range 7·6 to 10·2 on the relative amounts of different products formed from kaurenoic acid by P_{150} in the presence of S_{150} showed pH 8·4 to be optimal for the formation of gibberellin A_{14} (A-1). The relative amounts of other products did not vary appreciably over the range

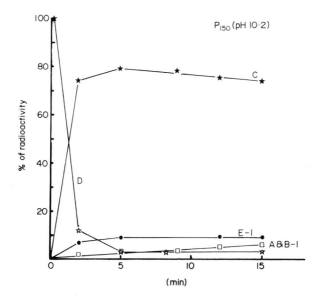

FIG. 8. Kinetics of production of radioactive fractions from [17-^{14}C]kaurenoic acid by P_{150} (pH 10·2) preparations of *F. moniliforme*. See Fig. 3 for the general conditions of incubation and analysis.

8·0 to 9·6. Above pH 9·6 the activity for 7β-hydroxykaurenoic acid (C) production remained high while very little of any other products was formed. Figure 8 shows the kinetics of production of different fractions from kaurenoic acid by an enzyme preparation consisting of P_{150} prepared in the usual fashion except that the crushed mycelia were resuspended in 0·10 M Tricine at pH 10·2 before high speed centrifugation. 7β-Hydroxykaurenoic acid is formed very rapidly by such preparations, and is only very slowly converted to other products. However, the combination of S_{150} prepared at pH 10·2 (which has only a low capacity itself for catalysing the oxidation of kaurenoic acid) with P_{150} prepared at pH 10·2 restores the overall activity including the formation of gibberellins A_{12} and A_{14}. Furthermore, the boiled S_{150} (pH 10·2) is as active as unheated S_{150} (pH 10·2) in this respect. One explanation of these results is that some factor(s) needed for the further metabolism of 7β-hydroxykaurenoic acid is transferred from the P_{150} to the S_{150} when these are prepared at pH 10·2.

Since 7β-hydroxykaurenoic acid may be the immediate substrate for ring contraction it is possible that this factor(s) plays a role in the ring contraction reaction.

Some preliminary attempts have been made to demonstrate the nature of the stimulatory factor(s) in the S_{150} (pH 10·2). The general test utilized has been to assess the ability of an added fraction or substance to restore the formation of Fraction A (mixture of gibberellins A_{12} and A_{14}) from kaurenoic acid in

TABLE IV

Effect of S_{150} preparations on kaurenoic acid metabolism in *F. moniliforme* P_{150} (pH 10·2)

Additions	% Radioactivity recovered in fraction[a]		
	$D^b + C^{b,\,c}$		A^b
None	88		1
$S_{150}{}^d$	41		27
S_{150}, boiled	31		43
S_{150}, dialysed	73		7
	D	C	A
None	2	78	4
S_{150}	2	21	44
S_{150}, extracted with ethyl acetate	2	4	56
S_{150}, treated with chelex	2	70	9
S_{150}, treated with XAD	5	28	34
S_{150} (pH 8)	2	82	3

[a] For general incubation conditions and analysis, see Fig. 3.
[b] D = kaurenoic acid; C = 7β-hydroxykaurenoic acid; A = gibberellins A_{12} and A_{14}.
[c] D and C were not well-resolved in this experiment.
[d] S_{150} (pH 10·2) except where otherwise indicated.

P_{150} (pH 10·2) suspensions. From Table IV it can be seen that the active factor(s) in S_{150} is stable to boiling and largely removed by dialysis, thus suggesting that it is a low molecular weight substance(s). Furthermore, it is not extracted by ethyl acetate, is partially removed by treatment with polystyrene beads (XAD) and is more completely removed by treatment with chelex resin, a polystyrene resin containing covalently bound metal ion chelating groups. The stimulating factor(s) is not found in S_{150} prepared at pH 8.

Another approach has been to test the ability of known substances to substitute for S_{150} in stimulating gibberellin formation in this system. This has led to a rather confusing picture. The data in Table V show that FAD at 10^{-5} M gives some stimulation of gibberellin formation; higher levels of FAD do not give further stimulation. Several metal ions at relatively high concentrations, including Fe^{2+}, Fe^{3+}, and Al^{3+}, are as effective or nearly as effective as S_{150}. Co^{2+} gives a slight stimulation over FAD alone while Ni^{2+} is ineffective. Cu^{2+} at 1 mM inhibits the conversion of kaurenoic acid to 7β-hydroxykaurenoic acid while Ca^{2+} at the high concentration of 50 mM appears to inhibit more selectively the further metabolism of 7β-hydroxykaurenoic acid. The effects of high concentrations of chelators in the absence of added metal ions is even more baffling (Table VI). Chelating agents which have a high association constant

TABLE V

Effect of FAD and metal ions on kaurenoic acid metabolism in *F. moniliforme* P_{150} (pH 10·2)

| Addition | 10^{-5} M FAD | % Radioactivity recovered in fraction[a] | | |
		D^b	C^b	A^b
None	−	3	71	7
S_{150} (pH 10·2)	−	4	17	45
None	+	5	55	16
Fe^{2+}, 10 mM	−	4	14	47
Fe^{2+}, 10 mM	+	4	7	50
Fe^{2+}, 1 mM	+	11	24	35
Fe^{2+}, 0·1 mM	+	13	44	18
Fe^{3+}, 10 mM	+	19	17	30
Fe^{3+}, 1 mM	+	9	24	34
Co^{2+}, 1 mM	+	4	44	25
Ni^{2+}, 1 mM	+	6	51	18
Cu^{2+}, 1 mM	+	86	7	1
Cu^{2+}, 0·1 mM	+	10	60	11
Ca^{2+}, 50 mM	+	9	67	6
Al^{3+}, 1 mM	+	3	14	48
Al^{3+}, 1 mM	−	8	48	21

[a] For general incubation conditions and analysis see Fig. 3.
[b] D = kaurenoic acid; C = 7β-hydroxykaurenoic acid; A = gibberellins A_{12} and A_{14}.

for Fe^{3+}, such as Tiron and "Fe^{3+}-specific", are as effective at high concentrations as S_{150} in stimulating gibberellin formation. EGTA is similarly effective, especially in the presence of added FAD. *o*-Phenanthroline and bipyridyl, on the other hand, act at high concentration to inhibit the conversion of kaurenoic acid to 7β-hydroxykaurenoic acid and appear to have no stimulatory effect at all.

These results give little indication of the nature of the presumably low molecular weight stimulatory factor(s) in S_{150} (pH 10·2). Part of the stimulation of S_{150} is probably provided by its FAD content, but this can not account for the total effect. There are some indications that a metal ion or some complex of a metal ion may be involved. The level of iron in S_{150} (pH 10·2) is 5×10^{-5} M, as determined by atomic absorption spectroscopy, so it clearly is not iron alone which provides the stimulation in view of the high levels of added Fe^{2+} or Fe^{3+}

TABLE VI

Effect of chelating agents on kaurenoic acid metabolism in *F. moniliforme* P_{150} (pH 10·2)

Additions	% Radioactivity recovered in fraction[a]		
	D^b	C^b	A^b
None	3	71	7
S_{150} (pH 10·2)	4	17	45
Tiron,[c] 25 mM	6	21	41
Tiron,[c] 8 mM	11	40	25
Tiron,[c] 0·8 mM	3	74	7
"Fe^{3+}-specific",[d] 25 mM	5	10	48
"Fe^{3+}-specific",[d] 8 mM	8	53	5
EGTA,[e] 25 mM	12	26	22
EGTA,[e] 25 mM; FAD, 0·05 mM	2	6	55
o-Phenanthroline, 25 mM	87	6	1
o-Phenanthroline, 1·5 mM	13	70	2
Bipyridyl, 25 mM	49	41	1
Bipyridyl, 1·5 mM	10	67	5

[a] For general incubation conditions and analysis, see Fig. 3.
[b] D = kaurenoic acid; C = 7β-hydroxykaurenoic acid; A = gibberellins A_{12} and A_{14}.
[c] Tiron = 4,5-dihydroxy-*m*-benzenedisulfonic acid, disodium salt.
[d] "Fe^{3+}-specific" = *N,N-bis*-(2-hydroxyethyl) glycine.
[e] EGTA = *bis*-(aminoethyl) glycolether-*N,N,N′,N′*-tetra-acetic acid.

needed for an equivalent effect. The manner in which certain of the chelating agents can act themselves to stimulate gibberellin formation is not at all obvious. The addition of low levels of Fe^{2+} or Fe^{3+}, which are not in themselves stimulatory, does not appear to alter the efficacy of the chelating agents. In spite of this confusing picture, it is nonetheless hoped that continued investigation of this system will provide some insight into the mechanism of the interesting ring contraction reaction.

It is clear that our knowledge of the enzymes catalysing reactions of gibberellin biosynthesis is still quite incomplete. However, progress is being made here as well as in our understanding of the pathway of gibberellin biosynthesis. There are some indications from this work of possible control points and regulatory mechanisms which may be operating and there is basis for hope that further work will bring still better understanding of the mechanisms of regulation of gibberellin biosynthesis and the resultant implications of these for the regulation of plant development.

ACKNOWLEDGEMENTS

Special thanks are due to Mr. Russell Frost, Mr. Dennis Nakata and Mr. William Schook whose experimental results are presented and discussed in the second part of this paper. These results will be presented as portions of Ph.D. dissertations in partial fulfillment of the requirements for the Ph.D. degree from the University of California, Los Angeles, for which each of these three is a candidate.

The research reported in this paper has been supported in part by National Institutes of Health Grant GM-07065 of the United States Public Health Service, Training Grant 5 T01 GM00463 of the United States Public Health Service, and the National Science Foundation.

REFERENCES

Birch, A. J., Rickards, R. W. and Smith, H. (1958). *Proc. chem. Soc.* 192–193.
Briggs, D. E. (1972). *In* "Biosynthesis and Its Control in Plants", Phytochemical Society Symposium No. 9 (B. V. Millborrow, ed.), p. 219. Academic Press, London and New York.
Cohen, B. S. and Estabrook, R. W. (1971). *Archs Biochem. Biophys.* **143**, 54–65.
Coolbaugh, R. C. and Moore, T. C. (1971). *Phytochemistry* **10**, 2401–2412.
Cross, B. E. (1968). *In* "Progress in Phytochemistry" (T. Reinhold and Y. Liwschitz, eds.), Vol. 1, pp. 195–222. Wiley and Sons, New York.
Cross, B. E., Galt, R. H. B. and Hanson, J. R. (1964). *J. chem. Soc.* 295–300.
Cross, B. E., Norton, K. and Stewart, J. C. (1968). *J. chem. Soc.* (C) 1054–1063.
Cross, B. E. and Stewart, J. C. (1970). *Phytochemistry* **9**, 1065–1071.
Dennis, D. T., Upper, C. D. and West, C. A. (1965). *Pl. Physiol., Lancaster* **40**, 948–952.
Dennis, D. T. and West, C. A. (1967). *J. biol. Chem.* **242**, 3293–3300.
Evans, R., Hanson, J. R. and White, A. F. (1970). *J. chem. Soc.* (C) 2601–2603.
Fall, R. R. and West, C. A. (1971). *J. biol. Chem.* **246**, 6913–6928.
Graebe, J. E., Bowen, D. H. and MacMillan, J. (1972). *Planta* **102**, 261–271.
Hanson, J. R. (1968). *In* "The Tetracyclic Diterpenes", pp. 114–121. Pergamon Press, London.
Hanson, J. R. and Hawker, J. (1971). *Chem. Commun.* 208.
Hanson, J. R. and White, A. F. (1969a). *J. chem. Soc.* (C) 981–985.
Hanson, J. R. and White, A. F. (1969b). *Chem. Commun.* 410–411.
Hedden, P. and MacMillan, J. (1971). *Tetrahedron Letters* 4939–4940.
Hower, C. D. and Batra, P. P. (1970). *Biochem. biophys. Acta* **222**, 174–179.
Jaffe, M. J. (1971). *Pl. Physiol.* (Supplement) **47**, 49.
Jefferies, P. R., Knox, J. R. and Ratajazak, T. (1970). *Tetrahedron Letters* 3229–3231.
Lang, A. (1970). *A. Rev. Pl. Physiol.* **21**, 537–570.
Lew, F. T. and West, C. A. (1971). *Phytochemistry* **10**, 2065–2076.
MacMillan, J. (1971). *In* "Aspects of Terpenoid Chemistry and Biochemistry" (T. W. Goodwin, ed.), pp. 153–180, Academic Press, New York.
Miller, W. L., Brady, D. R. and Gaylor, J. L. (1971). *J. biol. Chem.* **246**, 5147–5153.
Mitchell, J. W., Wirwille, J. W. and Weil, L. (1949). *Science, N.Y.* **110**, 252–254.
Murphy, P. J. and West, C. A. (1969). *Archs Biochem. Biophys.* **133**, 395–407.
Newhall, W. F. (1969). *Nature, Lond.* **223**, 965–966.
Oster, M. O. and West, C. A. (1968). *Archs Biochem. Biophys.* **127**, 112–123.
Piozzi, F., Passananti, S., Paternostro, M. P. and Spiro, V. (1971). *Phytochemistry* **10**, 1164–1166.
Pitel, D. W., Vining, L. C. and Arsenault, G. P. (1971). *Can. J. Biochem.* **49**, 194–200.
Reid, D. M. and Crozier, A. (1970). *Planta* **94**, 95–106.
Ruzicka, L. (1953). *Experientia* **9**, 357–367.
Shechter, I. and West, C. A. (1969). *J. biol. Chem.* **244**, 3200–3209.
Upper, C. D. and West, C. A. (1967). *J. biol. Chem.* **242**, 3285–3293.
Verbiscar, A. J., Cragg, G., Geissman, T. A. and Phinney, B. O. (1967). *Phytochemistry* **6**, 807–814.
Waller, G. R. and Burström, H. (1969). *Nature, Lond.* **222**, 576–578.

CHAPTER 8

Stereochemical Aspects of Enzyme Action

J. W. CORNFORTH

Shell Research Ltd., Milstead Laboratory of Chemical Enzymology, Sittingbourne Laboratories, Sittingbourne, Kent, England

One of Louis Pasteur's original methods (Pasteur, 1858, 1860) for optical resolution of racemic substances—and one that is still being used—is to allow an organism to feed on the racemic substance, when quite often only one of the two mirror-image forms is consumed. This was the first experimental observation of the stereospecificity of enzyme action, though at the time even the existence of enzymes was not recognized. Later, Emil Fischer was so impressed by the specificity of his crude enzyme preparations in selecting particular stereoisomers that he likened the relationship between enzyme and substrate to that between a lock and a key.

Since this time, and particularly since isotopic labelling was used for the solution of biochemical problems, the stereochemical control exerted by enzymes over the substrates that they will accept, and the reactions that they mediate, has been explored in much greater detail. Enzymes are easily the most specific of all known catalysts in the stereochemical sense, and it is worth while to wonder why specificity of this order is necessary to the processes of life.

Substrate specificity is for most purposes an obvious advantage. A living cell manages to keep an almost incredibly complex system of chemical changes under harmonious control with surprisingly little recourse to the physical separation of different cellular components from each other. The refusal of many enzymes to accept molecules—even molecules of similar chemical type—other than their "natural" substrates, is a powerful aid to orderly functioning. The less substrate-specific enzymes are generally those dealing with foreign substances that need to be removed from the cell.

Less obvious are the reasons for the detailed stereospecificity of the processes catalysed. Of course, if a symmetrical molecule is transformed enzymically into a molecule that can exist in two mirror-image forms, it is not surprising, given substrate stereospecificity of the enzymes which use the product, that only one of the two forms should be produced. If it were not, we should need another set of enzymes to deal with the opposite form; and application of this principle to all the asymmetric substrates in a cell would certainly lead to horrible metabolic chaos. Yet one finds that even when the product is what a chemist would call a symmetrical substance it has been produced by a strictly stereospecific, and invariably asymmetric, procedure; though the same substance could have been formed by a completely non-stereospecific catalysis. It is only during the past 25 years, and only then by the use of isotopic labelling, that this inherent and often apparently unnecessary stereochemical control has been discovered.

When a chemical reaction takes place in solution between symmetrical reagents, asymmetric molecules may be formed; but if they are, statistical laws require that right- and left-handed molecules shall be formed in equal numbers. In enzymic catalysis, however, the reaction appears to occur only after the binding of the substrates to a specific region, known as the active site, of the enzyme protein; and whatever else an enzyme is, it is certainly not symmetrical, being constructed from amino acids which all (except glycine) contain centres of asymmetry and occur in one particular stereochemical form. Not even the assembly of these amino acids has any recognizable element of symmetry. The environment in which reaction occurs is, then, so asymmetric that occurrence of equal numbers of the two mirror-image forms of a non-symmetric product is no longer to be expected.

Enzymes, however, go much further than this. A good example is provided by the enzyme isopentenyl pyrophosphate isomerase, and it happens that we have just completed (Clifford *et al.*, 1971) a study of the substrate stereochemistry of this enzyme (from pig liver). It catalyses the interconversion of isopentenyl pyrophosphate (IPP) and dimethylallyl pyrophosphate (DMAPP); and the combination of these two intermediates on another enzyme is the first step in the biosynthesis of polyisoprenoids.

Formally, this is a simple chemical process: a hydrogen (which comes from the aqueous medium in which the enzymic reaction proceeds) is added to an olefinic carbon atom and a second hydrogen is abstracted from a saturated carbon atom to form a double bond in a new position. There are some chemical reagents and catalysts that execute similar interconversions, though it would be difficult indeed to find one that would work in aqueous solution with this particular pair of isomers. But if it did, and if one allowed the isomerization to proceed in excess deuterium oxide for long enough to reach isotopic equilibrium, the two isomers would certainly have the following composition

Now although this experiment has not specifically been carried out with the enzyme as catalyst, there is no doubt from the available evidence that equilibrium would be reached when the substrates have not seven deuterium atoms each, but three, thus:

It is easy to see that the enzyme has discriminated not only between the two methyl groups of DMAPP but between the two hydrogens on the saturated methylene group of IPP.

There is another and even more subtle discrimination. If IPP specifically labelled with tritium at an olefinic hydrogen is used as a substrate in deuterium oxide, it can be shown that the new methyl group in the DMAPP is asymmetric and that its chirality (right- or left-handedness) depends on the position of the tritium in the IPP, thus:

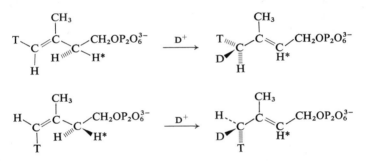

This means that addition of hydrogen (in this case, of deuterium) from the medium to the double bond in IPP is always to the same side of the double bond (in the projection shown, from above the plane of the paper).

Why is it necessary for the enzyme to exert such rigorous stereochemical control over a reaction of this kind? Neither of the isomers is an asymmetric molecule, and the result would be exactly the same if none of these three elements of discrimination was effective. I think that there are two basic reasons, not entirely independent of each other: the way in which the substrate is bound to the enzyme, and the way in which the enzyme catalyses the chemical reaction.

An energetically favourable combination of enzyme and substrate can arise from several types of interaction: electrostatic attraction between groups of opposite charge or polarization, van der Waals forces, and changes in solvation energy are among these. For the isomerase, one can see the possibilities for ionic binding: the substrates have an anionic pyrophosphate group and the enzyme, like the majority of those handling organic phosphates, requires magnesium and/or manganous cations as co-factors. This is not, however, sufficient; many other pyrophosphates are present in a living cell and it is necessary for the specific action of the enzyme that these should not be bound too efficiently. Thus the active site presumably includes hydrophobic locations which "recognize" and receive, say, the methyl group of IPP, and hence it is likely that the conformational movement possible to IPP in the enzyme-substrate complex is strictly limited. The freedom of DMAPP will be even more limited, since the central double bond already holds all its five carbon atoms in a single plane.

The chemical reaction catalysed by the isomerase is in the nature of a prototropic shift. There is no serious doubt that the hydrogen is added and abstracted in essentially the cationic form, since both hydrogens are freely exchangeable with the medium, water. Whichever way the isomerization is running a C—H bond therefore dissociates heterolytically, leaving the electrons with carbon.

Water is both a donor and a receiver of protons, and perhaps the simplest model that could be formulated for the isomerase would postulate hydrophilic regions, accessible to water molecules, around the upper side of the double bond in IPP, and around the underside of the double bond in DMAPP. Hydrophobic regions around other locations could ensure that water has no other access to the region where the chemical reaction occurs. On this simple view, a water molecule or derived oxonium ion collides with the bound IPP molecule, generating some species similar to a carbonium ion that loses a proton by collision with a water molecule in the other hydrophilic region. This would explain all the stereochemical findings.

Against this view it could be argued that at the neutral pH of the enzymic reaction the protonation of an olefinic double bond by water in aqueous solution is a highly improbable event and that there is no obvious reason why this particular environment should make it happen so easily. And although the relative stereochemistry of the added and abstracted protons is favourable for a *concerted* reaction—the new double bond being created as the old one is saturated—this also is an improbable process if it has to depend on two simultaneous collisions.

An alternative and perhaps more probable view supposes that the addition and abstraction of protons are under the control of two amphoteric groups (B) attached to the enzyme (Fig. 1). Suitable groups could be —SH/S$^-$, —NH$_3^+$/ —NH$_2$, —CO$_2$H/CO$_2^-$, =$\overset{|}{\text{N}}$H$^+$/=$\overset{|}{\text{N}}$: groups regularly found in all proteins. Note that one such group is insufficient for the purpose—with the known stereochemistry it would either have to be in two places at once or the substrate would have to move very considerably in relation to the active site.

In the enzyme-substrate complex, both these groups could be in a hydrophobic environment. This not only excludes competition by water for the acidic proton; the effective charge on both the acidic and basic groups is also not dissipated in a polarizable medium. Further, both groups can be sited at exactly the most favourable positions for addition and removal of protons.

This not only makes these events much more probable; it also permits the concerted mechanism mentioned above. The acid and the base do not have to find their way simultaneously to the correct positions: they are there already, and there all the time.

An enzyme molecule that had catalysed in this way the conversion of a molecule of IPP to one of DMAPP would, after dissociation of the product, have its amphoteric groups arranged to execute the reverse transformation of DMAPP to IPP. Presumably at this stage the aqueous medium can equilibrate with the two amphoteric groups—the ultimate source of the protons forming new C—H bonds is undoubtedly the medium—but it does not necessarily follow that this process always happens faster than the binding of another

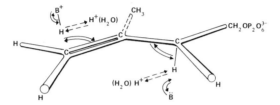

FIG. 1. Hypothetical concerted mechanism for isopentenyl pyrophosphate isomerase.

substrate molecule. Interestingly, this point could be tested experimentally, though the experiment would not be an easy one.

Simple reasoning has thus led us to the view that this enzyme catalyses a stereospecific process because it would be much more difficult to catalyse a stereochemically random process; and this is probably true of most enzymes. Unlike ourselves, they follow the strait and narrow path because it is the line of least resistance.

The work that led to elucidation of the stereochemistry of the isomerase completed a study, begun by Popják and the writer ten years ago, of the detailed stereochemistry of squalene biosynthesis from mevalonate (Popják and Cornforth, 1966). Squalene has fifty hydrogen atoms; of these, one is supplied from the pro-4S position of a reduced nicotinamide coenzyme, two come from the aqueous medium, five from the pro-5S hydrogen of mevalonate, six each from the pro-4R, pro-2R, pro-2S and pro-5R hydrogens of mevalonate, and eighteen from the methyl groups of mevalonate (Fig. 2). The disposition of these hydrogen atoms represents a choice of one stereospecific outcome from a total of 16 384 possibilities, and we have no evidence that any of the other 16 383 is operative to a significant extent, though the product—squalene—would be exactly the same if they were all operating at once.

Specificity, in the stereochemical sense, is so usual with enzymes that the exceptions become more interesting. It is, so far as I know, true that no enzyme has been found to deal with its natural substrate in other than a stereospecific manner; but liver alcohol dehydrogenase, for example, can reduce certain cyclic ketones so as to give both epimers of the corresponding alcohol (Graves *et al.*, 1965; and earlier references cited therein). Generally, these are formed at different rates. The enzyme mevaldate reductase, found in mammalian liver, will reduce mevaldate to mevalonate; but mevaldate is not an intermediate in the normal biosynthesis of mevalonate and the "natural" substrate of the reductase is not known. The enzyme reduces almost indifferently the *R* and *S* forms of mevaldate; but so far as the stereochemistry of reduction of the aldehyde group is concerned, this is strictly stereospecific, at least for the *R* form (Donninger and Popják, 1965). Similarly, propanediol dehydrase will accept indifferently *R* and *S* propanediol; but the rearrangement of each form

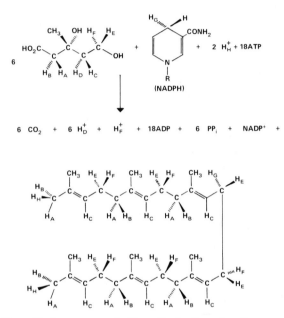

Fig. 2. Fate of hydrogen atoms in biosynthesis of squalene from mevalonate.

is highly stereospecific as has been shown in the laboratories of Abeles and Arigoni (review, Akhtar and Wilton, 1970). It is still undecided whether an enzyme (unless it happens to be a racemase!) ever produces, by operation on the same substrate, both possible enantiomeric or epimeric forms of a product. One can imagine circumstances in which this might happen: for example an enzymic reaction in which the final stage, not rate-limiting, was the acceptance of a molecule of water from the medium. This addition, perhaps, need not always occur in an environment where approach is possible from one direction only. A long-known and curious feature of plant biochemistry is the frequency with which terpenoids, particularly monoterpenoids, occur in partly racemized form. No doubt this is often due to the operation of two enzymes, producing molecules of opposite chirality, but I wonder if this is always so.

I have said that much of the information on the stereochemical control exerted by enzymes over their substrates has been gained by isotopic labelling. The question then must be faced: to what extent are isotopically labelled substrates "unnatural" to the enzyme?

Isotopically labelled substrates differ from their normal counterparts in three significant ways. First, the isotopic atoms do not have the same effective radii. This difference is probably insignificant for [^{13}C] and [^{14}C] but it is large enough in hydrogen isotopes to be detectable (Horeau and Nouaille, 1966) by Horeau's method for investigating absolute configurations: a method dependent on differential steric hindrance to a chemical reaction. Secondly, the energy required to make and break bonds with the isotopic atom is not the same as for the normal atom: the primary isotope effect. Thirdly, the presence of adjacent isotopic atoms may affect the energy of bond formation and breakage between normal atoms: the secondary isotope effect.

It is difficult to judge the importance of the first and the third factors in enzymic reactions. Perhaps the best indications are provided by our own work, with Eggerer and his colleagues, on enzymic reactions of the asymmetric methyl group (Cornforth et al., 1970). Here, methyl groups were made that contained one each of the three hydrogen isotopes, and they were put through enzymic reactions which chose one of these three atoms for removal. Often, a parallel experiment was done with methyl groups containing one tritium atom and two normal hydrogens. Now so far, we have always found that the removal of hydrogen follows the general rules for the primary isotope effect: hydrogen is removed more easily than deuterium, deuterium more easily than tritium, and when two hydrogens accompany one tritium on a methyl group they are abstracted to an equal extent, though in an asymmetric environment responsive to the smaller size of the tritium atom this need not be the case. Cautiously, therefore, we may conclude that the most important effect of isotopes on enzymic reactions is the primary effect, and that the stereochemical effects are less significant.

The selection of ordinary hydrogen rather than deuterium or tritium when all three isotopes are present in a single methyl group has, however, a stereochemical element. Two possible means of selection can be imagined: in one, the methyl group has space to rotate in the enzyme-substrate complex and a particular hydrogen can be chosen. In the other, the complex is rigid and the substrate may dissociate, becoming free to rotate, and become re-attached until the right hydrogen atom happens to be bound next to the group that removes it. Again, it is not impossible to think of experiments that will distinguish between these two possibilities.

In Milstead Laboratory's stereochemical investigations of enzyme action, our main motive has been to understand the nature of the catalytic process. Thus far, the principal application of our findings by others has been the elucidation of pathways of biosynthesis. We are well content that this should be so; for in the end, the reduction of a biosynthetic pathway to a sequence of well-understood enzymic processes is the outcome that all of us seek.

REFERENCES

Akhtar, M. and Wilton, D. C. (1970). *A. Rep. chem. Soc.* (B) 558.

Clifford, K., Cornforth, J. W., Mallaby, R. and Phillips, G. T. (1971). *Chem. Commun.* 1599.

Cornforth, J. W., Redmond, J. W., Eggerer, H., Buckel, W. and Gutschow, C. (1970). *Europ. J. Biochem.* **14**, 1.

Donninger, C. and Popják, G. (1965). *Proc. R. Soc.* (B) **163**, 465.

Graves, J. M. H., Clark, A. and Ringold, H. J. (1965). *Biochemistry* **4**, 2655.

Horeau, A. and Nouaille, A. (1966). *Tetrahedron Letters* 3953.

Pasteur, L. (1858). *C. r. hebd. Séanc. Acad. Sci., Paris* **46**, 615.

Pasteur, L. (1860). *C. r. hebd. Séanc. Acad. Sci., Paris* **51**, 298.

Popják, G. and Cornforth, J. W. (1966). *Biochem. J.* **101**, 553.

CHAPTER 9

The Control of Fatty Acid Biosynthesis in Plants

B. SEDGWICK

Milstead Laboratory of Chemical Enzymology, Shell Research Limited,
Sittingbourne Laboratories, Sittingbourne, Kent, England

I. INTRODUCTION

The general field of fatty acid biosynthesis has been the subject of widespread investigations over the past few decades, and as a result of this it is not possible to cover this topic comprehensively in the present chapter due to limitations of time and space. In addition, the title of this symposium imposes certain limitations on the scope of this article within which I feel obliged to remain. Accordingly, it is the aim of this chapter to examine the mechanism of fatty acid biosynthesis and to attempt to relate this to the control systems which operate in this pathway and, furthermore, to relate these to plant systems wherever possible. It is, however, necessary to refer to bacterial, yeast and animal systems as well in view of the important nature of investigations into fatty acid biosynthesis which have been performed using such organisms.

Aspects of the biosynthesis of fatty acids not dealt with in this chapter, such as the formation of non-conjugated ethylenic acids, acetylenic acids, hydroxyacids and other topics have been reviewed in a recently published monograph entitled "Plant Lipid Biochemistry" (Hitchcock and Nichols, 1971).

II. MECHANISM OF FATTY ACID BIOSYNTHESIS

A. *De novo* BIOSYNTHESIS

The *de novo* biosynthesis of even-numbered long chain fatty acids involves the sequential addition of C_2 units, in the form of acetate (Rittenberg and Bloch, 1944) by a series of reactions which have now been well characterized in a wide variety of plant and animal tissues (Wakil, 1961, 1963; Lynen, 1961, 1967; Lennarz *et al.*, 1962; Vagelos, 1964; Stumpf, 1969; Porter *et al.*, 1971). For reasons of specificity and energy requirements, the C_2 units do not react as such but as thiol esters of certain cofactors, or of the enzymes involves in the subsequent transformations (Baddiley, 1955; Lynen, 1961, 1967; Sauer *et al.*, 1964; Phillips *et al.*, 1970a, b). Furthermore, C_2 fragments which condense with the priming acyl thioester are first additionally activated to form a C_3 unit, malonate, which exists in its initial form as the thioester malonyl-CoA, from which the name "malonyl-CoA pathway" is derived to describe the *de novo* system of fatty acid biosynthesis (Gibson *et al.*, 1958a, b; Wakil, 1958; Wakil and Ganguly, 1959).

There then follows a sequence of reduction-dehydration-reduction reactions after each condensation of malonate. It is in the condensation reaction that the carbon added as bicarbonate to form the C_3 unit is lost as carbon dioxide.

The scheme shown in Fig. 1 illustrates the sequence of reactions catalysed by the best characterized yeast fatty acid synthetase complex, that of *Saccharomyces cerevisiae*, which are responsible for the formation of long chain fatty acids from acetyl-CoA and malonyl-CoA (Lynen, 1967). The carboxylation of acetyl-CoA to form malonyl-CoA is not shown in this scheme, and is discussed later. The mechanism of fatty acid synthesis shown in Fig. 1 is essentially the same for *de novo* fatty acid synthesis in a wide variety of organisms with the following exceptions.

The participation of flavin mononucleotide (FMN) in hydrogen transfer in the second reduction is peculiar to yeast and *Clostridium kluyveri*, and does not occur in other bacteria, higher plant or animal systems. Secondly, the termination reaction involving transfer of the completed fatty acyl chain to CoA does occur in certain other organisms such as dark-grown *Euglena*, but does not apparently exist in systems which produce free fatty acids or acyl-carrier protein esters as the major product (see Table I). Finally, the thiol

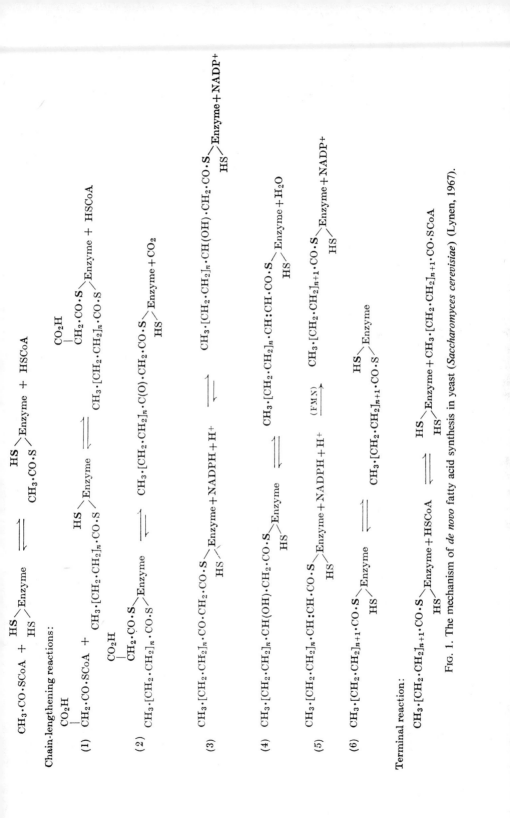

Fig. 1. The mechanism of *de novo* fatty acid synthesis in yeast (*Saccharomyces cerevisiae*) (Lynen, 1967).

TABLE I

Properties of fatty acid synthetase systems (Brindley et al., 1969)

Source	Molecular weight	Products	
Type I			
Yeast	$2 \cdot 3 \times 10^6$	14:0–18:0	CoA esters
Pigeon liver	$4 \cdot 5 \times 10^5$	16:0	Free acid
Rat liver	$5 \cdot 4 \times 10^5$	16:0	Free acid
Adipose tissue	—	14:0, 16:0	Free acid
Mammary gland	—	8:0–18:0	Free acid
Euglena gracilis (etiolated)	ca 1×10^6	16:0	CoA ester
Mycobacterium phlei	$1 \cdot 7 \times 10^6$	14:0–26:0	Esters
Type II			
E. coli		16:0, 18:0, 16:1	
Clostridium species		18:1, 3h-10:0	ACP esters
Pseudomonas species		3h-12:0, 3h-14:0	
Bacillus subtilis		C_{15} and C_{17} isoacids and anteisoacids	—
Avocado mesocarp		16:0, 18:0	—
Lettuce chloroplasts		16:0, 18:0	—
Spinach chloroplasts		16:0, 18:0	—
Euglena gracilis (photoauxotrophic)		18:0	ACP ester

groups which carry the reacting molecules are not always enzyme bound as indicated in Fig. 1.

1. Formation of Acetyl-CoA

The three reactions shown in Fig. 2 provide most of the acetyl-CoA required for fatty acid synthesis. Reaction A illustrates the activation of free acetate by acetate thiokinase or acetyl-CoA synthetase, a cytoplasmic enzyme, forming the CoA thioester in a classical magnesium dependent, ATP-driven kinase reaction, a mechanism for which has been proposed (Green and Allman, 1968). In higher plants, this enzyme has been purified from peanut (*Arachis hypogea*) cotyledons (Rebeiz et al., 1965) and from potato (*Solanum tuberosum*) tuber, in which tissue it has been shown to exist in multiple molecular forms, each with a molecular weight of about 60000 (Huang and Stumpf, 1970).

Equation B shows the formation of acetyl-CoA via the citrate cleavage enzyme, which also occurs in the soluble fraction of the cell. Acetyl-CoA for fatty acid synthesis may also arise from the glycolytic pathway as a result of the oxidative decarboxylation of pyruvate. However, the pyruvate dehydrogenase complex which catalyses this reaction is located in the mitochondria, hence acetyl-CoA formed there has to be transported to the cytoplasm via the carnitine transferase system of the mitochondrial membrane before it can be utilized for *de novo* synthesis. There is, however, a mitochondrial system for the elongation of preformed fatty acids which utilizes acetyl-CoA, and it is likely that acetyl-CoA produced in this manner is an important C_2 donor for this system.

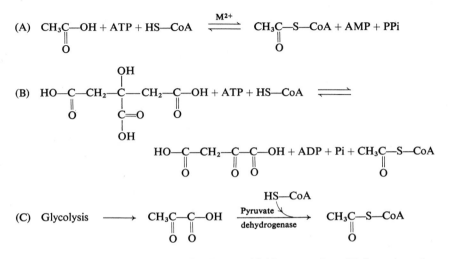

FIG. 2. The formation of acetyl-CoA. (A) Acetate thiokinase reaction; (B) from citrate by the action of the citrate cleavage enzyme; (C) from pyruvate by the action of the pyruvate dehydrogenase complex.

The β-oxidation of fatty acids is, of course, a major pathway for the formation of acetyl-CoA in the cell. However, the main purpose of this catabolic pathway is to provide a source of energy from fat storage tissues. Accordingly, this acetyl-CoA is channeled mainly into the citric acid cycle, or glyoxylate cycle, for maximum energy production, and under the metabolic conditions favouring β-oxidation it is unlikely that acetyl-CoA produced by degradation of long chain fatty acids will become available for their immediate resynthesis.

2. Formation of Malonyl-CoA

Generally speaking, the carboxylation of acetyl-CoA (shown in Fig. 3, equation a) catalysed by the enzyme acetyl-CoA carboxylase is the major source of malonyl-CoA for fatty acid synthesis in most organisms, and this process will be discussed in more detail. Equation b shows the malonate kinase reaction which has been demonstrated in a wide variety of plant tissues (Hatch and Stumpf, 1962; Hayaishi, 1955), where it appears to occur more frequently than in animal cells. Malonate kinase seems to be of major importance in tissues such as bush bean (*Phaseolus vulgaris*) roots where acetyl-CoA carboxyl-ase is apparently absent and where the oxidative decarboxylation of oxalo-acetate has been shown to give rise to free malonate (Shannon *et al.*, 1963; de Vellis *et al.*, 1963). A third pathway whereby malonyl-CoA can be produced is via a thiophorase reaction (equation c) in which CoA is transferred to malonate from either succinyl-CoA or acetoacetyl-CoA (Mahler and Cordes, 1966), but its importance in *de novo* synthesis of fatty acids must be questioned in terms of the availability of these two substrates, and also of malonate itself.

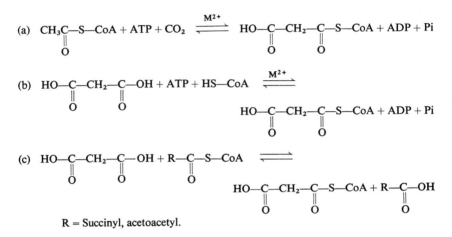

Fig. 3. The formation of malonyl-CoA. (a) Acetyl-CoA carboxylase reaction; (b) malonate thiokinase reaction; (c) from malonate via a thiophorase reaction.

3. *Acetyl-CoA Carboxylase*

Until 1958, it was generally assumed that the *de novo* biosynthesis of fatty acids took place by the reversal of β-oxidation. Then it was observed, using an avian liver system (Gibson *et al.*, 1958a, b), that ATP was required for optimum synthesis, and that use of bicarbonate buffer caused a tremendous stimulation of the incorporation of acetate into fatty acids. This lead to the discovery that the importance of bicarbonate and ATP lay in their involvement in the carboxy-lation of acetyl-CoA to form malonyl-CoA (Wakil, 1958) which was shown to be the actual C_2 donor in fatty acid synthesis in pigeon liver (Wakil and Ganguly, 1959), rat liver (Brady, 1958) and yeast (Lynen, 1959). The enzyme catalysing this carboxylation, acetyl-CoA carboxylase, was shown to be a biotin containing enzyme (Wakil and Gibson, 1960) and it was subsequently shown that the protein-bound prosthetic group becomes carboxylated (Lynen *et al.*, 1963; Matsuhashi *et al.*, 1964) to form 1'-*N*-carboxybiotin (Numa *et al.*, 1964), thus fulfilling the role of carbon dioxide carrier in this and several other biotin-dependent carboxylases (Lynen, 1967).

Acetyl-CoA carboxylase has been widely studied in animal tissues (see recent Review by Numa *et al.*, 1970) where it has been shown to be an extremely complex protein of high molecular weight, up to 8×10^6 in chicken liver (Gregolin *et al.*, 1966a, b), which consists of several subunits whose state of association can be controlled by substances such as citrate and isocitrate. Correlated with these changes in subunit structure was an enzymatic activation produced by these compounds. This effect was shown to be allosteric in nature and this, together with data on the apparently rate limiting effect of acetyl-CoA carboxylase on fatty acid synthesis from acetyl-CoA has led to the widely accepted view that this enzyme is the major controlling factor on the rate of fatty acid synthesis in animal tissues (see Numa *et al.*, 1970). However, in most yeasts, *Escherichia coli* and higher plants this activation phenomenon, with associated changed in the quaternary structure of the protein(s), does not occur. It thus seems questionable whether this enzyme exerts a significant controlling influence on fatty acid synthesis in these systems.

This enzyme has recently been studied in great detail in *E. coli* (Alberts and Vagelos, 1968; Alberts *et al.*, 1969a, b) where it has been shown to consist of three proteins, E_1, E_2 and E_3. The first step in the reaction involves the carboxylation of fraction E_2, the biotin carrier protein, catalysed by fraction E_1 or biotin carrier protein carboxylase, to form E_2-biotin-CO_2. The second part of the reaction involves transfer of the carbon dioxide from this complex to acetyl-CoA, forming malonyl-CoA, this being catalysed by E_3 or N-carboxybiotin: acetyl-CoA transcarboxylase. This cyclic process is shown schematically in Fig. 4 (Alberts *et al.*, 1969b). Fig. 5 shows the structure of biotin (R = enzyme protein, fraction E_2 in the *E. coli* system) and two possible mechanisms for its carboxylation, a concerted mechanism (a) (Lynen, 1967) and a stepwise process (b), involving a phosphorylated-biotin intermediate (Green and Allman, 1968). The mechanism of biotin action in enzymic systems has recently been reviewed by Knappe (1970).

In higher plants, this enzyme has been best characterized in wheat (*Triticum sativum*) germ (Hatch and Stumpf, 1961; Heinstein and Stumpf, 1969) and apparently consists of an $E_1 + E_2$ protein of molecular weight about 220000 and an E_3 protein of molecular weight 180000. Both fractions are capable of aggregation to form units of higher molecular weight, but this is not accom-

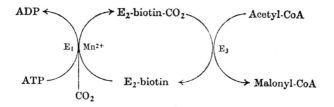

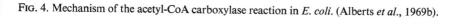

FIG. 4. Mechanism of the acetyl-CoA carboxylase reaction in *E. coli*. (Alberts *et al.*, 1969b).

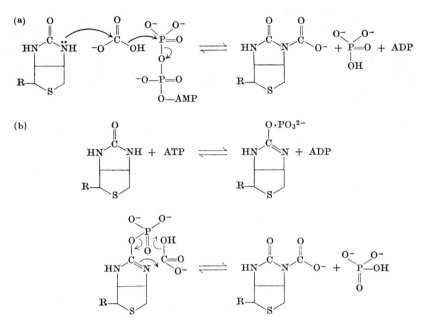

FIG. 5. Suggested mechanisms for the carboxylation of biotin. (Lynen, 1967). (a) A concerted mechanism; (b) a stepwise process involving a phosphorylated biotin intermediate.

panied by any activation processes. This enzyme and the carboxylase from lettuce (*Lactuca sativa*) chloroplasts (Burton and Stumpf, 1966) have been shown to be strongly inhibited by an unknown compound released from the disrupted chloroplasts, and it has been tentatively suggested that this substance may exert a regulatory effect *in vivo*.

4. *CoA and Acyl-carrier Protein*

It was thought for a long time that CoA thioesters were the reacting species in fatty acid synthesis, as in the degradative β-oxidation pathway. This idea was supported by the fact that acetyl-CoA and malonyl-CoA were produced in the cell by various reactions of intermediary metabolism, and, when added to enzymic systems producing fatty acids, were rapidly incorporated into these products. However, it has now been clearly established that in synthetic reactions of fatty acid metabolism, primary substrates and intermediates do not react as thioesters of CoA, but as protein-bound thioesters (Wakil *et al.*, 1964; Majerus *et al.*, 1964; Lynen, 1967; Phillips *et al.*, 1970a, b). Accordingly, the primary substrates acetyl-CoA and malonyl-CoA first undergo reactions catalysed by distinct and specific transacylase enzymes (Williamson and Wakil, 1965; Lynen, 1967; Phillips *et al.*, 1970b; Plate *et al.*, 1970; Joshi and Wakil, 1971) in which the acetate and malonate moieties are transferred to the thiol site of a carrier protein. In bacterial and most plant systems, this carrier is a discrete protein of low molecular weight called acyl-carrier protein or ACP for short. ACP from *E. coli* has been well characterized (Sauer *et al.*, 1964; Majerus *et al.*, 1964; Pugh and Wakil, 1965; Vanaman *et al.*, 1968a, b; Majerus 1967, 1968) and has been shown to be a highly acidic protein containing 77 amino acid residues and having a molecular weight of 8780 (Vanaman *et al.*,

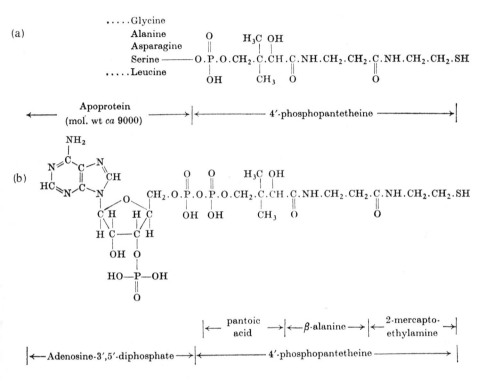

FIG. 6. The structures of acyl-carrier protein and coenzyme A. (a) Acyl-carrier protein from *E. coli*, showing the prosthetic group and adjacent amino acids of the *apo*-protein; (b) coenzyme A.

1968b). ACP from spinach (*Spinacia oleracea*) chloroplasts contains 88 residues, has a molecular weight of 9700 and has properties almost identical to ACP from *E. coli* (Simoni and Stumpf, 1969). Simoni *et al.* (1967) have also compared other ACPs. The dominant feature of these proteins is a 4'-phosphopantetheine side chain linked to a serine hydroxyl and terminating in the free thiol group of 2-mercaptoethylamine. Fig. 6 shows this structure and the adjacent amino acids in ACP from *E. coli*. Also shown is the structure of CoA, and it is seen that both have a 4'-phosphopantetheine side chain, and in fact this group of ACP is apparently derived from that of CoA (Alberts and Vagelos, 1966; Elovson and Vagelos, 1968).

5. *Fatty Acid Synthetase*

The ACPs referred to in the previous section are found in *de novo* fatty acid synthetases in which the constituent biosynthetic enzymes can be easily separated from one another and which can be termed dissociated systems. They are most commonly found in bacteria and higher plants, and the function of ACP in these systems is to provide a common carrier for the intermediates of fatty acid synthesis which can be recognized by the binding site or sites of all the enzymes involved. In systems reassembled *in vitro* from, for example, the separated component enzymes of the *E. coli* synthetase, addition of ACP is obligatory for fatty acid synthesis from acetyl-CoA and malonyl-CoA. In such an *in vitro* assay, it has not been possible to demonstrate any interaction between component enzymes of the synthetase, but this does not necessarily mean that none exists *in vivo*.

In the multienzyme complex type of system such as is found in yeasts and animal tissues, the component enzymes are bound extremely tightly by a variety of forces into a relatively high molecular weight, soluble multienzyme system which apparently contains all the enzyme activities found in the dissociated *E. coli* system. However, a discrete ACP has not been positively identified, although attempts have been made (Joshi *et al.*, 1968; Willecke *et al.*, 1969) to isolate a protein containing the covalently bound 4′-phosphopantetheine which is present and which has been shown to be the central carrier thiol in these multienzyme complexes (Lynen, 1967; Phillips *et al.*, 1970a, b; Nixon *et al.*, 1970; Joshi *et al.*, 1970). Figure 7 shows a tentative schematic representation of the structure of the synthetase complex from yeast (Lynen, 1967) which ties together many functional aspects of this complex, even down to its visualization by negative staining electron microscopy (Hagen and Hofschneider, 1964). This technique reveals an approximately spherical or disc-shaped body composed of several subunits and having a central "hole" or zone of low density.

Most of the knowledge concerning *de novo* synthesis of fatty acids in higher plants is due to the work of Stumpf and his collaborators. An interesting feature of these fatty acid synthetases is that, although showing most of the characteristics of the dissociated, bacterial type systems discussed above, many are associated with membrane structures, especially those in photosynthetic tissues. This may reflect the dependence of the enzyme in the intact cell on cofactor production by the chloroplasts, and in tissues such as lettuce chloroplasts both membrane and soluble fractions from osmotically disrupted organelles are required for *de novo* synthesis (Mudd and McManus, 1962). However, more violent disruptive processes such as the French press gave an entirely soluble system (Brooks and Stumpf, 1966) which was completely dependent on added ACP, NADPH and NADH for fatty acid synthesis, whereas the intact chloroplast did not require these added cofactors. In germinating seeds (McMahon and Stumpf, 1966) complete synthetase activity appears to exist in both particulate and soluble fractions of the cell, as is the case with avocado (*Persea*

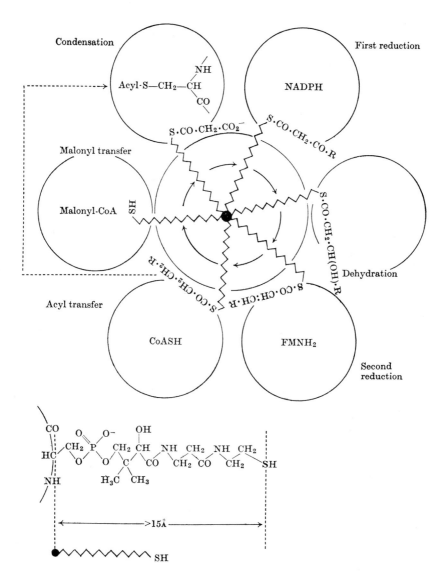

FIG. 7. Schematic representation of the multienzyme fatty acid synthetase complex of yeast (*Saccharomyces cerevisiae*) showing the structure and dimensions of the covalently bound 4′-phosphopantetheine group (Lynen, 1967).

gratissima) mesocarp preparations (Overath and Stumpf, 1964; Yang and Stumpf, 1965). These systems are interesting in that whereas the soluble fractions require ACP and are classical dissociated synthetases, the activity of the particles does not require ACP and is apparently tightly bound to the membranes. In other higher plant tissues such as developing soybean (*Glycine max*) (Rinne, 1969) and potato tubers (Huang and Stumpf, 1970), the *de novo* synthetase system is entirely soluble and of the dissociated type, requiring ACP.

B. CHAIN ELONGATION OF FATTY ACIDS

The synthetases discussed in Section A, whether dissociated or complex, only produce fatty acids up to a chain length of C_{18}, equivalent to the condensation of eight malonate units (see Fig. 1). However, longer chain acids are found in all tissues, especially so in plants. Their presence can be accounted for by two additional systems of fatty acid synthesis, which are termed chain elongation systems. These reactions do not occur in the cytoplasm, but in membranous organelles such as mitochondria and the microsomal system.

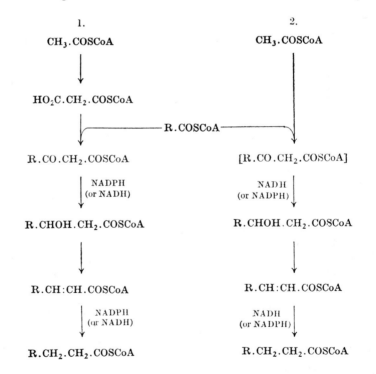

FIG. 8. Two pathways for the elongation of fatty acids. 1. "Microsomal" system utilizing malonyl-CoA and preferring NADPH. 2. "Mitochondrial" system utilizing acetyl-CoA and preferring NADH.

The major difference between the two elongation systems seems to be that in the microsomal system, malonyl-CoA is the C_2 donor (Nugteren, 1965; Guchait *et al.*, 1966) as in *de novo* synthesis, whereas in the mitochondria, acetyl-CoA is apparently the C_2 donor (Harlan and Wakil, 1963; Dahlen and Porter, 1968). Figure 8 shows the main features of these two systems in a generalized scheme which seems to be applicable to a variety of tissues. The intermediates and partial reactions are apparently the same as those in *de novo* synthesis, except that the CoA thioesters are the reacting species in most systems. However, there is some evidence, especially in higher plants, that ACP esters are utilized by the malonyl-CoA dependent elongation system (Harwood and Stumpf, 1971). Elongation reactions can occur with both saturated and unsaturated precursors, and in most tissues it is generally up to a maximum chain length of C_{22} or C_{24}. However, elongation up to C_{30} is quite common in many higher plants (Kolattukudy, 1966) where decarboxylation of this C_{30} fatty acid may give rise to the C_{29} paraffins characteristic of many plant waxes.

C. FORMATION OF UNSATURATED FATTY ACIDS

A study of the fatty acid composition of any organism reveals the presence of many unsaturated fatty acids, and this is particularly so in plants. It is necessary, therefore, to consider the formation of these acids in order to complete the biosynthetic picture.

Basically, two systems exist, the first of which is shown in Fig. 9, and is termed the anaerobic pathway in that it does not require oxygen. It is not active in higher plants and in animal tissues, but is not confined solely to anaerobic bacteria and has been demonstrated in aerobic bacteria, photosynthetic bacteria such as *Rhodopseudomonas* species, and pseudomonads (Hitchcock and Nichols, 1971). Knowledge concerning this pathway is largely due to the work of Bloch and his collaborators working on bacterial systems (Goldfine and Bloch, 1961; Scheuerbrandt *et al.*, 1961; Kass *et al.*, 1967; Bloch, 1969), and it is seen that the mechanism is due to a subtle variation in the basic scheme of *de novo* synthesis from malonyl-CoA. Enzyme i in this scheme is the classical 3-hydroxyacyl thioester dehydrase activity of the malonyl-CoA pathway (Mizugaki *et al.*, 1968a, b). However, it has been shown in *E. coli* that in addition there exists another dehydrase, which has an isomerase activity (shown as ii in Fig. 9) associated with it, and which is capable of forming *cis*-3-decenoyl thioesters from the normal *trans*-2-decenoyl thioester intermediate (Kass *et al.*, 1967; Brock *et al.*, 1967). As the subsequent reductase is specific for the *trans*-2-unsaturated species, the unsaturation present in the *cis*-3-decenoyl thioester is preserved during subsequent C_2 additions, thus giving rise to acids such as palmitoleate and *cis*-vaccenate. This partial decoupling of the normal reactions of *de novo* synthesis is controlled entirely by this specific dehydrase-isomerase enzyme, and as such this enzyme plays a key role in controlling monoenoic fatty acid synthesis in bacteria and similar species. In *E. coli* this enzyme is essentially specific for C_{10} acids, and has been shown to consist of two polypeptide chains, both of which are required for activity (Helmkamp and Bloch, 1969). However, in other bacteria the chain length

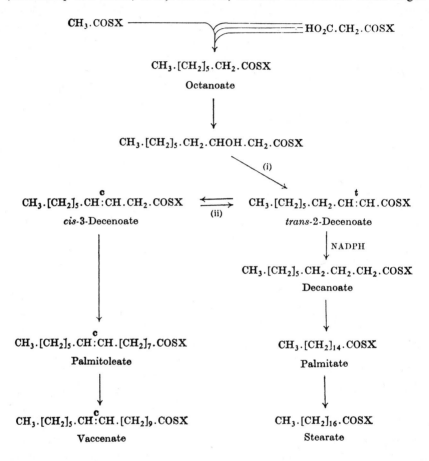

FIG. 9. The mechanism of fatty acid synthesis in *E. coli*, showing the pathway for the anaerobic formation of monounsaturated acids (Bloch, 1969).

specificity would appear not to be absolute, thus allowing the formation of the wide spectrum of monoenoic acids as shown in Fig. 10 (Scheuerbrandt and Bloch, 1962).

The second system for the formation of unsaturated fatty acids operates in yeasts, *Chlorella*, *Euglena* and higher plants as well as in animal tissues, and gives rise to polyenoic acids as well as monounsaturated acids. It is aerobic in nature, requiring oxygen and a reduced pyridine nucleotide to directly de-

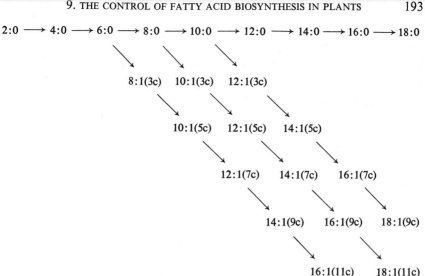

FIG. 10. General scheme for the anaerobic synthesis of monounsaturated fatty acids in bacteria (Scheuerbrandt and Bloch, 1962).

saturate a preformed long chain fatty acid (Bloomfield and Bloch, 1960), as shown in the following equation for desaturation of stearate to oleate.

$$CH_3(CH_2)_7CH_2CH_2(CH_2)_7COSCoA + O_2 + NADPH \rightarrow$$
$$CH_3(CH_2)_7CH{=}CH(CH_2)_7COSCoA + H_2O + NADP^+$$

The enzymes catalysing this type of reaction are called desaturases, are always intimately associated with membrane systems and require the participation of a flavoprotein mediated electron transport system. A system solubilized from membranes of etiolated *Euglena* cells (Nagai and Bloch, 1968) was shown to consist of at least three components, namely the desaturase enzyme, a flavin containing NADPH-oxidase and a ferredoxin protein. How these components may interact in the desaturase complex is illustrated in Fig. 11 (Bloch, 1969). In animal and yeast cells, the CoA thioesters are apparently the reacting

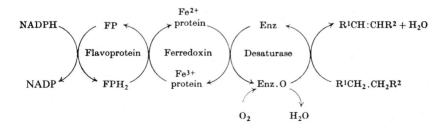

FIG. 11. Possible mechanism of action of fatty acid desaturase complexes (Bloch, 1969).

species, and whilst these substrates may also be utilized by higher plants, it would seem that ACP esters are the actual substrates in the cell (James *et al.*, 1968). Added CoA substrates are first transacylated to ACP before desaturation occurs, and this transfer is also required prior to elongation reactions in higher plants. Figure 12 shows possible pathways for the formation of monounsaturated acids by an aerobic system, and explains the mechanism of the so called "plant pathway" for the formation of unsaturated acids. This "plant

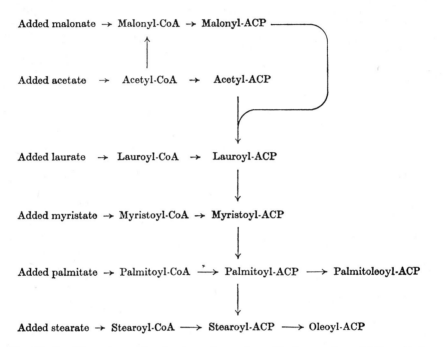

Fig. 12. Possible pathways for the elongation and aerobic desaturation of fatty acids in higher plants (after James *et al.*, 1968).

pathway" was originally thought to be a separate system for the aerobic desaturation of fatty acids, as it was observed in many higher plants (Stumpf and James, 1963), and also in green algae (Erwin and Bloch, 1964), that palmitate and stearate were apparently not directly desaturated to palmitoleate and oleate respectively, whereas shorter chain acids such as laurate and myristate were. The preferential utilization of palmitoyl-CoA and stearoyl-CoA in the synthesis of phospholipids and glycerides is thought to account for these observations in many systems, whilst the inhibition or absence of the specific CoA-ACP transferase enzymes in other tissues may also be a factor (Hitchcock and Nichols, 1971).

The formation of polyenoic acids occurs as a result of the further action of specific desaturase enzymes on monoenoic fatty acids. In animal and yeast cells the substrates again appear to be the CoA thioesters, whilst in higher plants the ACP esters are the actual substrates. However, there is evidence that phospholipid fatty acids may be the preferred substrate for the desaturation of oleate to linoleate in *Chlorella* (Gurr *et al.*, 1969). Whatever the exact mechanism of these desaturations, the generalized scheme shown in Fig. 13 illustrates how the sequential operation of desaturases, together with the elongation systems previously described, gives rise to the wide spectrum of unsaturated fatty acids found in animal and plant tissues.

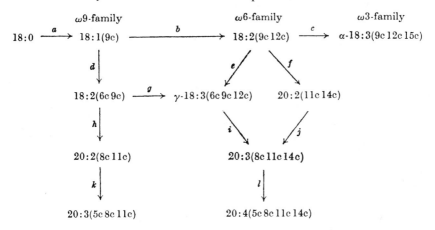

Fig. 13. Pathways for the biosynthesis of some polyenoic fatty acids (Hitchcock and Nichols, 1971). In higher plants the major route is a:b:c (involving α-linolenate). Route a:b:e:i:l is sometimes significant, especially in lower plants and algae. Routes d:h:k, f:j:l and e:i:l (involving γ-linolenate) are characteristic of animals, in which reactions b and g do not occur.

The exact mechanism of these desaturase enzymes, and the interrelationship of various classes of unsaturated fatty acids are far too complicated to consider in more detail in this chapter, but these subjects have been comprehensively discussed by Morris (1970) and by Hitchcock and Nichols (1971).

III. General Factors Affecting Fatty Acid Biosynthesis

A. SPECIES SPECIFICITY

Any perusal of the fatty acid composition of tissues immediately shows a marked species variation, and this may be attributed to the basic genetic differences between cells, or it may reflect the response of a particular cell or organism to environmental conditions. Obviously, these differences can ultimately be ascribed to variations in enzymatic activity in these cells, and this will be dealt with in Section IV. The general consideration of species variation in relation to fatty acid composition in higher plants is of particular importance commercially when a plant may be grown and harvested for food or other use, and it is here that the plant geneticist plays an important role. For example, it is desirable that Safflower (*Carthamus tinctorius*) seed oil should contain a high proportion of oleic acid so that it is liquid, but at the same time the amount of linoleic and linolenic acid must be kept at a minimum as these conjugated polyunsaturates are particularly susceptible to peroxidation reactions which lead to the formation of undesirable colour, odour and flavour in the oil. Genetic manipulation has made possible the isolation of strains of safflower, the seed oil of which meets these requirements (Putt *et al.*, 1969). At the same time, the final level of stearate in the seed oil has also been kept very low. Different genetic loci have been shown to control stearate formation and the formation of unsaturates (Knowles, 1969), as one would expect from the previous considerations concerning the separate nature of these biosynthetic pathways. It is apparent that the formation of high levels of oleate with low levels of polyunsaturates is caused by a lack of the desaturase responsible for introducing the Δ^{12} double bond into oleic acid (see Fig. 13). Alternatively, it may be better to grow a particular species, for example safflower instead of rape (*Brassica napus*), when variable climatic conditions prevail because this plant produces a more consistent seed oil composition at varying growth temperature.

The data presented in Table I show briefly the variations in the products of *de novo* synthesis of fatty acids in different species. This variation is seen not only in chain length, but also in the unsaturation and the nature of the product, as previously discussed. It should be stressed that the overall fatty acid composition of cells from these species may well differ significantly from that shown in Table I due to the subsequent activity of elongation and desaturase systems on the initial products of synthesis illustrated in this Table.

B. NUTRITIONAL STATUS

This heading is particularly applicable to animal tissues where it has been shown that starvation, for example, decreases the level of both fatty acid synthetase (Butterworth *et al.*, 1966) and acetyl-CoA carboxylase (Numa *et al.*, 1970) in the liver. Refeeding on a low fat diet produces an immediate increase not only in total amount of these enzyme proteins, but also a sharp increase in their specific activity. Similarly in plants, a poor nutritional level in the soil will lead to depleted metabolism and this will be reflected in a variety of pathways, including fatty acid synthesis. For example, deficiencies of iron (Newman, 1964), manganese (Bloch and Chang, 1964) and nitrogen (Wallace and Newman, 1965) have been shown to lead to a decrease in the level of polyunsaturated acids in photosynthetic tissues, and this has been correlated with a slow rate of chloroplast development. The fact that similar nutritional deficiencies caused only slight changes in the desaturation process in non-photosynthetic tissues such as the seeds of safflower (Yermanos *et al.*, 1964), cotton (*Gossypium hirsutum*) (van den Driessche, 1964) and rape (Appelqvist, 1968) suggests that this effect in leaves and similar tissues is directly linked with chloroplast synthesis and function. The level of carbon dioxide, a major nutritional factor for photosynthetic organisms, has been shown in *Chlorella fusca* to affect markedly the formation of both palmitic and oleic acid, there being a 50 % increase in overall fatty acid synthesis on increasing the carbon dioxide concentration from 1 % to 30 % (Dickson *et al.*, 1969). It is likely that this effect is mediated through the acetyl-CoA carboxylase enzyme, which becomes rate limiting at low levels of carbon dioxide.

C. GROWTH CONDITIONS

1. *Temperature*

The fact that temperature changes directly affect enzymatic rates and alter the availability of gases such as oxygen and carbon dioxide due to different solubilities makes interpretation of the effect of temperature on fatty acid synthesis extremely complicated. Furthermore, in photosynthetic tissues, oxygen production by this process further complicates the problem of oxygen availability. Thus it was initially observed (Harris and James, 1969) that in castor (*Ricinus communis*) seeds a decrease in temperature led to an increase in the formation of unsaturated acids. However, when the experiments were repeated under conditions of constant oxygen tension, the opposite effect was noted. In higher plants, the approach of winter has been shown to cause significant changes in cellular lipids (Gerloff *et al.*, 1966), including an increase in unsaturated fatty acid synthesis. In addition, variation in temperature has been shown (Hilditch and Williams, 1964) to cause changes in the plant and seed oils of many, but not all, species of higher plants. In this instance, lowered temperature increased the amount of polyunsatutated C_{18} acids compared with oleic acid.

Temperature has been shown to play a key role in the induction of specific desaturase enzymes, and in bacilli it has been observed (Fulco, 1970) that decreasing the growth temperature from 30° to 20° causes the induction of a Δ^5 desaturase responsible for the formation of 5-hexadecenoate from palmitate. What is of special interest is that this desaturase appears to be active at 20°, but not at 30°. This provides an example of the effect of external environment on the control of fatty acid synthesis at the molecular level. A similar stimulation of desaturase activity has been observed in yeasts grown at lowered temperatures (Meyer and Bloch, 1963).

The apparent reason for these modifications has been indicated by the recent studies on the fatty acid composition of membrane phospholipids in *E. coli* (Haest *et al.*, 1969; Esfahani *et al.*, 1971). The increase in unsaturation of these acids accompanying a decrease in growth temperature has been correlated with transition temperatures of the liquid crystal structures of isolated phospholipids having varying fatty acid composition, and with changes in the X-ray diffraction pattern of intact membrane fragments from cells grown at differing

temperatures. The observed changes tend to preserve the physical characteristics of the cell membranes so that membrane functions such as transport and
biosynthesis are minimally affected by these temperature changes.

2. Light

It has been indicated in Section II that in photosynthetic tissues, the chloroplasts appear to be the main site of fatty acid synthesis, and this process is
apparently intimately connected with photosynthesis. For example, fatty acid
synthesis is stimulated by light and inhibited by photophosphorylation inhibitors. Although it has been shown (Stumpf et al., 1963) that non-cyclic
photophosphorylation, producing cofactors such as ATP and NADPH,
occurs together with fatty acid synthesis in these tissues, the effect of light in
stimulating fatty acid synthesis could not be accounted for in terms of cofactor
production alone. It was suggested (Stumpf et al., 1967) that another effect of
light may be to cause the photoreduction of certain disulphides to form the
active thiols, for example ACP.

The conditions of illumination of Chlorella and Euglena have also been shown
to affect unsaturated fatty acid biosynthesis in these cells (Nagai and Bloch,
1965, 1968) in two ways. For example, growth of illuminated Chlorella in
carbon dioxide causes a stimulation of unsaturated fatty acid biosynthesis,
though this may be related to increased membrane synthesis in the chloroplasts
under these conditions. Secondly, illumination of Euglena cells apparently
causes the induction of the CoA-ACP transferase enzyme which enables
stearate to be directly desaturated to form oleate (see Fig. 12).

Finally, light may affect not only the rate and type of fatty acid biosynthesis,
but also the conformation of the fatty acid synthetase itself, as shown in Table I.
Dark grown Euglena cells have a type I or complex synthetase which produces
palmitoyl-CoA as the major product, resembling the yeast system. Illumination of these cells alters the nature of the synthetase to a type II or dissociated
type, characteristic of the soluble systems found in higher plants. There is an
associated synthesis of the corresponding ACP which is required for activity,
and the product of this system is stearate as the ACP ester (Delo et al., 1971;
Ernst-Fonberg and Bloch, 1971).

D. OTHER ENVIRONMENTAL FACTORS

1. Chemical Factors

Included under this heading are such materials as pesticides and herbicides,
and other metabolic blocking agents such as arsenite and fluoride. Of particular
interest recently are the carbamate weedkillers which have been shown to affect
wax formation in vivo in higher plants, such as the pea (Harwood and Stumpf,
1971). When certain of these carbamates were used in in vitro experiments using

pea tissue, they caused a decrease in the formation of C_{20}–C_{24} acids, although total synthesis of fatty acids was slightly raised. It is these longer chain acids which are important in wax formation, and this finding agrees with the results obtained *in vivo* and suggests a specific inhibition of the elongation system in these tissues, leaving the *de novo* synthetase unaffected. Arsenite has been shown in the pea to block the elongation of palmitate to stearate although the formation of longer chain acids from stearate was not affected (Harwood and Stumpf, 1971). This suggests that two separate elongation systems occur in this tissue. In safflower seeds, arsenite reduced the formation of all acids longer than C_{16} and also prevented linoleate formation, though this latter effect may be due to the lack of a suitable C_{18} precursor. Fluoride blocked stearate elongation in the pea, but not palmitate elongation, and in this respect its action resembled that of arsenite. However, in safflower seeds elongation was unaffected by fluoride, but desaturation was specifically inhibited. Finally, in the avocado system arsenite did not affect the *de novo* formation of palmitate, but blocked the elongation of this acid to stearate (Harwood and Stumpf, 1972), as was found in the pea (*Pisum sativum*) and safflower seed. It is thus seen that the effect of chemical modifiers on fatty acid synthesis in higher plants is quite complex and shows some species variation, and it is possible that use could be made of these relatively simple chemicals to alter selectively the synthesis of certain fatty acids so as to make the control of certain species in this manner commercially useful.

2. *Physical Damage*

Physical damage to a tissue leads to a cycle of events or repair processes which tend to restore the tissue to its condition prior to damage. Inevitably, membrane synthesis is involved in these processes, and as a result of this it is not surprising to find a stimulation in the synthesis of fatty acids. For example, it has been observed that in freshly sliced potato (Willemot and Stumpf, 1967) there is produced an "ageing" effect that seems to be associated with a de-repression phenomenon, involving an increase in RNA and protein synthesis, and in membrane formation. This type of damage has been shown to cause a ten-fold increase in the *de novo* synthesis of fatty acids, and shortly following this and paralleling an increase in mitochondrial formation was an increase in polyunsaturated acids and longer chain acids up to C_{24}, which were not formed by the undamaged tissue. These effects are almost certainly caused by hormone action and are directed at repairing the cell damage caused by slicing.

E. STAGE OF DEVELOPMENT

This heading includes seed development and germination, organogenesis and maturation of tissues, chloroplast development and "greening" of tissues,

the "steady-state" metabolism of these tissues, followed by ageing and senescence processes. Hitchcock and Nichols (1971) have recently considered
these changes in relation to fatty acid synthesis, and the following summarizes
briefly the general changes which occur.

In growing green tissues there is an initial increase in overall fatty acid synthesis, including the rapid formation of unsaturated acids, which can be
correlated with growth and the synthesis of membrane systems. Then, as the
tissue matures (as distinct from "ageing") there is a gradual decrease in the
proportion of unsaturated acids, and the rate of overall synthesis declines.

A similar cycle of synthesis occurs in the storage tissue of developing seeds.
At the onset of germination, all synthesis in the endosperm ceases abruptly,
and fatty acid synthesis commences in the developing cotyledons, using the
storage tissue lipids (and/or starch) as a source of material. It is interesting
that in the pea, this initial synthesis of fatty acids in the cotyledons appears to
be catalysed by enzymes already synthesized and present in these tissues
(Harwood and Stumpf, 1970) which become activated as a result of an increase
in water content.

IV. FACTORS CONTROLLING FATTY ACID BIOSYNTHESIS
AT THE MOLECULAR LEVEL

The previous section has dealt with some of the general factors which have
been shown to affect fatty acid synthesis, and in some cases these effects were
related to events at the molecular level. It is the purpose of this section to consider in more detail factors at the molecular level which control fatty acid
biosynthesis, and to relate these to the mechanisms of synthesis discussed in
Section II.

When one talks about control of fatty acid synthesis, one can consider this
under various headings, and these are·shown in Table II, which conveniently
summarizes many of the points concerning the mechanism and control of
synthesis which have already been mentioned in this chapter. Table III
summarizes, under six main headings, the major factors which control the
synthesis of fatty acids at the enzymic or molecular level, and these will now
be considered in more detail.

A. SYNTHESIS OF ENZYMES

In addition to the obvious requirements of nucleic acid and protein synthesis
necessary for the synthesis of individual enzymes, the complex nature of many
of the enzymes involved in fatty acid synthesis raises the additional question
of their structural organization into multienzyme units. If one considers the
multienzyme complex type of *de novo* synthetase, there is the problem of
whether the individual enzymes are synthesized in a specific order and inserted

TABLE II

Aspects of the control of fatty acid synthesis

1. Site of synthesis	(a)	Soluble fraction
	(b)	Mitochondria
	(c)	Microsomes
2. Type of synthesis	(a)	*de novo*
	(b)	Elongation
	(c)	Desaturation
3. Rate of synthesis Dependent on	(a)	Substrates
	(b)	Enzyme activity
4. Nature of product	(a)	Chain length
	(b)	Unbranched or branched
	(c)	Saturated
	(d)	Unsaturated-Monoenoic, dienoic, polyenoic
	(e)	Unsaturated-Acetylenic
	(f)	Keto-, hydroxy-, epoxy- or cyclopropane acid
	(g)	Free acid, CoA-thioester, ACP thioester, enzyme bound thioester, *O*-ester

TABLE III

Factors controlling fatty acid synthesis at the molecular level

A. Synthesis of the enzymes of fatty acid synthesis		
B. Synthesis of cofactors involved in fatty acid synthesis		
C. Availability and synthesis of primary substrates		
D. Regulation of primary substrates	(1)	Transport
	(2)	Competing reactions
	(3)	Ionic effects
E. Regulation of biosynthetic enzymes	(1)	Kinetic parameters
	(2)	Allosteric control
	(3)	Subunit dissociation
	(4)	Substrate inhibition
	(5)	Substrate protection
	(6)	Product inhibition
	(7)	Ionic effects
F. Termination processes		

sequentially into a multienzyme unit by a specific mechanism, or whether the enzymes are synthesized in random order and assembled into the complex (quaternary structure) by random interactions favoured by the tertiary structure of the individual proteins. The fact that the complexes from pigeon liver (Butterworth *et al.*, 1967) and yeast (Sumper *et al.*, 1969b) can be dissociated into subunits and reassembled to form the intact functional complex suggest that possibly the second alternative is correct. The subunits are inactive with respect to overall fatty acid synthesis, but exhibit many, but not all, of the partial reactions shown in Fig. 1. However, it has not been possible to dissociate either of these complexes so as to release the individual, active enzymes, let alone reassemble such a system. One thing that is certain is that the integrity of the complex is not functionally determined. In other words, it is not dependent on the activity of the constituent enzymes. Recently, a thermosensitive yeast mutant requiring palmitate for growth has been isolated (Meyer and Schweizer, 1972) in which the purified fatty acid synthetase complex was apparently identical with that from normal cells except that it did not synthesize palmitate. On analysis for the partial reactions shown in Fig. 1 it was found that only the condensing enzyme activity was lacking, although the enzyme complex apparently contained this inactive protein and had been assembled in the usual manner, indicating that this type of complex does not depend on function for assembly.

Mention has already been made of the two types of synthetase from *Euglena*, shown in Table I, and of the effect of light on the nature of the synthetase in these cells. However, there is no evidence that the two are interconvertible in any way and in fact it would seem that they are synthesized independently (Ernst-Fonberg and Bloch, 1971).

Also shown in Table I is the existence of a unique bacterial synthetase of type I, the multienzyme complex of *Mycobacterium phlei*. It contains covalently bound 4′-phosphopantetheine and is similar to the yeast enzyme in many respects, except that the complex (molecular weight 1.7×10^6) is very unstable and dissociates into inactive subunits of molecular weight less than 45000. There exists in *M. phlei* a typical ACP, but this is not required for the functioning of the complex. However, there is evidence for an ACP dependent synthetase of molecular weight about 225000, and it has been suggested (Brindley *et al.*, 1969) that this may be related to the type I complex, though it has not been possible to interconvert them *in vitro*. There has been a recent preliminary report (Knoche and Koths, 1972) that *Corynebacterium diptheria*, which is phylogenetically related to the *Mycobacteria*, also contains a type I or complex synthetase which is apparently membrane bound *in vivo*. The multienzyme complex can be solubilized by sonication and requires NADH, NADPH and FMN for the synthesis of palmitic acid, which only occurs *in vitro* in buffers of high ionic strenth (>0·5 M). The existence of these atypical fatty acid synthetase systems, and the membranous nature of certain higher plant synthetases, suggests that perhaps all synthetases *in vivo* exist as multienzyme complexes, some of which break up to yield the dissociated type on disruption of the cells.

B. SYNTHESIS OF COFACTORS

Under this heading can be considered CoA, ACP, ATP and the pyridine nucleotides. Provided the tissue is in good nutritional status, levels of CoA are not usually rate limiting because only catalytic quantities are required and this coenzyme is readily recycled. The same can be said with regard to the availability of ATP for the thiokinases and acetyl-CoA carboxylase. In the photosynthetic tissues photophosphorylation reactions are an additional source of ATP and in these cells the enzymes of fatty acid synthesis are located in the chloroplasts and thus have immediate access to the ATP produced in this manner.

It has previously been mentioned that CoA is the source of the 4'-phosphopantetheine prosthetic group of ACP, and in *E. coli* this has been shown to be an important control on ACP levels. The protein moiety of ACP, called *apo*-ACP, is first synthesized without the prosthetic group, which is later transferred to it from CoA by an enzyme called *holo*-ACP synthetase to form the active ACP (Alberts and Vagelos, 1966; Elovson and Vagelos, 1968). There is also present an enzyme called *holo*-ACP hydrolase which cleaves the prosthetic group from ACP to form *apo*-ACP, and it is thought that the balance between this enzyme and the *holo*-ACP synthetase may control ACP levels, and hence fatty acid synthesis, in cells which possess a discrete ACP and a dissociated synthetase.

The level of NAD and NADP in the cell is not usually rate limiting in healthy tissues, so their availability for the formation of their reduced forms, NADH and NADPH, required for fatty acid synthesis, is not rate limiting. As NADH and NADPH are really primary substrates for the synthetase, their formation will be considered in more detail below.

C. AVAILABILITY AND SYNTHESIS OF PRIMARY SUBSTRATES

The term primary substrates in fatty acid synthesis applies to acetyl-CoA, malonyl-CoA and the reduced pyridine nucleotides, though in dissociated synthetases one could say that acetyl-ACP and malonyl-ACP are the primary substrates and not the corresponding CoA thioesters. The synthesis of the acetyl and malonyl substrates has already been dealt with in some detail in Section II, but one reaction merits further consideration in this section. The citrate cleavage enzyme (see Fig. 2, equation B) is an important source of acetyl-CoA and may play a regulatory role in fatty acid synthesis. However, the availability of citrate in the cytoplasm is perhaps more important in this context. The cytoplasmic level of citrate is controlled not only by the functioning of the tricarboxylic acid cycle in the mitochondria, but also by its transport from there to the cytoplasm. Once in the soluble fraction, it is available for the formation of acetyl-CoA by the cleavage enzyme, and in cells which possess an activatable acetyl-CoA carboxylase it also controls the formation of malonyl-CoA because of its affect on the acetyl-CoA carboxylase. Even in tissues in which the carboxylase is not under this allosteric control, such as the potato (Huang and Stumpf, 1970) and yeast (Sumper et al., 1969a), the level of malonyl-CoA has been shown to have a profound influence on the rate and nature of fatty acid synthesis.

The availability of reduced pyridine nucleotides depends primarily on the mitochondrial formation of NADH. However, extramitochondrial reactions are usually mediated by NADP linked reductases, and it is NADPH which is the main substrate in fatty acid synthesis. However, although NADPH alone is satisfactory for de novo fatty acid synthesis in most tissues, optimum rates in other systems also require the presence of small amounts of NADH (White et al., 1971). This requirement seems to be associated with the second reduction of fatty acid synthesis catalysed by the enoyl reductase, and in fact this enzyme has been isolated from E. coli in a form which requires NADH for activity (Weeks and Wakil, 1968). In addition, mitochondrial elongation enzymes prefer NADH to NADPH. It thus seems likely that in the intact cell the respective levels of these two nucleotides and their interconversion by a well characterized transhydrogenase enzyme (Middleditch and Chung, 1971) could exert an important control on the rate of fatty acid synthesis, and on the nature of the product.

D. REGULATION OF PRIMARY SUBSTRATES

Following their synthesis, the availability of primary substrates for fatty acid synthesis may be affected by other regulatory processes such as cellular transport mechanisms, competing reactions and ionic effects. Transport has been generally discussed, but the importance of the carnitine acyl transferase system in transporting acyl groups across the mitochondrial membrane should be specifically mentioned in the context of making mitochondrial acetyl-CoA available for cytoplasmic fatty acid synthesis (Bressler and Katz, 1965).

The large number of reductive reactions in the cell which utilize reduced pyridine nucleotides are self-evident, and compete with fatty acid synthesis for these substrates. The utilization of acetyl-CoA by the citric acid cycle or in the synthesis of mevalonic acid are two major pathways competing for acetyl-CoA in the cell. Finally, there are hydrolytic enzymes, termed thioesterases, which split the thioester bond of acetyl-CoA and malonyl-CoA, thus removing these activated units from the pool available for fatty acid synthesis. These thioesterases are present in all cells, and are frequently troublesome when studying fatty acid synthesis in crude cell homogenates and partially purified enzyme preparations because of their high specific activity.

Specific and non-specific ionic effects play an important role in many protein-protein interactions, and these have recently been shown to affect the interaction of ACP substrates with biosynthetic enzymes in fatty acid synthesis. The affinity of the β-keto reductase enzyme from $E.$ $coli$ for acetoacetyl-ACP has been shown (Schulz et al., 1969) to be markedly affected by divalent cations, which were shown to bind to the many acidic groups present in ACP (Schulz, 1972), thus facilitating the formation of the enzyme-substrate complex. Although the structures of ACP from $E.$ $coli$ and spinach are known to be very similar (Simoni et al., 1967), the spinach ACP forms poor substrates for the $E.$ $coli$ synthetase, and causes the slow accumulation of C_{12}-C_{18} hydroxy acids (Overath and Stumpf, 1964). This is probably due to the fact that subtle differences in the ionization of certain groups in the spinach ACP make the corresponding β-hydroxyacyl thioester interact weakly with the $E.$ $coli$ dehydrase.

E. REGULATION OF BIOSYNTHETIC ENZYMES

1. Kinetic Parameters

Kinetic parameters such as pH, substrate concentrations and temperature are primarily applicable to in vitro assays, though obviously changes in these parameters will usually produce a similar affect in vivo. The pH/activity curves for overall fatty acid synthesis in a variety of tissues show a maximum between 6·5 and 7·5, and the pH optima of individual enzymic activities usually lie within this range.

2. Allosteric Control

The allosteric activation of acetyl-CoA carboxylase has already been discussed, and mention has been made of the fact that this affect seems to be confined to animal cells. It has recently been shown in the chicken liver enzyme that citrate exerts its effect by controlling the conformation of the E_2-biotin-CO_2 complex prior to the transcarboxylation reaction catalysed by E_3 (Hashimoto et al., 1971).

In certain cells, the synthetase may also be under allosteric control, as studies on the pigeon liver enzyme have indicated (Plate et al., 1968). In this tissue, malonyl-CoA was shown to be inhibitory under certain conditions, and kinetic studies on the nature of this inhibition and its removal by fructose-1,6-diphosphate, suggested binding at a regulatory site on the enzyme.

Specific allosteric controls have not been demonstrated in dissociated synthetases such as those found in higher plants and thus the significance of this type of control of fatty acid synthesis in these tissues must be questioned.

3. *Subunit Dissociation*

This may play an important role in controlling the activity of complex synthetases and two factors are important in this respect. The first of these is ionic strength, and the second is the state of reduction of sulphydryl groups. It has been shown (Butterworth *et al.*, 1967; Kumar and Porter, 1971) that low ionic strength buffers, especially of high pH, cause the dissociation of the pigeon synthetase into subunits which are inactive in the overall synthesis of palmitate, but which show most of the partial reactions. However, the condensation reaction is not active, and this is apparently due to the inability of the cysteine sulfhydryl group at the active site of this enzyme to bind acetyl groups from acetyl-CoA. This dissociation can also be brought about by disulphides such as cystine, and can be reversed partially by high ionic strength buffer and completely by thiols such as dithiothreitol (DTT). This dissociation can also be prevented by NADPH, but not by NADH, and this effect is related to binding sites other than those at the active centres of constituent enzymes.

4. *Substrate Inhibition*

The effect of malonyl-CoA on the pigeon synthetase has been discussed. In addition, free CoA is a potent inhibitor in the pigeon and in other systems, and may play a role in the metabolic regulation of these synthetases (Kumar and Porter, 1971).

5. *Substrate Protection*

The protective effect of NADPH in controlling the dissociation of the pigeon synthetase complex was mentioned previously. A similar protective effect of either NADPH or NADH against the denaturation of the purified β-ketoreductase from *E. coli* has recently been reported (Schulz and Wakil, 1971), but in this instance the protective effect seems to be due to binding at the active site of the enzyme.

Preincubation of the pigeon synthetase with acetyl-CoA or malonyl-CoA protected the enzyme from inhibition by thiol poisons, and this appeared to be due to protection of the transacylase enzymes of this complex (Plate *et al.*, 1970).

6. *Product Inhibition*

This probably plays an important role in a feedback control mechanism in both complex and dissociated synthetases, and the inhibitory effect of palmitoyl CoA has been demonstrated in both animal and yeast systems. However, this compound has strong detergent properties and has been shown to dissociate the pigeon enzyme into subunits (Butterworth *et al.*, 1967). Thus, any effect long chain acyl-CoA derivatives may have on fatty acid synthesis must be considered in terms of the physicochemical properties of these compounds as well as in specific feedback inhibition theories.

7. Ionic Effects

These are basically of two types, and apply to both complex and dissociated synthetases. Firstly specific effects such as interactions with substrates and enzymes to facilitate substrate binding have been considered in the regulation of primary substrates. Divalent metal ion requirements by thiokinases and carboxylases have also been discussed. Secondly, non-specific ionic effects governing the maintenance of the general configuration of protein chains in an aqueous medium (tertiary structure) are important for all proteins, but in fatty acid synthetases they apparently play an additional role in controlling the configuration of the multienzyme complex (quaternary structure) in this type of synthetase.

F. TERMINATION REACTIONS

These processes play an important role in determining the chain length, nature of the product, and to a lesser degree the rate of synthesis, of fatty acids as is shown in Table I.

It would seem that the primary product of all type II or dissociated synthetases is the ACP ester. However, in *E. coli* there is an extremely active thioesterase associated with the synthetase enzymes (Barnes and Wakil, 1968) which is highly specific for C_{16} ACP esters, and to a lesser extent C_{18}, and which is responsible for termination in this system *in vivo*. The specificity of this enzyme explains why palmitate, palmitoleate and *cis*-vaccenate are the major fatty acids synthesized by the *E. coli* system. In higher plants, the products of fatty acid synthesis *in vitro* consist of a mixture of free acids and the ACP-esters, which again suggest that the activity of a thioesterase is important in determining chain length.

A similar thioesterase activity is present in the synthetase complexes of rat and pigeon liver, which explains why free acids are the only detectable products in these systems.

The yeast synthetase is interesting in that the major products are CoA esters, and there is a specific termination reaction in this complex involving a transfer of the C_{16} or C_{18} chain from an enzyme thiol to CoA (see Fig. 1). However, it would seem that it is not the chain length specificity of this enzyme which is the primary factor in terminating *de novo* synthesis at a chain length of C_{16} or C_{18} in yeast. Apparently, the sulfhydryl binding site of the condensing enzyme which accepts the lengthening acyl chain during synthesis tends to have greater affinity for binding a new C_2 unit from acetyl-CoA as the lengthening acyl chain approaches C_{16} (Sumper *et al.*, 1969a). It has been suggested that this site is in a hydrophilic region of the protein, and that the hydrophobic nature of C_{16} fatty acyl chains causes them to be preferentially transferred to the CoA-transferase enzyme of the synthetase.

Although termination reactions may play a role in determining the overall rate of fatty acid synthesis, examination of the rates of individual partial reactions in yeast (Lynen, 1969) and other cells (Kumar et al., 1972) do not indicate that the termination process is rate limiting. In most cells, the condensation step is rate limiting, and in view of the key role this enzyme plays in de novo synthesis it seems likely that it is this enzyme which controls the overall activity of the fatty acid synthetase.

V. The Stereochemistry of Reactions in Fatty Acid Synthesis

A. GENERAL CONSIDERATIONS

It has long been known that enzymic reactions involving molecules containing centres of asymmetry are, with few exceptions, stereospecific. In other words, an enzyme tends to be specific for one optical isomer of a compound which is its substrate, or to form one stereoisomer of its product, but exactly why this is so is not clear (possible reasons for this are discussed by Professor Cornforth on p. 173 of this volume). This general stereospecificity of enzymic reactions imposes an inherent form of control on these reactions, and in the metabolism of fatty acids the most commonly quoted example of this concerns the configuration of the 3-hydroxyacyl intermediates of fatty acid synthesis and oxidation. It is known that in the synthetic pathway, this intermediate has the D(−) configuration, whereas in the degradative pathway it is the L(+) stereoisomer. In this chapter, the absolute R and S terminology (Cahn et al., 1966) will be used in place of the D(−) and L(+) nomenclature.

B. ACETYL-COA CARBOXYLASE

Although the absolute stereochemistry of the product of this enzyme has not been determined, a closely related, biotin dependent carboxylase, propionyl-CoA carboxylase, has been examined in this context (Rétey and Lynen, 1965; Arigoni et al., 1966; Prescott and Rabinowitz, 1968). The mechanism of the proton replacement on the α-carbon is shown in Fig. 14, and this process appears to be a concerted mechanism, involving simultaneous removal of the pro-R hydrogen and insertion of carbon dioxide with retention of configuration. The hydrogen isotope effect in the abstraction of the proton was observed to be small when [(2R)-2-^{3}H$_1$]propionyl-CoA was used. This is in contrast to the large tritium isotope effect observed in another biotin dependent carboxylase, pyruvate carboxylase (Rose, 1970). It has been inferred (Lynen, 1967, 1970) that because acetyl-CoA carboxylase and propionyl-CoA carboxylase appear to have similar biotin dependent mechanisms for the carboxylation of their substrate, it is likely that the acetyl-CoA carboxylase reaction has the same stereochemistry. While this may be proved to be so, it must be borne in mind that in the carboxylation of propionyl-CoA, the enzyme is distinguishing between the two protons of a methylene group positioned between a methyl carbon and a carbonyl carbon, whereas in the carboxylation of acetyl-CoA the enzyme abstracts a proton from a terminal methyl group.

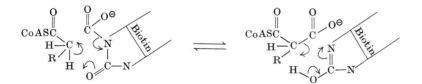

Fig. 14. The mechanism of transcarboxylation in biotin dependent carboxylases (Lynen, 1967).

Attempts to gain some insight into this problem have been made (Sedgwick, 1972) making use of the availability of the stereospecifically labelled chiral acetates (Cornforth et al., 1970).

The corresponding R- and S-acetyl-CoA thioesters were prepared enzymically from these acetates, then incorporated into long chain fatty acids using a combined acetyl-CoA carboxylase-fatty acid synthetase enzyme system. The carboxylase was purified from chicken liver (Goto et al., 1967) and the synthetase was prepared from either chicken liver (Hsu and Yun, 1970) or yeast (Lynen, 1969). Table IV shows the results obtained in this type of experiment in which the doubly labelled acetyl-CoA thioesters were used as substrates, and the isotope ratio in the palmitic acid product was determined to compare the relative tritium retention from the three acetates. The greater retention of tritium from the S-substrate compared with the R-substrate, the non-chiral [2-^{3}H$_1$]-acetyl-CoA giving an intermediate result, indicates an overall stereospecificity of the carboxylase-synthetase reactions. In this experiment, the chirality of the malonyl-CoA formed by carboxylation of the R- and S-acetyl-CoA substrates is dependent on the magnitude of the hydrogen isotope effect. This effect in the acetyl-CoA carboxylase reaction appears to be small, which is in agreement with the effect observed for the propionyl-CoA carboxylase (Rétey and Lynen, 1965). This, together with some non-specific proton exchange which occurs, possibly at the C-2 methylene of the β-keto-acyl thioester product of the condensation reaction, provide the most likely explanation of the small differentiation indicated in Table IV. Because the absolute stereochemistry of all the synthetase enzyme reactions is unknown, it is not possible to determine the stereochemistry of the carboxylase reaction from this experiment alone. However, comparison of this result with that obtained using

TABLE IV

The incorporation of radioactivity from double labelled acetyl-CoA preparations into palmitic acid by combined acetyl-CoA carboxylase/fatty acid synthetase systems

Substrate	$[^3H]/[^{14}C]$ ratio	Chicken liver synthetase Palmitic acid product		Yeast synthetase Palmitic acid product	
		$[^3H]/[^{14}C]$ ratio	% $[^3H]$ retention	$[^3H]/[^{14}C]$ ratio	% $[^3H]$ retention
R-acetyl-CoA	6·94	2·11 ± 0·04	30·4 ± 0·60	2·54 ± 0·08	36·6 ± 1·1
S-acetyl-CoA	6·90	2·37 ± 0·01	34·3 ± 0·15	2·78 ± 0·06	39·9 ± 0·5
$[^3H]$-acetyl-CoA	5·38	1·74 ± 0·04	32·3 ± 0·70	2·01 ± 0·07	37·5 ± 1·2

$[2\text{-}^3H_1]$malonyl-CoA of known chirality and the same synthetase enzyme will enable the stereochemistry of the acetyl-CoA carboxylase to be definitely assigned.

C. REACTIONS OF THE FATTY ACID SYNTHETASE

1. Condensation Reaction

Although a mechanism for this reaction, shown in Fig. 15, has been proposed (Lynen, 1967) the stereochemistry of this reaction is not known. Direct analysis of the absolute configuration of the protons at C-2 in the β-keto acyl product of this reaction is not possible due to their rapid equilibration with the water of the reaction medium. Thus, any approach to this problem using acetyl-CoA or malonyl-CoA of known chirality involves taking this intermediate through to saturated fatty acids in a similar way to that described for the carboxylase. Again, any definite assignment of mechanism in the condensation reaction is dependent on knowledge of the stereochemistry of the other individual reactions catalysed by the synthetase. The hypothetical stereochemistry of the reactions catalysed by the synthetase shown in Fig. 16 indicates retention of configuration at C-2 during the condensation reaction.

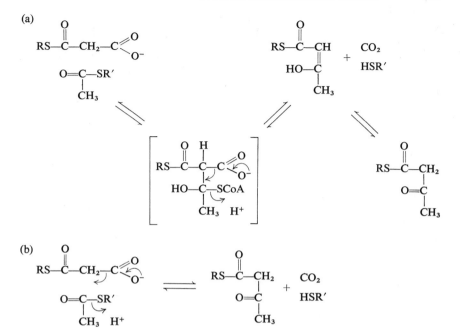

FIG. 15. Possible mechanisms for the formation of β-ketoacyl thioesters by the condensing enzyme (Lynen, 1967). (a) Stepwise mechanism involving the formation of an enol intermediate. (b) Concerted mechanism.

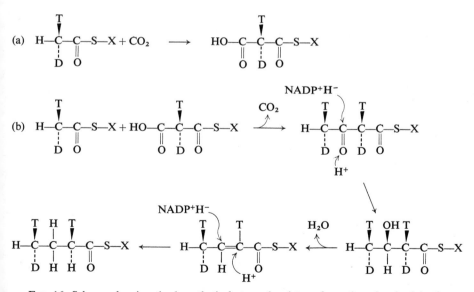

FIG. 16. Scheme showing the hypothetical stereochemistry of reactions involved in the *de novo* synthesis of fatty acids. (a) The acetyl-CoA carboxylase reaction. (b) Reactions of the fatty acid synthetase.

2. *The First Reduction Reaction*

As mentioned previously, reduction of the β-keto acyl product of the condensing enzyme results in the formation of the D($-$) or 3R-hydroxyacyl intermediate (Lynen, 1961; Wakil and Bressler, 1962). In this reduction, the hydroxyl group which is formed contains a proton from the reaction medium, and the hydrogen attached directly to C-3 comes from the hydride ion of NADPH, as illustrated in Fig. 16. The stereospecificity of hydrogen transfer from NADPH during this first reduction has been examined, and it has been shown (Dugan *et al.*, 1970) that in this reaction it is the hydrogen on the B side of the nicotinamide ring which is donated, that is the pro-4S hydrogen.

3. *Dehydration Reaction*

Elimination of the elements of water from the 3R-hydroxyacyl intermediate results in the formation of the *trans*-2,3-enoyl derivative, as shown in Fig. 16. An exception to this is the β-hydroxydecanoyl thioester dehydrase of *E. coli* (Brock *et al.*, 1967) which also possesses isomerase activity and forms a mixture of *trans*-2,3-decenoyl and *cis*-3,4-decenoyl intermediates and is responsible for the formation of *cis*-vaccenic acid by this organism, as previously discussed. However, it is not known whether it is the pro-R or pro-S hydrogen at C-2 which is removed together with the C-3 hydroxyl in this reaction. It is necessary to determine the stereochemistry of this proton removal before one can fully define the overall stereochemistry of the synthetase reactions. The hypothetical stereochemistry shown in Fig. 16 shows the removal of the pro-R hydrogen, i.e. a *trans* elimination.

If the synthetase is incubated with an L($+$) or 3S-hydroxyacyl thioester model substrate, dehydration will not occur unless the enzyme preparation is contaminated with racemase activity capable of forming the 3R-hydroxyacyl substrate. As mentioned previously, the 3S-hydroxyacyl intermediate is the normal product of crotonase, the enzyme of the β-oxidation pathway responsible for hydrating the *trans*-2,3-enoyl intermediate in this pathway. Crotonase will also act on isocrotonyl-CoA, that is the *cis*-2,3-isomer, but at only about one third of the rate at which crotonyl-CoA is utilized (Stern and del Campillo, 1956), and in this instance the product is the 3R-hydroxybutyryl-CoA.* However, in both these cases the stereochemistry of proton addition at C-2 is not known (whether a *re*- or *si*- addition according to the system of Hanson, 1966), so the possible stereochemistry of proton removal at C-2 catalysed by the dehydrase activity of the synthetase can not be inferred by analogy.

4. *The Second Reduction*

This reaction results in the formation of the saturated acyl-thioester intermediate as shown in Fig. 16. The mechanism involves an attack at C-3 by the

* See Note Added In Proof, p. 217.

hydride ion of a reduced pyridine nucleotide, the saturation at C-2 being completed by the addition of a proton from the reaction medium. The stereospecificity of hydrogen transfer from NADPH to C-3 of the *trans*-2,3-enoyl intermediate has been examined, and it has been shown (Dugan *et al.*, 1970) that it is the hydrogen on the A side of the nicotinamide ring, the pro-4*R* hydrogen, which is donated in this reaction. However, again one does not know the absolute stereochemistry of the proton additions at C-3 and at C-2 which occur during this second reduction. In the terminology of Hanson (1966), the additions at these two carbons can be one of four possibilities namely *re-re-*, *re-si-*, *si-re-* or *si-si-*. In terms of interpreting the stereochemistry of other reactions of the synthetase based on results using chiral acetyl-CoA or malonyl-CoA and analysis of the fatty acid produced, it is only the proton addition at C-2 during this second reduction which is of importance as it is this carbon which carried the asymmetry in the original primary substrate. Preliminary results based on analysis of the optical rotatory dispersion pattern of palmitic acid (as the methyl ester) formed by the yeast synthetase from a malonyl thioester fully deuterated at C-2 have indicated that this material has the *R* configuration at C-2, this conclusion being based on a comparison with the O.R.D. spectrum of synthetic $[(2R)\text{-}2\text{-}^2H_1]$palmitic acid (Sedgwick, 1972). If this result is substantiated by further experiments, it implies that reduction at C-2 catalysed by the enoyl reductase enzyme of the yeast synthetase occurs by means of a *si*-addition of a proton from the reaction medium, irrespective of which proton at C-2 is removed during the dehydration reaction discussed above.

It has been reported by a private communication (see Dugan *et al.*, 1970) that the crotonyl-CoA reductase enzyme from yeast (not the same enzyme as that present in the synthetase) transfers hydrogen from NADPH to C-3 of crotonyl-CoA to give the pro-*R* configuration, that is, a *re*-addition. However, it has not been reported whether the same stereochemistry at C-3 applied to the enoyl-reductase enzyme of the fatty acid synthetase.

REFERENCES

Alberts, A. W. and Vagelos, P. R. (1966). *J. biol. Chem.* **241**, 5201.
Alberts, A. W. and Vagelos, P. R. (1968). *Proc. natn. Acad. Sci. U.S.A.* **59**, 561.
Alberts, A. W., Majerus, P. W. and Vagelos, P. R. (1969a). *In* "Methods in Enzymology" (J. M. Lowenstein, ed.), Vol. XIV, pp. 50–66. Academic Press, New York and London.
Alberts, A. W., Nervi, A. M. and Vagelos, P. R. (1969b). *Proc. natn. Acad. Sci. U.S.A.* **63**, 1319.
Appelqvist, L.-A. (1968). *Physiol. Pl.* **21**, 455.
Arigoni, D., Lynen, F. and Rétey, J. (1966). *Helv. Chim. Acta* **49**, 311.
Baddiley, J. (1955). *Adv. Enzymol.* **16**, 1.
Barnes, E. M. and Wakil, S. J. (1968). *J. biol. Chem.* **243**, 2955.
Bloch, K. (1969). *Acc. Chem. Res.* **2**, 193.

Bloch, K. and Chang, S. B. (1964). *Science, N. Y.* **144**, 560.
Bloomfield, D. K. and Bloch, K. (1960). *J. biol. Chem.* **235**, 337.
Brady, R. O. (1958). *Proc. natn. Acad. Sci. U.S.A.* **44**, 993.
Bressler, R. and Katz, R. I. (1965). *J. biol. Chem.* **240**, 622.
Brindley, D. N., Matsumura, S. and Bloch, K. (1969). *Nature, Lond.* **224**, 666.
Brock, D. J. H., Kass, L. R. and Bloch, K. (1967). *J. biol. Chem.* **242**, 4432.
Brooks, J. L. and Stumpf, P. K. (1966). *Archs Biochem. Biophys.* **116**, 108.
Burton, D. and Stumpf, P. K. (1966). *Archs Biochem. Biophys.* **117**, 604.
Butterworth, P. H. W., Guchait, R. B., Baum, H., Olson, E. B., Margolis, S. A. and Porter, J. W. (1966). *Archs Biochem. Biophys.* **116**, 453.
Butterworth, P. H. W., Yang, P. C., Bock, R. M. and Porter, J. W. (1967). *J. biol. Chem.* **242**, 3508.
Cahn, R. S., Ingold, C. K. and Prelog, V. (1966). *Angew. Chem. Int. Ed. Engl.* **78**, 413.
Cornforth, J. W., Redmond, J. W., Eggerer, H., Buckel, W. and Gutschow, C. (1970). *Europ. J. Biochem.* **14**, 1.
Dahlen, J. V. and Porter, J. W. (1968). *Archs Biochem. Biophys.* **127**, 207.
Delo, J., Ernst-Fonberg, M. L. and Bloch, K. (1971). *Archs Biochem. Biophys.* **143**, 384.
Dickson, L. G., Galloway, R. A. and Patterson, G. W. (1969). *Pl. Physiol., Lancaster* **44**, 1413.
Dugan, R. E., Slakey, L. L. and Porter, J. W. (1970). *J. biol. Chem.* **254**, 6312.
Elovson, J. and Vagelos, P. R. (1968). *J. biol. Chem.* **243**, 3603.
Ernst-Fonberg, M. L. and Bloch, K. (1971). *Archs Biochem. Biophys.* **143**, 392.
Erwin, J. and Bloch, K. (1964). *Science, N. Y.* **143**, 1006.
Esfahani, M., Ioneda, T. and Wakil, S. J. (1971). *J. biol. Chem.* **246**, 50.
Fulco, A. J. (1970). *J. biol. Chem.* **245**, 2985.
Gerloff, E. D., Richardson, T. and Stahmann, M. A. (1966). *Pl. Physiol., Lancaster* **41**, 1280.
Gibson, D. M., Titchener, E. B. and Wakil, S. J. (1958a). *J. Am. Chem. Soc.* **80**, 2908.
Gibson, D. M., Titchener, E. B. and Wakil, S. J. (1958b). *Biochim. biophys. Acta.* **30**, 376.
Goldfine, H. and Bloch, K. (1961). *J. biol. Chem.* **236**, 2596.
Goto, T., Ringelman, E., Riedel, B. and Numa, S. (1967). *Life Sci.* **6**, 785.
Green, D. E. and Allman, D. W. (1968). *In* "Metabolic Pathways" (D. M. Greenberg, ed.), Vol. II, 3rd Edn. pp. 1–36. Academic Press, New York and London.
Gregolin, C., Ryder, E., Kleinschmidt, A. K., Warner, R. C. and Lane, M. D. (1966a). *Proc. natn. Acad. Sci. U.S.A.* **56**, 148.
Gregolin, C., Ryder, E., Warner, R. C., Kleinschmidt, A. K. and Lane, M. D. (1966b). *Proc. natn. Acad. Sci. U.S.A.* **56**, 1751.
Guchait, R. B., Putz, G. R. and Porter, J. W. (1966). *Archs Biochem. Biophys.* **117**, 541.
Gurr, M. I., Robinson, M. P. and James, A. T. (1969). *Europ. J. Biochem.* **9**, 70.
Haest, C. W. M., De Gier, J. and van Deenen, L. L. M. (1969). *Chem. Phys. Lip.* **3**, 413.
Hagen, A. and Hofschneider, P. H. (1964). "Proceedings of the 3rd European Regional Conference on Electron Microscopy, Prague, Vol. B, p. 69. Prague: Czechoslovak Academy of Sciences.
Hanson, K. R. (1966). *J. Am. chem. Soc.* **88**, 2731.
Harlan, W. R. and Wakil, S. J. (1963). *J. biol. Chem.* **238**, 3216.
Harris, P. and James, A. T. (1969). *Biochim. biophys. Acta* **187**, 13.
Harwood, J. L. and Stumpf, P. K. (1970). *Pl. Physiol., Lancaster* **46**, 500.
Harwood, J. L. and Stumpf, P. K. (1971). *Archs Biochem. Biophys.* **142**, 281.
Harwood, J. L. and Stumpf, P. K. (1972). *Archs Biochem. Biophys.* **148**, 282.

Hashimoto, T., Isano, H., Iritani, N. and Numa, S. (1971). *Europ. J. Biochem.* **24**, 128.
Hatch, M. D. and Stumpf, P. K. (1961). *J. biol. Chem.* **236**, 2879.
Hatch, M. D. and Stumpf, P. K. (1962). *Pl. Physiol., Lancaster* **37**, 121.
Hayaishi, O. (1955). *J. biol. Chem.* **215**, 125.
Heinstein, P. F. and Stumpf, P. K. (1969). *J. biol. Chem.* **244**, 5374.
Helmkamp, G. M. and Bloch, K. (1969). *J. biol. Chem.* **244**, 6041.
Hilditch, T. P. and Williams, P. N. (1964). "The Chemical Constitution of the Natural Fats", 4th Edn. Chapman and Hall, London.
Hitchcock, C. and Nichols, B. W. (1971). "Plant Lipid Biochemistry." Academic Press, London and New York.
Hsu, R. Y. and Yun, S.-L. (1970). *Biochemistry*, **9**, 239.
Huang, K. P. and Stumpf, P. K. (1970). *Archs Biochem. Biophys.* **140**, 158.
James, A. T., Harris, P. and Bezard, J. (1968). *Europ. J. Biochem.* **3**, 318.
Joshi, V. C. and Wakil, S. J. (1971). *Archs Biochem. Biophys.* **143**, 493.
Joshi, V. C., Plate, C. A. and Wakil, S. J. (1970). *J. biol. Chem.* **245**, 2857.
Joshi, V. C., Sedgwick, B., Plate, C. A. and Wakil, S. J. (1968). *Fedn Proc.* **27**, 817.
Kass, L. R., Brock, D. J. H. and Bloch, K. (1967). *J. biol. Chem.* **242**, 4418.
Knappe, J. (1970). *A. Rev. Biochem.* **39**, 757.
Knoche, H. W. and Koths, K. E. (1972). *Fedn Proc.* **31**, (No. 2), 476.
Knowles, P. F. (1969). *J. Am. Oil Chem. Soc.* **46**, 130.
Kolattukudy, P. E. (1966). *Biochemistry, Easton* **5**, 2265.
Kumar, S. and Porter, J. W. (1971). *J. biol. Chem.* **246**, 7780.
Kumar, S., Phillips, G. T. and Porter, J. W. (1972). *Int. J. Biochem.* **3**, 15.
Lennarz, W. J., Light, R. J. and Bloch, K. (1962). *Proc. natn. Acad. Sci. U.S.A.* **48**, 840.
Lynen, F. (1959). *J. Cell. Comp. Physiol.* **54** (Suppl. 1), 33.
Lynen, F. (1961). *Fedn Proc.* **20**, 941.
Lynen, F. (1967). *Biochem. J.* **102**, 381.
Lynen, F. (1969). *In* "Methods in Enzymology" (J. M. Lowenstein, ed.), Vol. XIV, pp. 17–33. Academic Press, New York and London.
Lynen, F. (1970). *Biochem. J.* **117**, 47.
Lynen, F., Matsuhashi, M., Numa, S. and Schweizer, E. (1963). *In* "The Control of Lipid Metabolism" (J. K. Grant, ed.), pp. 43–56. Academic Press, New York and London.
Mahler, H. R. and Cordes, E. H. (1966). "Biological Chemistry" p. 627. Harper & Row, New York, Evanston and London, and John Weatherhill, Inc., Tokyo.
Majerus, P. W. (1967). *J. biol. Chem.* **242**, 2325.
Majerus, P. W. (1968). *Science, N.Y.* **159**, 428.
Majerus, P. W., Alberts, A. W. and Vagelos, P. R. (1964). *Proc. natn. Acad. Sci. U.S.A.* **51**, 1231.
Matsuhashi, M., Matsuhashi, S. and Lynen, F. (1964). *Biochem. Z.* **340**, 263.
McMahon, V. and Stumpf, P. K. (1966). *Pl. Physiol., Lancaster* **41**, 148.
Meyer, F. and Bloch, K. (1963). *Biochim. biophys. Acta* **77**, 671.
Meyer, K. H. and Schweizer, E. (1972). *Biochem. biophys. Res. Commun.* **46**, 1674.
Middleditch, L. E. and Chung, A. E. (1971). *Archs Biochem. Biophys.* **146**, 449.
Mizugaki, M., Swindell, A. C. and Wakil, S. J. (1968a). *Biochem. biophys. Res. Commun.* **33**, 520.
Mizugaki, M., Weeks, G., Toomey, R. E. and Wakil, S. J. (1968b). *J. biol. Chem.* **243**, 3661.
Morris, L. J. (1970). *Biochem. J.* **118**, 681.
Mudd, J. B. and McManus, T. T. (1962). *J. biol. Chem.* **237**, 2057.

Nagai, J. and Bloch, K. (1965). *J. biol. Chem.* **240**, 3702.

Nagai, J. and Bloch, K. (1968). *J. biol. Chem.* **243**, 4626.

Newman, D. W. (1964). *J. exp. Bot.* **15**, 525.

Nixon, J. E., Phillips, G. T., Abramovitz, A. S. and Porter, J. W. (1970). *Archs Biochem. Biophys.* **138**, 372.

Nugteren, D. H. (1965). *Biochim. biophys. Acta* **106**, 280.

Numa, S., Nakanishi, S., Hashimoto, T., Iritani, N. and Ozaki, T. (1970). *Vitams Horms* **28**, 213.

Numa, S., Ringelman, E. and Lynen, F. (1964). *Biochem. Z.* **340**, 228.

Overath, P. and Stumpf, P. K. (1964). *J. biol. Chem.* **239**, 4103.

Phillips, G. T., Nixon, J. E., Abramovitz, J. W. and Porter, J. W. (1970a). *Archs Biochem. Biophys.* **138**, 357.

Phillips, G. T., Nixon, J. E., Dorsey, J. A., Butterworth, P. H. W., Chesterton, C. J. and Porter, J. W. (1970b). *Archs Biochem. Biophys.* **138**, 380.

Plate, C. A., Joshi, V. C. and Wakil, S. J. (1970). *J. biol. Chem.* **245**, 2868.

Plate, C. A., Joshi, V. C., Sedgwick, B. and Wakil, S. J. (1968). *J. biol. Chem.* **243**, 5439.

Porter, J. W., Kumar, S. and Dugan, R. E. (1971). *In* "Progress in Biochemical Pharmacology" (W. L. Holmes and W. M. Bortz, eds), Vol. 6, pp. 1–101. Karger, Basel.

Prescott, D. J. and Rabinowitz, J. L. (1968). *J. biol. Chem.* **243**, 1551.

Pugh, E. L. and Wakil, S. J. (1965). *J. biol. Chem.* **240**, 4727.

Putt, E. D., Craig, B. M. and Carson, R. B. (1969). *J. Am. Oil. Chem. Soc.* **46**, 126.

Rebeiz, C. A., Castelfranco, P. and Breidenbach, R. W. (1965). *Pl. Physiol., Lancaster* **40**, 286.

Rétey, J. and Lynen, F. (1965). *Biochem. Z.* **242**, 256.

Rinne, R. W. (1969). *Pl. Physiol., Lancaster* **44**, 89.

Rittenberg, D. and Bloch, K. (1944). *J. biol. Chem.* **154**, 311.

Rose, I. A. (1970). *J. biol. Chem.* **245**, 6052.

Sauer, F., Pugh, E. L., Wakil, S. J., Delaney, R. and Hill, R. L. (1964). *Proc. natn. Acad. Sci. U.S.A.* **52**, 1360.

Scheuerbrandt, G. and Bloch, K. (1962). *J. biol. Chem.* **237**, 2064.

Scheuerbrandt, G., Goldfine, H., Baronowsky, F. E. and Bloch, K. (1961). *J. biol. Chem.*, **236**, PC70–PC71.

Schulz, H. (1972). *Biochem. biophys. Res. Commun.* **46**, 1446.

Schulz, H. and Wakil, S. J. (1971). *J. biol. Chem.* **246**, 1895.

Schulz, H., Weeks, G., Toomey, R. E., Shapiro, M. and Wakil, S. J. (1969). *J. biol. Chem.* **244**, 6577.

Sedgwick, B. (1972). Unpublished Results.

Shannon, L. M., de Vellis, J. and Lew, J. Y. (1963). *Pl. Physiol., Lancaster* **38**, 691.

Simoni, R. D. and Stumpf, P. K. (1969). *In* "Methods in Enzymology" (J. M. Lowenstein, ed.), Vol. XIV, pp. 84–88. Academic Press, New York and London.

Simoni, R. D., Criddle, R. S. and Stumpf, P. K. (1967). *J. biol. Chem.* **242**, 573.

Stern, J. R. (1961). *In* "The Enzymes" (P. D. Boyer, H. Lardy and K. Myrbäck, eds), Vol. V, pp. 511–529. Academic Press, New York and London.

Stern, J. R. and del Campillo, A. (1956). *J. biol. Chem.* **218**, 985.

Stumpf, P. K. (1969). *A. Rev. Biochem.* **38**, 159.

Stumpf, P. K. and James, A. T. (1963). *Biochim. biophys. Acta* **70**, 20.

Stumpf, P. K., Bove, J. M. and Goffeau, A. (1963). *Biochim. biophys. Acta* **70**, 260.

Stumpf, P. K., Brooks, J. L., Galliard, T., Hawke, J. C. and Simoni, R. (1967). *In* "Biochemistry of Chloroplasts" (T. Goodwin, ed.), Vol. 11, pp. 213–219. Academic Press, London and New York.

Sumper, M., Oesterhelt, D., Riepertinger, C. and Lynen, F. (1969a). *Europ. J. Biochem.* **10**, 377.
Sumper, M., Riepertinger, C. and Lynen, F. (1969b). FEBS *Letters* **5**, 45.
Vagelos, P. R. (1964). *A. Rev. Biochem.* **33**, 139.
Vanaman, T. C., Wakil, S. J. and Hill, R. L. (1968a). *J. biol. Chem.* **243**, 6409.
Vanaman, T. C., Wakil, S. J. and Hill, R. L. (1968b). *J. biol. Chem.* **243**, 6420.
Van den Driessche, T. (1964). *Annls Physiol. vég., Brux.* **9**, 13.
Vellis, J. de, Shannon, L. M. and Lew, J. Y. (1963). *Pl. Physiol., Lancaster* **38**, 686.
Wakil, S. J. (1958). *J. Am. chem. Soc.* **80**, 6465.
Wakil, S. J. (1961). *J. Lipid Res.* **2**, 1.
Wakil, S. J. (1963). *A. Rev. Biochem.* **32**, 369.
Wakil, S. J. and Bressler, R. (1962). *J. biol. Chem.* **237**, 687.
Wakil, S. J. and Ganguly, J. (1959). *J. Am. chem. Soc.* **81**, 2597.
Wakil, S. J. and Gibson, D. M. (1960). *Biochim. biophys. Acta* **41**, 122.
Wakil, S. J., Pugh, E. L. and Sauer, F. (1964). *Proc. natn. Acad. Sci. U.S.A.* **52**, 106.
Wallace, J. W. and Newman, D. W. (1965). *Phytochemistry* **4**, 43.
Weeks, G. and Wakil, S. J. (1968). *J. biol. Chem.* **243**, 1180.
White, H. B., Mitsuhashi, O. and Bloch, K. (1971). *J. biol. Chem.* **246**, 4751.
Willecke, K., Ritter, E. and Lynen, F. (1969). *Europ. J. Biochem.* **8**, 503.
Willemot, C. and Stumpf, P. K. (1967). *Pl. Physiol., Lancaster* **42**, 391.
Williamson, I. P. and Wakil, S. J. (1965). *J. biol. Chem.* **241**, 2326.
Yang, S. F. and Stumpf, P. K. (1965). *Biochim. biophys. Acta* **98**, 27.
Yermanos, D. M., Hall, B. J. and Burge, W. (1964). *Agron. J.* **56**, 582.

NOTE ADDED IN PROOF

In a subsequent publication (Stern, 1961) it has been suggested that the major product formed by the action of crotonase on isocrotonyl-CoA is 3*S*-hydroxybutyryl-CoA.

CHAPTER 10

Hormones and Carbohydrate Metabolism in Germinating Cereal Grains

D. E. BRIGGS

*Department of Biochemistry, University of Birmingham,
Edgbaston, Birmingham, England*

I. INTRODUCTION

A. GENERAL CONSIDERATIONS

The scope of this review is limited to considerations of some aspects of the mobilization of reserve carbohydrates in germinating barley (*Hordeum vulgare*) corns. In general terms the changes undergone by the germinating grains of the Gramineae are very similar (Lehmann and Aichele, 1931). Some references will be made to other cereals as from the limited data available it seem that the types of mechanism that operate in mobilizing carbohydrate reserves in barley operate also in the other varieties of British corn, i.e. wheat

(*Triticum* spp.), rye (*Secale cereale*) and oats (*Avena* spp.). However, the detailed chemistry of these cereals is clearly distinct and the more remotely related cereals seem to be even more different.

Attempts to understand the chemistry of germinating barley using recognizable scientific techniques were certainly being made in the United Kingdom about the year 1800, the object being to raise as much tax as possible on barley malt. The enzyme chemistry of starch conversion dates from about this period when Kirchhoff (1815) recognized that the "albumen" of malt had the ability to saccharify starch. Dubrunfaut, in 1830, prepared an extract of malted barley that could catalyse the hydrolysis of starch to sugars and three years later Payen and Persoz used alcohol precipitation to prepare a stable powder of a malt extract, called diastase, that retained its amylolytic power (Bayliss, 1919). Maltsters and their customers continue to support related investigations to the present day. However, the extensive and well-documented part of the literature regarding this work seems to be little known to biochemists at large.

B. THE STRUCTURE OF THE BARLEY GRAIN

A knowledge of the structure of cereal grains is an absolute prerequisite for understanding the control mechanisms operating during germination. Failure to take account of the different natures of the parts of the grain, and their interactions during germination, has sharply reduced the value of many studies. The detailed morphology of the barley grain is shown in Fig. 1. The kernel is an indehiscent fruit or caryopsis enclosed in a husk (Bergal and Clemencet, 1962; Carson and Horne, 1962). In most European barleys the husk strongly adheres to the caryopsis, but in some naked varieties, as in the more familiar varieties of wheat and rye, the husk separates during threshing. Within the husk, at the base of the grain, situated at each side of the embryo and partly covering it, is a pair of lodicules. These serve to lever the palea and lemma apart at anthesis, but their function, if any, during germination is unknown. The residue of the ovary wall, the pericarp, consists of several layers of microscopically distinct but dead cells which invest the whole seed and adhere to it. The husk and pericarp together comprise some 9–12% (dry wt) of the grain and seem to restrict the swelling of the grain in steeping but not to take part in the chemistry of germination (Hough et al., 1971). The outermost structure of the true seed, the thin, membraneous testa, is made of two cuticularized, impermeable layers slightly separated by the remains of cell walls and it is from there that the pericarp and husk separate during decortication. In the whole grain the pericarp also affects the penetration by solutes (Collins, 1918; Tharp, 1935). In the furrow the testa merges with the "pigment strand", the dark brown, suberized remains of a vascular bundle that runs the length of each grain. The outermost layers of even the most healthy grains are always colonized by a wide variety of microorganisms, fungi, yeasts, actinomycetes and bacteria. The testa and pigment strand prevent the penetration of these organisms into undamaged grain.

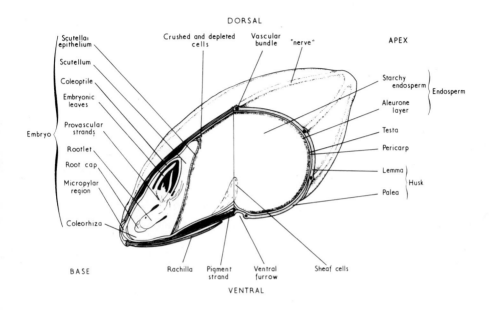

FIG. 1. Idealized diagram of a barley grain, with a quarter section removed, to show the interior.

At the base of the grain, within the testa is the diploid embryonic body (2–5% dry wt grain) which may be considered to have at least two functional parts, the axis of the embryonic plant and the scutellum. The base of the axis is differentiated into several embryonic roots, often five in number, set to grow towards the base of the grain, and they are enclosed in a root-sheath or coleorhiza. The apex of the embryonic axis, which will grow along the dorsal side of the grain, is the acrospire or blade, botanically, it is the coleoptile together with the enclosed embryonic leaves and the primary shoot meristem. The embryonic axis is joined to the scutellum at the primary node and here provascular strands run between the embryonic structures and then ramify through the scutellum. The inner face of the scutellum ("shield", *Lat.*), is recessed into the starchy endosperm. At the interface the scutellum is covered by a columnar epithelium. All the cells of the embryonic body appear to be fully viable. They contain dense cytoplasm with well defined nuclei and numerous inclusions. Occasionally a few grains of starch are present.

The greater part of the grain, by weight, is made up of the remainder of the nucellar tissue and the endosperm tissues. The nucellus is reduced during the formation of the grain to a thin, hyaline layer situated between the aleurone layer and the testa, where it can be distinguished in sections swollen with alkali, and the so-called "sheaf cells". The triploid endosperm is genetically distinct from the diploid embryo and is differentiated into two clearly distinct tissues, the aleurone layer (a thin skin of 3 or 4 layers of cuboidal cells) and the starchy endosperm. The aleurone, thinned down to one layer of tiny, flattened cells, overlaps the edge of the scutellum before becoming indistinguishable from the other layers covering the embryonic tissues (Fig. 2). The cells of the

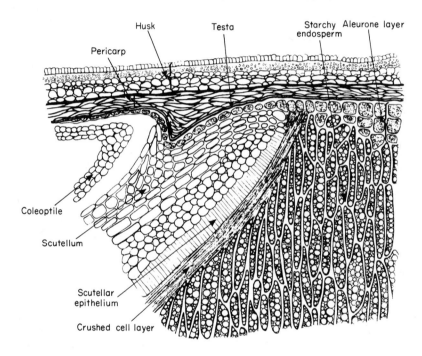

FIG. 2. Detail of the structure of the grain, vertical section, where the apex of the scutellum meets the aleurone layer.

aleurone layer are living, and respire, but they do not grow or divide. They have thick cell walls, dense cytoplasm rich in protein, lipids and numerous organelles, but no starch (Section III, c). The starchy endosperm, which constitutes the rest of the grain, roughly 75% (dry wt), is a parenchymatous tissue. The cells appear to be devoid of life as judged by several criteria including their failure to reduce tetrazolium salts. The cell walls are thin and the most apparent cell contents are massive numbers of closely packed starch grains. In the mature, ungerminated grain starch is practically confined to this tissue. The starch grains are embedded in a protein matrix. Sometimes in immature grain, or grain that has not accumulated a full complement of starch, the distorted and broken remains of nuclei may be detected in these cells. The starchy endosperm is not a homogeneous tissue. A transverse section of the grain shows that the cell walls are closest together near to, and appear to radiate out from, the nucellar "sheaf cells" (Fig. 3.)

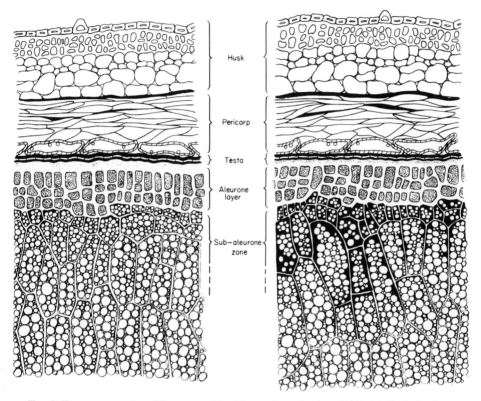

FIG. 3. Transverse sections (diagrammatic) of the grain on the dorsal side. (a) Grain having a low nitrogen content, of good malting quality. (b) Grain with a high nitrogen and protein content.

To anticipate: The starchy endosperm serves as a dead storehouse of nutrients that support the growth of the embryonic axis. The breakdown of these reserves is achieved by the activity of the aleurone layer and the scutellum, and the products are taken up by the scutellum and transported to the embryo. Neither the scutellum nor the aleurone layer grow, nor do their cells divide.

C. EXPERIMENTAL MATERIALS AND CONDITIONS OF GERMINATION

Samples of barley vary in their germinative and chemical characteristics, and these alter slowly during storage. Even after dressing, however, samples contain grains of differing shapes, sizes and chemical compositions; further-more, a proportion of the grains will certainly have started to sprout in the ear (Gordon, 1970) if harvested in a maritime climate. For reliable biochemical experimentation it is essential to select good grain and to sterilize the surface. The following routine procedure for preparing grain is preferred, (Briggs, 1968a). Uniformly-sized grain is selected by means of slotted sieves. Husked grain cannot be sterilized or adequately inspected for damaged or pregerminated grains, so the husk, lodicules and most of the pericarp is removed with 50 % sulphuric acid, leaving the true seed bounded by the testa (Pollock et al., 1955). These decorticated grains are washed in cold water, initially containing suspended Ca CO_3. Then it is briefly immersed in a dilute solution of iodine in excess potassium iodate, rinsed, and air dried. The iodine is mainly present as the I_3^- ion, which is excluded from intact grains. Damaged or pregerminated grains allow the I_3^- ion to penetrate and this gives a blue-black colour with the starch of the endosperm, such grains are discarded. The cream-coloured "mealy" or "floury" grains are used in preference to the harder, grey "flinty" or "steely" grains that usually have a higher nitrogen content and which differ in their metabolism (Essery et al., 1956; Briggs, 1968b). The sulphuric acid treatment does not sterilize the surface of the grain, sterilization is effected with a solution of sodium hypochlorite, and subsequent manipulations are carried out under aseptic conditions. All the steps have been found to be essential if uniform results are to be obtained.

The conditions in which the grains germinate have very marked effects on the biochemical changes that occur. We use two basic conditions of germination (Briggs, 1968a; Groat and Briggs, 1969). (a) Sterile, decorticated grains are germinated in the dark at 25°C, in continuous contact with a filter-paper substratum, wetted with a regulated amount of water. Thus conditions may partly approximate to the germination of grain below ground. Considerable growth of the embryo occurs and the depletion of the starchy endosperm is very considerable and results in a large loss in dry weight and the liquefaction of the interior. (b) In the second method of germination the sequence approxi-mates to that used in temperature-controlled floor malting. The sterile grain is immersed in water, for two consecutive 24 h periods, at 14·4°C. After a period of draining the grain is formed into a layer and germinated, at 14·4°C with occasional turning, in air with a high humidity. Under these conditions the growth of the embryo is minimized by a lack of water and the endosperm remains solid, although it becomes softer and its contents are partly depleted.

II. The Carbohydrates of the Barley Grains

A. SOLUBLE CARBOHYDRATES, GUMS AND HEMICELLULOSES AND ENZYMES THAT DEGRADE THEM

The carbohydrates of the barley grain are very diverse (Harris, 1962). For the present purpose attention will be concentrated on those substances

TABLE I

Carbohydrate composition of ungerminated
barley (% dry weight)

Cellulose	4–5
Hemicelluloses	8–10
Gums:	
Total, soluble at 40°C	1–1·5
β-Glucan, soluble at 40°C	0·6–1·0
β-Glucan, soluble at 65°C	1·6–2·6
Starch	58–65
Amylose/Amylopectin	
(normal barley varieties)	20–30/80–70
Soluble reducing sugars	
Glucose	0·03–0·09
Fructose	0·03–0·16
Maltose	0·006–0·135
Soluble non-reducing sugars	
Sucrose	0·34–1·69
Raffinose	0·14–0·83
"Glucodifructose"	0·10–0·43
(Kestose + isokestose)	
Higher fructosans	0·5 (approx.)

Data a from Bourne and Pierce, 1970; Harris,
1962; Pullock, 1962.

occurring interior to the testa, chiefly those of the starchy endosperm (Table I).

Free pentoses are not usually detected among the water-soluble carbohydrates in barley. Glucose (1) and fructose (2) occur, apparently concentrated
in the embryo and aleurone layer, as are sucrose (3) and some higher fructosans.
The fructosans, varying from kestose (4) and iso-kestose (5) ("gluco-
difructose"), to molecules containing glucose combined with thirty or more
fructose residues, occur as two series. In each case the molecules contain

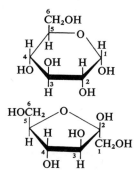

(1) Glucose α-D-glucopyranose

(2) Fructose β-D-fructofuranose

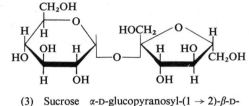

(3)   Sucrose   α-D-glucopyranosyl-(1 → 2)-β-D-fructofuranose.

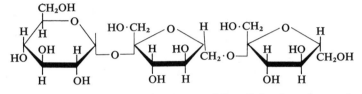

(4)   Kestose (6-kestose) α-D-glucopyranosyl-(1 → 2)-β-D- fructofuranosyl-(6 → 2)-β-D-fructofuranoside

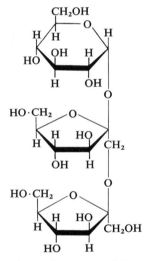

(5)   Isokestose (1-kestose) α-D-glucopyranosyl-(1 → 2)-β-D-fructofuranosyl-(1 → 2)-β-D-fructofuramoside.

a sucrose unit to which are attached chains of β-(1 → 2)-fructofuranosidic units in the kritesin or inulin series and β-(2 → 6)-fructofuranose units in the phlein or hordeacin series (6). Not much is known of the enzymes catalysing the metabolism of fructosans in barley (Schlubach, 1965; MacLeod, 1953). Raffinose (7) occurs in ungerminated barley, it disappears early in germination and does not reappear.

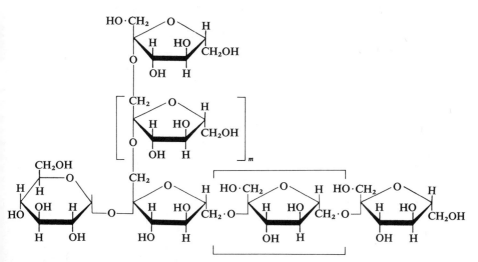

(6) The probable basic unit of the fructosan series. The kritesin (inulin) series is formed when $m = 1$ or more, $x = 0$—the chain is extended with a series of fructosfuranosidic residues joined by $(1 \rightarrow 2)$-β-links.

The hordeacin (phlein) series of fructosans is formed when $m = 0$, and $x = 1$ or more—the chain is extended by β-fructofuranosidic residues linked $(2 \rightarrow 6)$.

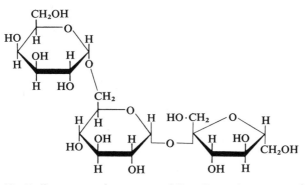

(7) Raffinose α-D-galactopyranosyl-$(1 \rightarrow 6)$-α-D-glucopyranosyl-$(1 \rightarrow 2)$-β-D-fructofuranoside.

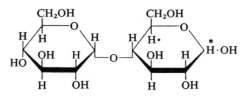

(8) Maltose α-D-glucopyranosyl-$(1 \rightarrow 4)$-D-glucopyranose.

Maltose (8), when it occurs, does so mainly in the sub-aleurone zone of the starchy endosperm, where the highest concentration of β-amylase occurs (Engel, 1947; Linderstrøm-Lang and Engel, 1937; Section III). In malting grain a minimum level of soluble sugars occurs at about the second day of germination, when respiration has utilized much of the "preformed" sugar, and before it has been replenished by the degradation of insoluble reserves (Pollock, 1962; Harris and MacWilliam, 1954). This "minimum" may be involved in triggering the processes leading to the breakdown of the endosperm reserves.

Warm-water extracts of barley contain a range of polysaccharide substances. If solubilized starch and maltodextrins are specifically degraded with α-amylase and the degradation products are removed the remaining substances are termed "gums". The yield of gum depends on the sample of barley and the temperature of extraction and the method of extraction used. Thus grist (ground up or milled grain) pre-treated with boiling 80% ethanol to inactivate gum-degrading enzymes, yields more gum if treated with activated papain or alcoholic alkali—a result explained by assuming that the gum is partly enmeshed in insoluble protein. The bulk of the gum is thought to come from the starchy endosperm. Fractionation with ammonium sulphate will at least partly separate gums into (a) β-glucans (9), (b) araboxylans (10), and (c) uncharacterized materials containing arabinose (11), xylose (12), glucose (1), mannose (13) and galactose (14). β-Glucan (9), when isolated in an undegraded state, is highly viscous in solution indicating molecular weights of about 200000 (Djurtoft and Rasmussen, 1955). It consists of linear chains of D-glucopyranose residues joined by random β-(1 → 3) and β-(1 → 4) linkages in the ratios of 1:1 to 1:4.

Araboxylans vary in composition depending on their site of origin in the grain (Harris, 1962). The araboxylans of the starchy endosperm (10) consists of linear chains of β-(1 → 4)-D-xylopyranose units (12) to which L-arabinofuranose residues (11) are attached at positions 2 and 3 on the xylose residues. The ratio of arabinose to xylose differs between different ammonium sulphate fractions. The cell walls of the starchy endosperm are thought to be similar in composition to the water soluble gums, but of higher molecular weight. Hemicelluloses extracted with dilute sodium hydroxide contain glucose, arabinose and xylose (Preece and Hobkirk, 1954). Uronic acids are virtually absent in materials from the starchy endosperm (but not in husk materials).

G$\,$(G·)$_n$·G···C—C··C··C·(·C)$_o$G—C···G—C··G$\,$(G·)$_p$C—G *

(9)   Barley β- glucan; Lichenin
G = D-glucopyranose residue
— = β-(1 → 4) link
··· = β-(1 → 3) link

Repeating units, cellobiose G—G (18), and laminaribiose, G ··· G (19)

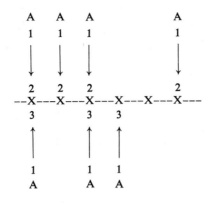

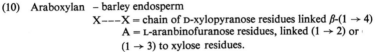

(10)   Araboxylan – barley endosperm
X---X = chain of D-xylopyranose residues linked β-(1 → 4)
A = L-aranbinofuranose residues, linked (1 → 2) or
(1 → 3) to xylose residues.

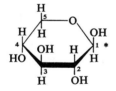

(11)   Arabinose   α-L-arabinofuranose.

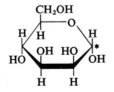

(12)   Xylose   β-D-xylopyranose.

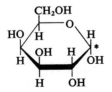

(13)   Mannose   α-D-mannopyranose.

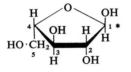

(14)   Galactose   α-D-galactopyranose.

Other experiments indicate that the cell-walls of the starchy endosperm are entirely hemicellulose in nature, and that true "cellulose" is confined to the husk in the barley grain (MacLeod and Napier, 1959). Lignin appears to be confined to the husk in ungerminated grain.

A wide range of enzymes that catalyse the hydrolysis of polysaccharides has been found in germinating barley. Some of the enzymes degrade polysaccharides of types that are not known to occur in the grain. Enzymes that hydrolyse cellodextrins (15), soluble derivatives of cellulose [e.g. carboxymethyl cellulose (16)], laminarin, lichenin, barley β-glucan, araboxylan, xylan and pectin have been detected (Harris, 1962). Studies with crude extracts of resting and germinating barley suggested that araboxylan is attacked by the following hydrolytic enzymes (Preece *et al.*, 1954; Preece and Hobkirk, 1955; Preece and MacDougall, 1958):

(a) An arabinosidase, which removed the arabinose residues to yield a xylan.
(b) An endo-xylanase that attacked the xylan-to-xylan links at random, producing araboxylan and xylan "dextrins".
(c) An exoxylanase which hydrolysed the penultimate links of xylan chains to release xylobiose (17) and,
(d) a xylobiase which hydrolysed the xylobiose to xylose and which possibly released xylose directly from the ends of the xylan chain. Thus it is proposed that breakdown occurs according to the scheme shown (Fig. 4).

Similar experiments indicated that the β-glucans were attacked by (a) an endo-β-glucanase that attacked the glucan chains at random, giving rise to a range of β-glucan dextrins. (b) An exo-glucanase which attacked the penulitimate glucose links giving rise to cellobiose (18) (β-D-glucopyranosyl-(1 $\rightarrow$ 4)-D-glucopyranose), and laminaribiose (19) (β-D-glucopyranosyl-(1 $\rightarrow$ 3)-D-glucopyranose), (c) β-glucosidase(s) which hydrolyse cellobiose and laminaribiose to glucose, and possibly attack the non-reducing end of the β-glucan chains to release glucose (Preece and Hoggan, 1956; Harris, 1962; Gelman, 1969). Transglucosidase activity has also been detected by incubating barley extracts with cellobiose (18), and detecting the formation of gentibiose (β-D-glucopyranosyl-(1 $\rightarrow$ 6)-D-glucopyranose), cellotriose, laminaribiose (19) and 6^2-glucosylcellobiose (e.g. Anderson and Manners, 1959; Luchsinger and Richards, 1968). Most studies on the development of gum-degrading enzymes have been made using crude extracts and inhibitors and the results have been interpreted according to this scheme. For example it was often assumed that the breakdown of substituted celluloses was carried by the *endo-β*-glucanase but at present, judged mainly on results obtained with enzymes separated using gel chromatography and continuous electrophoresis (Manners and Marshall, 1969), the situation seems to be as follows: (a) the presence of an exo-β-glucanase has not been demonstrated; (b) two β-glucosidases have been

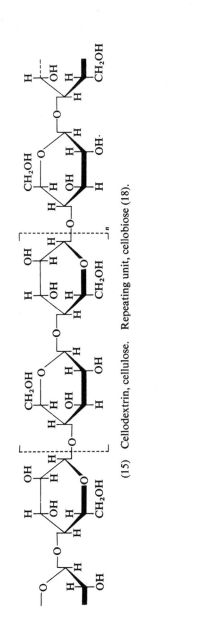

(15) Cellodextrin, cellulose.   Repeating unit, cellobiose (18).

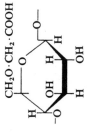

(16) Substituted glucose residue, as occurs in carboxymethyl cellulose.

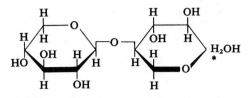

(17)   Xylobiose   β-D-xylopyranosyl-(1 → 4)-D-xylopyranose.

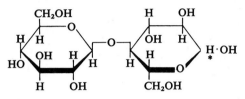

(18)   Cellobiose   β-D-glucopyranosyl-(1 → 4)-D-glucopyranose.

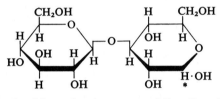

(19)   Laminaribiose   β-D-glucopyranosyl-(1 → 3)-D-glucopyranose.

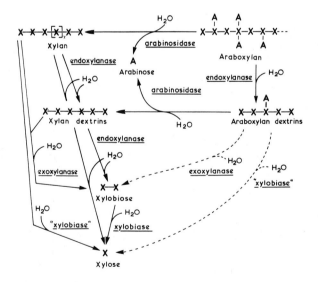

FIG. 4. Possible routes by which enzymes catalyse the breakdown of the araboxylan of the endosperm. A = Arabinofuranose residues; X = Xylopyranose residues.

distinguished. They both hydrolyse laminaribiose (19) and cellobiose (18), and also catalyse transglucosylation reactions. They also release glucose from the non-reducing ends of the molecules of β-barley glucan, lichenin, laminarin, pachyman and cellodextrins (15) (laminarin and pachyman are β-(1 → 3)-linked glucans). The quantities of these enzymes do not alter much in malting. (c) Three β-glucanases have been distinguished. (i) An endo-β-(1 → 3)-glucanase or "endo-laminarinase" which appears to be specific in that it attacks laminarin and pachyman, with little release of free glucose, but appears not to attack β-glucans containing β-(1 → 4) as well as β-(1 → 3) links. (ii) and (iii). The other barley endo-β-glucanases both degrade β-barley glucan (9), lichenin, laminarin, pachyman, cellodextrins (15) and laminaribiose (19). They are distinct in that during the breakdown of polysaccharides one produces a relatively greater ratio of glucose to oligosaccharides than does the other.

Most of the estimates that have been made of gum hydrolysing activity in grain have been based on determinations of the changes in viscosity or reducing power of substrates caused by crude grain extracts. Such activities must usually be brought about by the co-operative action of several enzymes and are, therefore, impossible to interpret in terms of changes in level of one specific protein. The overall enzyme activities alter independently to some extent and so an understanding of the changes in levels of individual enzymes is essential before control-mechanisms can be reliably unravelled. This must await the development and use of specific assays for the individual enzymes.

B. STARCH

Starch is the largest single constituent of barley grains, forming 58–65% of the dry weight (Harris, 1962). In the ungerminated grain it occurs practically exclusively in the starchy endosperm. Generally, the starch grains fall into populations of two size ranges, having an overall average diameter of 20–30 μ. Th carbohydrate can be separated into two fractions, (a) amylose (20), and (b) amylopectin (21), where the α-(1 → 4) linked chains are branched through α-(1 → 6) links. European barleys have amylose/amylopectin ratios of about 20–30/80–70. In germinating grain the starch is readily degraded; nevertheless, isolated, intact starch grains are surprisingly resistant to enzymic hydrolysis and there is no generally accepted explanation for this discrepancy. The activities of starch-degrading enzymes (diastase) are generally determined on soluble starch or its derivatives. Many results that have been ascribed to particular enzymes have been obtained with non-specific assays and are, in fact, due to the concerted activity of several enzymes. Phosphorylase catalyses the cleavage, by inorganic phosphate, of α-(1 → 4)-links between the non-reducing terminal D-glucopyranose residues of starch, producing glucose-1-phosphate and a shortened starch chain. The enzyme acts on amylose (20) to degrade it completely, but action on amylopectin (21) ceases as the enzyme

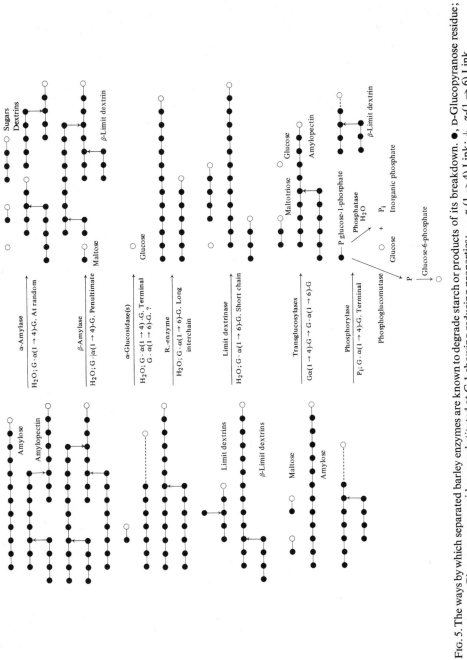

Fig. 5. The ways by which separated barley enzymes are known to degrade starch or products of its breakdown. ●, D-Glucopyranose residue; ○, D-Glucopyranose residue, unsubstituted at C-1 showing reducing properties; —, α-(1 → 4) Link; ↓, α-(1 → 6) Link.

approaches the α-(1 → 6)-branch points so that from this substrate the ultimate products are glucose-1-phosphate and β-limit dextrin (22) (Fig. 5). Barley is rich in phosphatases, and in their presence the glucose-1-phosphate will be hydrolysed to inorganic phosphate and glucose and the overall reaction will appear to be due to the action of a phosphate-stimulated α-glucosidase.

β-Amylase occurs in the starchy endosperm in bound and free forms, and isozymes are thought to exist in at least some varieties (Section III; Harris, 1962).

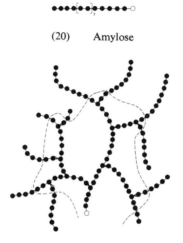

(20)   Amylose

(21)   Amylopectin.   The dotted line indicates the approximate β-amylolysis limit.

(22)   β-Limit dextrin—the molecule within these limits.

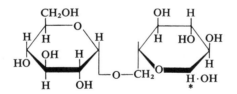

(23)   Isomaltose   α-D-glucopyranosyl-(1 → 6)-D-glucopyranoside

α-Amylase, in contrast, is formed entirely during germination (Graber and Daussant, 1964). The enzyme found in some samples of barley is almost certainly that formed in pregerminated grains (Greenwood and MacGregor, 1965; Gordon, 1970) or in contaminating microorganisms, and it creates more "chain ends" that can be degraded by β-amylase or phosphorylase. It has been suggested that even purified samples of enzyme carry traces of transglucosylase (French and Abdullah, 1966) but it is also possible that the transglucosylase activity is a property of the α-amylase itself. It requires Ca^{2+} for activity and in the presence of Ca^{2+} at pH 6·0 it is usually stable at 70°C for 20 min. Some varieties have one heat-stable α-amylase isozyme, others have two or possibly even five (Frydenberg and Nielsen, 1965; Van Onckelen and Verbeek, 1969). Malt α-amylase has been crystallized several times but, as noted, it is uncertain whether the crystalline product was pure. (Schwimmer and Balls, 1949; Fischer and Haselbach, 1951.)

An α-glucosidase, which is found in resting and germinating barley, catalyses the hydrolysis of maltose (8), isomaltose (23) (Jørgensen, 1963; 1964; 1965; Jørgensen and Jørgensen, 1963), and at least 40% of soluble starch. It also exhibits transglucosidase activity so that, for example, it forms isomaltose (23), panose (4-O-α-isomaltosyl-D-glucose), and maltotiriose (4-O-α-maltosyl-D-glucose), from maltose. Because, in studies on partly purified extracts treated in various ways, it was found that these enzymic activities occurred in the same ratio, it was thought that barley produced only one α-glucosidase (Jørgensen, 1963), however, the report of an acid α-glucosidase, having a pH optimum of 4, may be the first direct evidence for a new type of α-glucosidase in barley (Marshall and Taylor, 1971). Histochemical studies indicate that most of the α-glucosidase activity is confined to the embryo and aleurone layer and probably remains there during germination.

(24)   Nigeran, (mycodextran)

●	α-D-glucopyranose residue
o	Terminal, reducing D-glucopyranose residue
—	α-(1→4)-link
·	α-(1→3)-link
↓	α-(1→6)-link

(25)   Pullulan (idealized)

Germinating barley is known to contain two "debranching" enzymes that catalyse the hydrolysis of the α-(1 → 6)-glucosidic links in starch or its degradation products (MacWilliam and Harris, 1959). R-enzyme (amylopectin 6-glucanohydrolase) catalyses the hydrolysis of the outermost interchain links in amylopectin and some untypical glycogens but not pullulan (25) or β-limit dextrin (22) (Manners, 1971; Manners and Sparra, 1966; Manners and Rowe, 1971).

Limit dextrinase will not attack amylopectin but hydrolyses the outer α-(1 → 6) bonds in β-limit dextrin (22) and α-limit dextrins, and successively hydrolyses these links in pullulan (25). During germination its activity increases by a factor of about twenty (Manners and Yellowlees, 1970; 1971; Manners et al., 1971).

C. THE ESTIMATION OF ENZYMES DEGRADING STARCH

Diastase is a complex mixture of enzymes even if no account is taken of possible isozymes. The measurement of starch hydrolysis involves either the decline in viscosity or measuring the increase in the reducing power of incubation mixtures or following the decline in colouration given with iodine. No approach gives a specific assay of any one enzyme. Indeed, owing to the "cooperative" action of some of these enzymes their apparent activities are more than not additive (e.g. Briggs, 1964a; Meredith, 1965). However, it is possible to obtain near-specific assays of α-amylase if adequate precautions are taken. As this enzyme is the basis of a number of assays in popular use it is essential for the interpretation of experimental results to compare several of these methods and to indicate some of their shortcomings. Extraction of the enzyme from macerated grain is likely to be incomplete unless adequate concentrations of salts are present in the medium. Further, the enzyme will be unstable unless the extract is buffered at about pH 6 and contains calcium ions (e.g. Wiener and Hopkins, 1952; Preece, 1947). Any alteration in starch structure alters the colour yield and often the absorption spectrum produced with iodine and so all starch degrading enzymes will have effects on these parameters to a greater or lesser extent. The colour yields given by many reagents for reducing groups are altered by, for example, metal ions and other constituents of plant extracts (Briggs, 1967).

To overcome the lack of specificity a new assay was developed, in which α-amylase was extracted from ground germinated barley and simultaneously subjected to a selective heat inactivation process to destroy most of the interfering enzymes (Briggs, 1967). The activity of the heat-treated enzyme is followed using the decline in starch-iodine colour. Under these conditions enzyme activity is only slightly stimulated by high levels of salts indicating that, as expected from the known heat-lability of this enzyme, limit dextrinase activity is not significant. It would be expected that a proportion of R-enzyme would survive the heat-inactivation treatment and by its degrading action lead to a rise in the starch-iodine colour during incubation, but the contribution by this enzyme must be negligibly small as molybdate ions, which inhibit R-enzyme, did not affect the results of the assay.

The other widely used assay (Shuster and Gifford, 1962; Chrispeels and Varner, 1967a) is in no sense specific for α-amylase. We have confirmed that homogenized aleurone layer, totally lacking α-amylase, contains an α-glucosidase (V. J. Clutterbuck and D. E. Briggs, unpublished); these homogenates show "amylase" activity by the Shuster and Gifford method. The lack of specificity of these widely used assays makes it necessary to reinterpret much published work; for example, results obtained with de-embryonated grain (rich in β-amylase) or aleurone layers (poor in β-amylase but rich in α-gluco-sidase) cannot be quantitatively comparable. As far as possible results obtained with the "diastase" techniques of Shuster and Gifford or Chrispeels and Varner will be referred to as α-amylase-D.

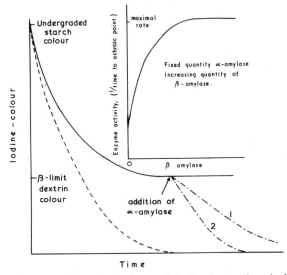

FIG. 6. Scheme illustrating colours given with iodine by starch or its breakdown products during degradation of α-amylase, β-amylase or mixtures of these enzymes. Main diagram: ——, β-amylase added to starch, time 0; ––––, α-amylase added to starch, time 0; ·—·—, α-amylase (two levels) breaking down β-limit dextrin, in the presence of β-amylase. Rate of breakdown of starch given by a fixed quantity of α-amylase and variable quantities of β-amylase. The basis of the Sandstedt *et al.* (1939) assay of cereal α-amylase.

III. GRAIN GERMINATION AND CARBOHYDRATE MOBILIZATION

A. THE BACKGROUND

When hydrated to an adequate extent (i.e. to more than 37% moisture, fresh wt) and placed in air or oxygen at a suitable temperature, the oxygen uptake of barley grains rises and they begin to germinate. In husked grain the first visible sign of germination is the appearance of the coleorhiza ("chit"), then the primary roots emerge. At the same time some new enzymes appear and, supplementing those already present, catalyse the breakdown of the contents of the starchy endosperm, and hydrolytic products accumulate. These are taken up by the scutellum and serve as nutrients for the developing embryo. The breakdown begins immediately below the scutellum and spreads progressively down the grain to the apex. The breakdown of the cell-walls precedes attack on the starch grains they enclose (Dickson and Shands,

1941). Some parts of the starchy endosperm dissolve completely, leaving fluid filled spaces (Briggs, 1968a). The exact pattern of endosperm breakdown, the so-called "modification process", is discussed later.

The embryo used to be regarded as a parasite growing at the expense of the endosperm, a dead reserve of food. However, while embryos could be transplanted successfully onto untreated, de-embryonated grains (even of different varieties or different species), the growth of transplanted embryos was much reduced on "killed" de-embryonated grains. Further, when united with living embryos, the "killed" endosperm did not break down in the usual way (Brown and Escombe, 1898; Stingl, 1907). Microscopic investigations with germinating rye grains convinced Haberlandt (1890) that the aleurone layer was an enzyme secreting tissue, and various other observations indicated that "vital processes" in the aleurone were essential for modification. Thus, the aleurone of sterile, de-embryonated barley could be peeled from the endosperm after a few days soaking (Brown and Escombe, 1898; Stoward, 1911). This was prevented by anaesthetics. If damp de-embryonated grains were incubated in air, then, after two or three weeks, starch grains were corroded, and the starchy endosperm liquefied. If the incubation was made under anaerobic conditions, or in the presence of chloroform, or if the aleurone were filed away, no liquefaction occurred. On the other hand, if small pieces of aleurone remained then these sank into the adjacent starchy endosperm showing a degradation of this tissue immediately beneath the aleurone fragments. Histochemical studies subsequently indicated that during grain germination the total diastase increased in amount in the endosperm adjacent to the aleurone layer and that this could be due to the release of α-amylase from this tissue, providing further evidence for its "glandular" nature (Linderstrøm-Lang and Engel, 1937).

B. MORE RECENT EXPERIMENTS: GIBBERELLINS AND α-AMYLASE

The modification of the starchy endosperm in whole grain is dependant on the growth of a viable embryo; factors which prevent embryo growth prevent modification of the endosperm. Agents which overcome dormancy accelerate the process, for example hyperbatic oxygen, H_2O_2, Fe^{2+}, Cu^{2+}, Hg^{2+} ions, some thiols, nitrous acid, formaldehyde, gibberellic acid, and a cocktail of antibiotics (Hough et al., 1971; Pollock, 1962; Gaber and Roberts, 1969; Morris, 1958). The activity of such diverse agents makes it difficult to interpret the slight stimulatory effects of some plant hormones.

It was known that "modification" of the endosperm spreads from the embryo and this process has been studied by Kirsop and Pollock (1958) who removed embryos from samples of germinating, decorticated barley on successive days and kept the samples of mutilated grain in "malting" conditions for 7 days. Modification was incomplete if the embryos were removed on germination days 1 or 2, but on day three, or subsequently, the continued presence of the embryo was not necessary. Thus something, guessed at that time to be enzymes, was released from the embryo by the third germination day in amounts sufficient to cause normal modification. It was at about this time that the effects of gibberellins on grain metabolism became known outside Japan. Hayashi (1940) had accelerated the malting process by adding gibberellic acid, and it was also found to break barley dormancy, and enhance seedling growth, respiration and the activities of hydrolytic enzymes. Embryos and de-embryonated grains cultivated together in liquid produced more α-amylase-SKB than was produced by the parts cultures separately or when either was dead (Yomo, 1958; Briggs, 1963), showing an interaction between these parts. The embryo produced a substance, termed AFX, that when applied to de-embryonated grain, induced the appearance of α-amylase-SKB and increased hydrolytic activities. AFX was partly purified and was shown to contain dialysable, weakly acidic, gibberellin-like compound. From 100 kg of green malt, 3 mg AFX were obtained and this was estimated by bioassay to be a 44% yield. Gibberellic acid could replace AFX in causing modification of de-embryonated grain (Yomo, 1960c; Paleg, 1960a, b; 1961). In liquid culture media gibberellic acid triggered almost complete dissolution of the starchy endosperm of de-embryonated grain, resulting in a massive drop in dry weight and a release into the culture medium of enzymes, reducing sugars, amino acids and inorganic phosphate (Yomo, 1960; Paleg, 1960a, b; 1961; MacLeod and Millar, 1962; Briggs, 1963).

In the presence of amino acids (but not in their absence) isolated embryos produce about twice as much α-amylase-SKB when incubated with GA_3 as without. Addition of GA_3 to the whole grain increases the amount of true α-amylase found in the embryo (Briggs, 1964b; 1968a, b, c). Results for α-amylase-SKB obtained from dissected grains cultured with amino acids indicated that at the end of malting 7% of the enzyme is located in the embryo and 93% in the endosperm. Of that in the endosperm 6·5% originated in the embryo and 86·5% in the aleurone (Briggs, 1964b). Subsequently it has been shown that, in grain growing under different conditions, true α-amylase occurs in the scutellum, but not in the acrospire or roots. It appears that α-amylase is produced in the aleurone layer in response to gibberellins synthesized in the embryo and not in response to other factors. If gibberellins are not the only substances capable of inducing α-amylase formation and endosperm modification, are the other substances physiologically important in the grain? It has been known since 1908 that there are substances which will activate the pre-existent β-amylase in the endosperm (Ford and Guthrie, 1908; Harris, 1962). In the presence of high salt concentrations β-amylase will slowly degrade starch granules. Also high salt solutions bring α-glucosidase into solution (Jørgensen and Jørgensen, 1963). Thus sugar release as in non-specific "diastase" assays may be due to the activation of preformed enzymes or the de novo synthesis of enzymes. High levels of sugar release have been elicited from isolated endosperms of barley by the fungal metabolite helminthosporol (27), and related compounds, various organic solvent residues, as well as gibberellins (e.g. Briggs, 1966a, b; Coombe, 1971; Crozier *et al.*, 1970; Paleg *et al.*, 1964). Using α-amylase formation as the test we can find no enzyme induction with cyclic-3' 5'-AMP, glutamine, ATP, phorone, isophorone, phenobarbital, kinetin, or benzylaminopurine (V. J. Clutterbuck and D. E. Briggs, unpublished), although reports to the contrary have appeared. However, some of these substances did alter the rate at which α-amylase was released from isolated aleurone layers induced with GA_3. We can find evidence for the induction of α-amylase by gibberellins and the fungal products helminthosporol and helminthosporic acid only. Neutral inducers of α-amylase have not been detected in germinating barley (Groat and Briggs, 1969), and kinetic studies seem to show that gibberellins are indeed the controlling factors in the grain.

1. *Subcellular Changes Occurring in the Scutellar Epithelium and the Aleurone Layer During Enzyme Production*

The changes undergone by the cells of the scutellar epithelium and the aleurone layer during grain germination, and of the aleurone layer when isolated and incubated *in vitro*, have been investigated by light and electron microscopy and are consistent with their supposed functions—i.e. enzyme secretion and, in the case of the scutellar epithelium, nutrient absorption, (e.g. Godineau, 1962; *vide infra*).

The scutellar epithelium consists of a single layer of columnar cells, separated by thin cell walls that are perforated with plasmodesmata. As germination begins the cytoplasm becomes more coarsely granular or cloudy as seen under the light microscope (Fig. 7). The aleurone grains swell and gradually the aleurone proteins disappear, followed by the lipids, at the same time the organelles develop and the lateral walls of the cells separate, and the formerly polygonal cells elongate. Later the cytoplasm clears and vacuolates, organelles become disorganized and the cells appear to empty. In this state the area exposed to the reserves of the starchy endosperm is increased.

The cells of the aleurone layer have well defined nuclei and dense granular cytoplasm, but in contrast to the cells of the scutellar epithelia they are approximately cuboidal and are bounded by thick cell walls, perforated by plasmodesmata. In the ungerminated grain the aleurone grains are dense and the mitochondria have few cristae. As the tissue hydrates the aleurone grains swell, and the endoplasmic reticulum and mitochondria develop.

In the absence of gibberellin little further visible change occurs but in the presence of gibberellin the spherosomes tend to rearrange themselves around the aleurone grains. The endoplasmic reticulum becomes "rough" with ribosomes and organized into lamellae, polysomes are seen, cisternae that have appeared reach their maximal dilation and the mitochondria develop

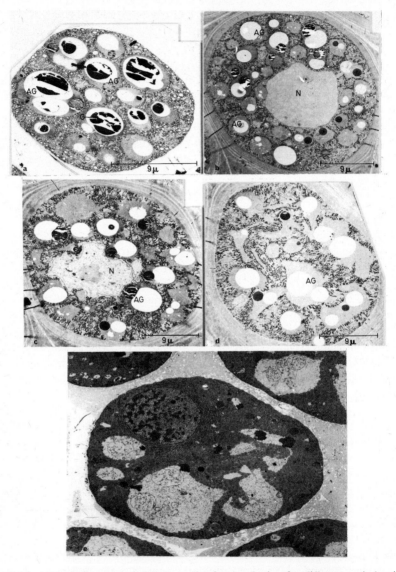

FIG. 7. Electron micrographs of aleurone cells after incubation for different periods with
GA₃, at 25°C. (Jones, 1969; Jones and Price, 1970, with permission.) a–d, fixed with KMnO₄;
(e), fixed with glutaraldehyde. (a) Water-imbibed. Exposed to GA₃ for (b) 2 h, (c) 6 h, (d) 10 h,
(e) 24 h. AG-aleurone grains. N-nucleus. G-globoid within aleurone grain. In (e) note the
vacuoles and stacked e.r. near the nucleus.

cristae. The aleurone grains then begin to degenerate and lose their contents. Ultimately they vacuolate, "spiny vesicles" appear at the plasmalemma, "P-protein" appears to be secreted while RNase leaks from the cell. The contents of the cells degenerate further, the cell-walls are eroded [the break-down occurs fastest close to the starchy endosperm (Taiz and Jones, 1970)].

2. Enzyme Formation in the Embryo

Separated barley embryos produced "diastase" that is different from the "diastase" found in the endosperm of resting grains. The production of this "diastase" which is enhanced by feeding some nitrogenous substances (Brown and Escombe, 1898; Stoward, 1911) is prevented by feeding sugars to the embryos (Massart, 1955). Embryos incubated after dissection from the whole grain contain and release hydrolytic enzymes which attack a range of substrates (Briggs, 1962). Isolated embryos were also shown to produce α-amylase-SKB (Briggs, 1964b) and this was prevented by various metabolic poisons. Thus the enzyme was thought to be synthesized de novo as in the aleurone layer, in response to endogenous gibberellins. However, in contrast to the aleurone, the production of enzyme was enhanced by feeding amino acids (a mixture being better than any one tested singly) and depressed by feeding sugars (Briggs, 1964b).

Enzyme production in the scutellum has subsequently been attributed to the thin layer of "reduced" aleurone tissue adhering round the rim, and separated with it during dissection (MacLeod and Palmer, 1966). The numerical value of the results on which this attribution is based may be queried as they were obtained in the absence of added nutrients when, as already seen, enzyme production is sharply limited. In the author's view the scutellar epithelium does synthesize α-amylase because (a) scraping the scutellar surface reduces the quantity of "diastase" produced by dissected embryos, (b) the histological changes undergone by aleurone and scutellar epithelial cells are so similar, (c) α-amylase is produced by plugs of scutellar tissue + epithelium (entirely devoid of "rim" and any attached aleurone), when incubated with GA_3 and amino acids (Table II). The different effects of nutrients on the production of amylase by the embryo and the aleurone indicate that there are differences between the tissues. At present it is impossible to say what proportions of α-amylase are produced by the scutellar epithelium and what by the "residual aleurone" clinging round the rim.

3. Biochemical Changes and Enzyme Formation in the Aleurone Layer

The release of α-amylases by barley endosperms treated with GA_3 are reduced or prevented by heat, acetate, $HgCl_2$, $CuSO_4$, KCN, fluoroacetate, iodoacetic acid, p-chloromercuribenzoic acid, anaerobiosis and 2,4-dini-trophenol (Briggs, 1963; Paleg, 1960a; Chandra and Varner, 1963; Varner,

TABLE II

Amylase production by dissected grains incubated in liquid medium containing GA$_3$ (1 mg/l) (unpublished results of Clutterbuck and Briggs)

| | α-Amylase in preparation (S.I.C./fraction) | | | |
	Medium 0–48 h	Medium 48–72 h	Grain part 0–72 h	Total (all fractions)
Embryonic axis + scutellum	2·0 2·5	2·2 2·3	1·5 1·5	5·7 6·3
Edges peeled. Protein hydrolysate in medium	1·8	3·1	1·9	6·8
Aleurone peelings and edges of scutellum	5·9 5·0 / 3·5	6·1 4·2 / 2·8	9·0 8·4 / 9·6	21·0 17·7 / 16·0
Remainder of pleurone and starchy endosperm	37·6 33·8 / 39.5	50·6 48·0 / 61.8	34·1 33·6 / 33·2	122·3 115·4 / 134·5
Halved grains	45·6 41·4 / 45·0	59·0 54·4 / 67·8	45·4 43·6 / 44·8	150·0 139·4 / 157·6

1964; Yomo and Iinuma, 1964a, b). Enzyme production is also blocked by puromycin, chloramphenicol and cycloheximide. Partial tryptic digestion of purified α-amylase and "fingerprinting" the peptides so obtained demonstrated that [^{14}C]-amino acids were incorporated throughout the enzyme molecule (Varner and Chandra, 1964), and [^{18}O] from $H_2^{18}O$ was also incorporated into the α-amylase-D, as determined by isopycnic centrifugation (Filner and Varner, 1967). All this suggested that de novo protein synthesis occurs (Briggs, 1963; Yomo and Iinuma, 1964a, b; Varner, 1964; Varner et al. 1965), again showing that GA$_3$ stimulated synthesis and did not activate a preformed enzyme zymogen, or release a preformed enzyme (Paleg, 1960b; MacLeod and Millar, 1962).

That gibberellic acid acted on the aleurone layer, and not the endosperm as a whole, was shown by (a) dissecting endosperm slices and showing that dissolution in response to GA$_3$ occurred only when the aleurone was present (MacLeod and Millar, 1962); (b) noting the pattern of breakdown of de-germed grain incubated with gibberellic acid (e.g. Briggs, 1963); (c) demonstrating enzyme production by separated aleurone layers (Briggs, 1964a, b; Paleg, 1964). Consistent with this finding is the fact that the respiration of the degermed grain is confined to the aleurone layer (Paleg, 1964; Briggs, 1964a, b; Varner et al., 1965). It is thought that enzyme synthesis is dependent on a supply of amino acids released by the hydrolysis of protein reserves in the aleurone cells. By blocking this mobilization of amino acids with the protease inhibitor $KBrO_3$ the production of α-amylase-D becomes stimulatable by an external supply of amino acids.

The formation of α-amylase-D in the aleurone layer, in response to gibberellic acid, is prevented by actinomycin-D which, in other organisms, inhibits the activity of DNA-directed RNA polymerase (Paleg, 1964; Varner and Chandra, 1964; Chrispeels and Varner, 1967a, b). Base analogues such as 6-mercaptopurine, 6-azaguanine, 8-azaadenine and 5-bromouracil reduce the aleurone response, but the effectiveness of these agents depends on whether the experiments are carried out on separated aleurone layers or if the starchy endosperms are also present (Varner et al., 1965; Chrispeels and Varner, 1967b; Yomo and Iinuma, 1964; Briggs, 1963; Paleg, 1964). In response to GA$_3$ the turnover of RNA increases, as shown by the incorporation of [^{32}P]-inorganic phosphate and [^{14}C]-uridine and adenine into the RNA (Chandra and Varner, 1965; Varner et al., 1965). Label from the uridine is also incorporated into cytidine and from adenine into guanine, incidentally demonstrating the presence of enzymes interconverting these bases.

Interpretation of the results is confused by the GA$_3$-induced increase in ribonuclease which accelerates the breakdown of cellular RNA, and hence accelerates the rate of RNA turnover. Like α-amylase, enhanced RNA formation is dependent on the continuous presence of hormone. Very low doses of GA$_3$ stimulate RNA synthesis without provoking maximal protein synthesis. Separation of the RNA on M.A.K. columns has demonstrated the appearance of a new, rapidly labelled fraction designated RNA 28S II that comprises only a small proportion of the total (Chandra and Duynstee, 1968). This is not formed in the absence of GA_3 and its formation is inhibited by actinomyocin-D. Hence it may be connected with α-amylase synthesis, or the synthesis of other enzymes. GA_3 treatment doubles the incorporation of [^{14}C]-adenine into cyclic-3, 5-AMP (Pollard, 1970a, Pollard and Venere, 1970). It also enhances the methylation of t-RNA and heavy r-RNA by [^{14}C-Me]-methionine. Abscisic acid blocks α-amylase-D formation and the methylation of RNA in the aleurone layer suggesting that there may be a link between the two events (Chandra, 1971). The demonstration of increased incorporation of [^{14}C]-choline into the endoplasmic reticulum and enhanced phospholipid biosynthesis gives a quantitative measure of the increase in complexity seen in the electron microscope and supports the view that GA_3 organizes the cell for enhanced protein synthesis (Evins and Varner, 1971; Koehler and Varner, 1971; Johnson and Kende, 1971). Reports of a subcellular preparation of aleurone tissue that incorporated limited quantities of radioactive amino acids into α-amylase raises the hope that further characterization of the enzyme formation sequence can be achieved (Duffus, 1967; Momotani and Kato, 1971).

α-Amylase accumulates in the aleurone cells before it appears in the medium (Chrispeels and Varner, 1967a; V. J. Clutterbuck and D. E. Briggs, unpublished) so a secretory or release mechanism may be involved.

Abscisic acid (ABA) inhibits the aleurone responses (Section III). ABA and actinomycin D inhibit α-amylase-D formation if added with GA_3. However, if they are added subsequently in low doses, enzyme formation is either not affected or is effected after a lag. Enzyme release is prevented, suggesting that the release mechanism is biosynthesized some hours after the enzyme (Chrispeels and Varner, 1967a, b; Varner and Johri, 1968). Not all hydrolytic enzymes required for endosperm dissolution are synthesized in response to gibberellins. However, it is known that activities ascribed to α-amylase, "endo-β-glucanase", "endo-pentosanase", ribonuclease, protease and phosphatase rise rapidly in germinating grain and in de-embryonated grain treated with gibberellic acid. Further, the protease, at least, is synthesized *de novo* (Jacobsen and Varner, 1967). Thus the changes that occur in GA_3 treated aleurone layers must be involved with many other events besides the production and the release of α-amylase. It might be assumed that the synthesis of these enzymes co-occurs, and is due to the simultaneous de-repression of a group of genes by gibberellin. While the data shows slight discrepancies (no doubt because "glucanase", "protease" are mixtures), yet it seems that some of these enzymes appear in the culture media in sequence, and not simultaneously as might be expected (MacLeod *et al.*, 1964; Jones and Price, 1970; Chrispeels and Varner, 1967a; Jacobsen and Varner, 1967). There are several possible explanations. For example, enzymes may induce the formation of their own release mechanisms at different rates; or gibberellins may de-repress different genes at different rates. Alternatively, it is known that the aleurone layer accumulates GA_1 (Pollard, 1970b; Musgrave et al., 1972), and converts it to more polar metabolities. The accumulation is evidently actively dependent on metabolism since it is prevented by dinitrophenol or NaF, possibly it is the gibberellin metabolites that act as inducers of the last-formed enzymes. It may be relevant that the most polar gibberellin known also has the greatest activity on the aleurone layer (Coombe, 1971).

The aleurone layer has a basal respiration, turnover of proteins and forms cyclic-3,5-AMP. Even in the resting state the aleurone contains hydrolytic enzymes, including carbohydrases. Thus the tissue appears to have a basal metabolism, which gibberellins alter to the production and release of hydrolytic enzymes.

Some of the early effects of gibberellins on separated aleurone layers, the release of a β-$(1 \rightarrow 3)$ glucan, inorganic phosphate and so on, may perhaps be due to enhanced leakage of metabolites and unrelated to the production of new enzymes (Pollard 1969, 1970b; Pollard and Nelson, 1971; Pollard and Singh, 1968). The elaboration of the mitochondria is consistent with enhanced respiration and energy production, while the increase in rough endoplasmic reticulum and the appearance of lamellae and polysomes suggests that protein synthesis increases. The build up of protein-containing vesicles and the apparent release of their contents at the plasmalemma is a possible mechanism for the controlled release of the newly synthesized enzyme. However, several enzymes are released in sequence and the visual observations of the emptying of the vesicles coincides with the release of ribonuclease only (Fig. 8).

While on balance the evidence favours de novo enzyme synthesis in the scutellum as well as in the aleurone layer, yet the metabolic differences betweeen the triploid aleurone layer and the diploid scutellar epithelium, with its dramatic changes in cell shape, should not be disregarded. Further, microscopic studies on the aleurone have been more extensive and refined than those on the scutellar epithelium, and so differences may have been overlooked. Even so, the general similarity between the tissues is striking (Fig. 9).

C. ENDOGENOUS GIBBERELLIN-LIKE SUBSTANCES AND α-AMYLASE FORMATION

From the data presented it seems clear that the embryo produces material that triggers the production of α-amylase by the aleurone layer. However, the data does not allow a decision as to whether or not gibberellins are the *only* factors generated by the embryo that are involved in stimulating enzyme-formation. Nor is it clear if the aleurone response to gibberellins (with or without other embryo-derived factors) is modified in the intact grain by substances from the starchy endosperm. To test if gibberellins are the only embryo derived trigger-substances studies were made of the changes in

FIG. 8. Changes in the cells of the scutellar epidermis of barley, during germination at 11°C.

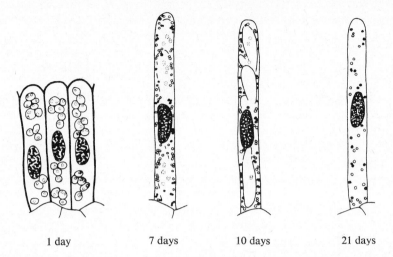

| 1 day | 7 days | 10 days | 21 days |

There is a progressive decline in reserve material throughout the period but starch grains appear in the leucoplasts by day 3 and then disappear; meanwhile the cells double in length. The endoplasmic reticulum increases in quantity until about day 7, then declines and vacuoles appear and enlarge. Mitochondria develop cristae and both leucoplasts and dictyosomes become progressively more complex until they begin to decline, approximately after day 10. The aleurone grains lose their protein matrix which is replaced by vacuoles.

α-amylase and gibberellin content of germinating barley, and the rate and extent of formation of α-amylase by de-embryonated grains in response to gibberellic acid (Groat and Briggs, 1969). Gibberellins were extracted from an acidified solution with ethyl acetate and after separation from inhibitory substances by paper chromatography they were estimated, as GA_3 equivalents, by the lettuce-hypocotyl extention assay (Frankland and Wareing, 1960). Tests showed that all gibberellin-like biological activity was to be found in this acid fraction. The α-amylase content and gibberellin content of decorticated grain was followed during seven days germination. A comparison of the rate of α-amylase formation and gibberellin content showed that the gibberellin content reached a peak about 14–18 h before the maximal rate of α-amylase synthesis. As it was known that α-amylase-D synthesis continues only in the presence of gibberellin, and it is believed that α-amylase is stable in malting barley, the result seemed consistent with enzyme formation being dependant on the hormone. The lag was presumed to be due to (a) the time taken for the gibberellin to diffuse from the embryo into the endosperm, and (b) the time taken for enzyme synthesis to occur. After about 24 hours the gibberellin content fell, and so subsequently did the rate of α-amylase formation; these declines occurred roughly in parallel. Later the gibberellin content remained constant although amylase formation had ceased (Fig. 10). This may have been due to the grain becoming dry and inhibiting both enzyme

FIG. 9. Diagrammatic representation of the subcellular changes in separated aleurone layers incubated at 25°C with GA₃ after 3 days imbibition. sph, spherosomes; ag, aleurone grain; r.e.r., rough e.r.; m, mitochondrion; v, vacuole.

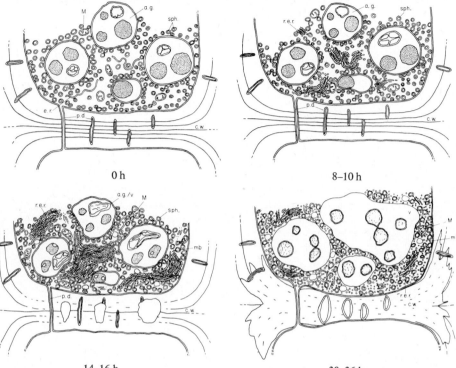

0 h

8–10 h

14–16 h   30–36 h

There is a pronounced increase in rough e.r. at 8 h and many polysomes are visible by 12 h. Vacuolation begins at 16 h and the small vacuoles fuse to form one large one by 36 h. Spherosomes are dispersed in the cell at first but become apressed to the aleurone grain membranes at 4 h. Leucoplasts gradually increase in numbers throughout the period while microbodies increase and then decrease. The cristae of mitochondria also increase until about 10 h, then degenerate. Cell walls show a progressive breakdown after about 8 h.

formation and gibberellin destruction or it may have been due to gibberellin being formed and retained in the embryo.

This was further investigated by removing embryos from malting, decorticated grains at different times and determining the α-amylase contents. (Fig. 11.) Embryo removal at zero time germination, prevented the subsequent appearance of α-amylase, and this was true for some hours more. But suddenly, when embryos were removed about 26 h from steep-out the residual, degermed grain was found to have gained the ability to form near maximal quantities of α-amylase on subsequent incubation for the remainder of the seven-day germination period. This could possibly be due to the arrival of a "dose" of stimulating substances from the embryo or the induced ability of the endosperm to make gibberellin. The first traces of the hormones arrival were found

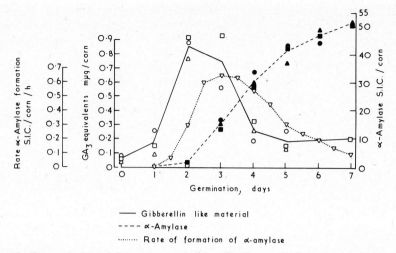

FIG. 10. Changes in levels of α-amylase and gibberellin-like materials in decorticated barley malted at 14·4°C. Results of three separate experiments. —○—□—△—, Gibberellin-like material; —●—■—▲—, α-amylase; - - - -▽- - - -, rate of formation of α-amylase (calculated) (Groat and Briggs, 1969).

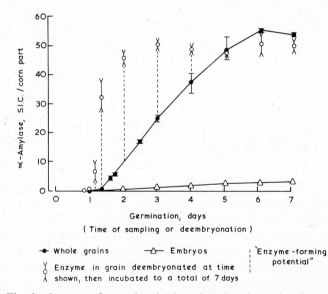

FIG. 11. The development of α-amylase in decorticated grains and embryos (malting at 14·4°C) compared with the α-amylase formed in grains degermed at various times and "malted" for a further period. "Enzyme forming potential", the vertical dashed lines, indicate the difference between the α-amylase present in the endosperm at the time of degerming and that found at the end of the 7 d malting period. (Groat and Briggs, 1969.)

in grains degermed after about 18 h, but the bulk of the rise occurred at about 24 h. When the "enzyme forming potential", as defined in the figure, that occurred at different times is plotted with measured gibberellin content it is seen that the rise in the two curves are approximately coincident (Fig. 12). Thus the result is consistent with gibberellins being at least a part of the stimulating substance(s) and arriving simultaneously with it at the aleurone layer. The apparent arrival of nearly all of the presumed gibberellin in the endosperm part of

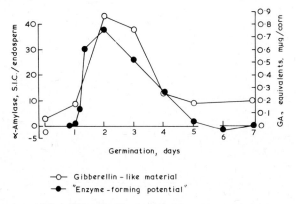

FIG. 12. Comparison of the gibberellin-like material found in whole grain with the enzyme-forming potential of degermed grain (see Fig. 11). (Groat and Briggs, 1969.)

the grain in a period of a few hours suggested that the application of individual doses of gibberellin to the de-embryonated grain might mimic the effect of the "dose" of substances from the embryo. If grains were de-germed, hydrated with buffer and then dosed with measured micro-drops of a solution of GA_3 applied to the depression left by the removal of the germ, then α-amylase formation did not occur for 19–26 h. Such a long lag may have been due to the early removal of the embryo, so preventing the movement of the hypothetical "embryo factor" into the endosperm (Section III). However, when grains were steeped normally and then degermed and individually dosed with gibberellin 24 h after embryo removal, at about the time the natural dose of stimulant arrived from the embryo, then α-amylase formation began in 14–18 h, a lag about equal to that between the rise of gibberellin content and the rise in α-amylase formation found in whole grains. Thus the time sequence could be explained in terms of most of the endogenous gibberellin arriving in the endosperm after 24 h germination.

In another experiment, fully steeped grains were degermed then, 24 h later, were dosed with various quantities of gibberellic acid. Subsequently α-amylase developed, and after a further 72 h it was seen that a sigmoid relationship existed between the quantity of α-amylase found and the log of the dose of GA_3 applied (Fig. 13). In different tests the quantity of α-amylase found in endosperms of entire grain, malted in parallel, equalled that induced by

0·3ng and 0·6ng GA_3 applied to degermed grains. The measured peak amount of gibberellins in endosperm was 0·6ng GA_3 equivalents, a value sufficiently close to suggest that in quantitative terms endogenous gibberellins are the major embryo-derived trigger substances.

While conclusions based on such an approach seem fully justified with regard to timing, the good quantitative agreement achieved could be accidental. It is known that several gibberellins occur in barley (Section III)

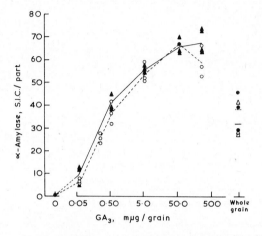

Grains deembryonated after steeping, and individually
dosed with 1 μl drops GA_3 solutions after 24 h
then incubated for a further 72 h

FIG. 13. α-Amylase in endosperms either steeped, deembryonated and dosed with GA_3, or prepared from whole grain at the end of the germination period. Results of two experiments carried out under malting conditions at 14·4°C. (Groat and Briggs, 1969). Experiment (a)—dosed endosperms, —▲—; endosperms from whole grains, —△. Experiment (b) —dosed endosperms, --○--; endosperms from whole grain, --●. (Groat and Briggs, 1969.)

namely GA_1, GA_3, GA_4 and GA_7, and these have different potencies in inducing responses in degermed barley (Table III). Further, their effectiveness in the lettuce hypocotyl assay differs from that on germed grains. For example, GA_3 has about 25 times the activity of GA_1 in the lettuce bioassay that we employed. The nearness of the quantitative results suggests that the hormone reaching the aleurone may be a mixture of gibberellins in constant proportions, or is predominantly GA_3. The extracted gibberellin occurring at the "peak" chromatographed on paper more like GA_3 than GA_1.

In grain germinated in water at 25°C the situation is different and more complicated. Under these conditions α-amylase synthesis and destruction occur (Briggs, 1968a, b, c; Groat and Briggs, 1969). The levels of α-amylase and gibberellins rise and fall *together* in this grain so that peak rate of enzyme increase *precedes* the peak of gibberellin content. Gibberellins may occur in different proportions in the grain grown in this fashion from those in which

TABLE III

The order of activity of several gibberellins, in a number of tests

Test assay	Order of activity	Reference
Lettuce hypocotyl extension	$A_7 > A_3 > A_4 > A_1$ (dose/response curve of A_4 and A_7 steeper than others) A_8 inactive, or slight activity	Crozier et al. (1970).
Lettuce hypocotyl extension	$A_4 > A_3 > A_1$. A_8 inactive (A_4 more efficient at lower doses, less good at high doses)	Paleg et al. (1964)
Rice seedling test	$A_3 > A_1 > A_7 > A_4 > \ldots \gg A_8$	Ogawa (1967)
Stimulation of barley germination	$A_1 > A_7 > A_4 > A_3 > A_8$	Griffiths et al. (1964)
Barley endosperm, acid phosphatase	$A_1 = A_3 > A_4 = A_7 \ldots \gg A_8$	Jones, K. C. (1969)
Barley endosperm, sugar release	$A_1 \approx A_3 > A_4 \approx A_7 \ldots \gg A_8$ (dose/response curves—slopes A_1 and A_7 steeper than /A_3 and A_4)	Paleg et al. (1964) Radley (1968)
Rice endosperm, diastase (?)	$A_1 \approx A_3 \approx A_4 \nsim A_7 \ldots \gg A_8$	Ogawa (1967)
Barley endosperm, α-amylase–diastase	$A_3 > A_1 > A_7 \approx A_4$	Jones and Varner (1967)
Barley endosperm, α-amylase–diastase	$A_1 > A_3 > A_7 > A_4 \ldots \gg A_8$	Crozier et al. (1970)
Barley endosperm, heated, autolysis	$A_1 = A_3 = A_4 = A_7 \ldots \gg A_8$	Griffiths et al. (1964)

they occur in malted grain. The pattern of growth is different in the two cases and the well-developed seedlings would be expected to contain gibberellins concerned with growth. The plants were grown in the dark, but were exposed to light during sampling, and light is known to cause a rapid rise in gibberellins in barley and wheat leaves (Reid and Clements, 1968; Loveys and Wareing, 1971). Also, roots were better developed than in malted grain. Various workers have failed to find bound forms of gibberellins in ungerminated barley embryos, yet in embryos cultivated *in vitro* gibberellins were released continuously for 60 h and the total amount released exceeded that initially present, evidently synthesis was occurring (Cohen and Paleg, 1967; Radley, 1967; Yomo and Iinuma, 1966). The gibberellin appeared to resemble GA_3 according to Cohen and Paleg (1967) and Groat and Briggs (1969), GA_1 according to Radley (1968).

1. *Gibberellin Formation and Distribution in the Embryo*

Some embryo growth seems to be obligatory before gibberellins can be detected. The attempts to find which parts of the embryo are involved have given conflicting results. MacLeod and Palmer (1966) removed pieces of the barley embryo and followed enzyme production and modification in the endosperm, assuming that this was dependant on a supply of gibberellins from the embryo. They concluded that gibberellins probably arose at the embryonic node, particularly from that region where the subsidiary rootlets arise.

On the other hand, experiments made with dissected embryos cultivated at 25°C on agar gave gibberellin levels in the embryo and agar which led to different conclusions (Radley, 1967). For the first two days gibberellin production occurred chiefly in the scutellum, but by the third day production in the scutellum had ceased, but was occurring in the embryonic axis. In contrast to other results (Cohen and Paleg, 1967) GA_1 seemed to predominate over GA_3. When α-amylase production was measured in grain in which the embryo had been damaged it was found that inhibition of coleoptile growth was not invariably linked to reduced enzyme production and, by inference, gibberellin production (Briggs, 1968b). The scutella of germinating barley embryos produced gibberellins in intact grains if the axis was present but not if the axis was removed (Radley, 1968, 1969). However, separated scutella, grown on agar, produced gibberellin whether or not embryonic axes were still attached, unless sugar was present, in which case they behaved as if attached to the endosperm, which of course acts as a source of sugars for the growing embryo. Thus it appeared that sugars repressed the synthesis of gibberellins in the scutellum, while the embryonic axis acted as a "sink", utilizing sugars and so reducing their concentration in the scutellum and allowing hormone synthesis to proceed (Radley, 1968, 1969). Such a result has important consequences regarding the regulation of gibberellin production in the initial stages of grain germination. Possibly this does not begin until the level of endogenous sugars, such as raffinose, has been reduced by embryonic metabolism. The later production by the axis could well account for the gibberellin that appeared unrelated to α-amylase synthesis. Clearly, experiments involving separated embryos must be interpreted with caution and indeed it has been shown repeatedly that they behave differently from embryos attached to the grain. Similarly, attempts to discover the pattern in which gibberellins spread through the grain by investigating the effects of damage (MacLeod and Millar, 1962; Briggs, 1968b) cannot be interpreted with any confidence. Slight damage to the aleurone layer causes a disproportionate reduction in the response to factors from the embryo or to saturating doses of gibberellin (Briggs, 1968b; Sparrow, 1965) suggesting that an aleurone layer behaves as an integrated tissue and not as a collection of independent cells.

Modification (cell wall degradation), the formation of α-amylase and other changes spread from the base of the starchy endosperm towards the apex, beginning under the scutellum. MacLeod and Palmer (1966) suggest that gibberellins are released from the apex of the scutellum and move towards the apex of the grain. However, other authors disagree.

A description given by Mann and Harlan (1916) indicates that modification slowly proceeds directly into the starchy endosperm from the scutellum, more rapidly adjacent to the aleurone layer. Radley's (1967, 1968, 1969) concept of gibberellin production occurring in the scutellum, not a highly differentiated tissue, suggests that release should occur generally. If this were true then modification should move down the grain round the entire circumference, except at the furrow where there is no aleurone tissue, and parallel to the scutellum. From none of the published work was it possible to reconstruct the changes; we have, therefore, attempted to map modification in malting, decorticated barley. Dorsal or ventral, proximal or distal pieces of aleurone are equally effective at breaking down the adjacent starchy endosperm and releasing sugars when incubated in liquid containing gibberellic acid (Briggs, 1964b). Furthermore, if de-embryonated, decorticated grain is incubated with gibberellic acid, the periphery of the starchy endosperm is rapidly and completely dissolved, presumably by enzymes from the aleurone layer, leaving a "core". Thus sequential modification must be due to differential distribution of the gibberellins. The α-amylase contents of the parts of samples of malting grains were determined; in each case α-amylase appeared first at the proximal end of the starchy endosperm, later distally and, as expected, approximately equally on the dorsal and central sides (D. E. Briggs, unpublished). The starch contents of discs punched from the dorsal and ventral sides of sections (50 μ) cut parallel to the scutellum were assayed enzymatically. Despite the scatter of results it appeared that the starch content declined faster on the dorsal than the ventral side in the vertical sections, but in the sections parallel to the scutellum the starch appeared to be removed at equal rates on the dorsal and ventral sides. Microscopical investigations of modification confirmed this although the process usually progressed a little faster close to the aleurone layer (Fig. 14). In decorticated grain endosperm breakdown was more complete, much of the endosperm liquefied, and the pattern of liquefaction could be followed by freezing the grain and dissecting it while frozen, the liquid

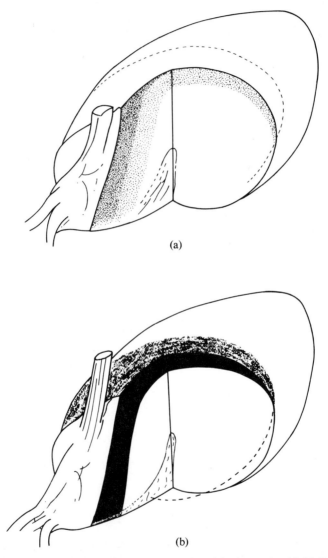

(a)

(b)

FIG. 14. The pattern of modification in decorticated barley grain. (a) Malted at 14·4°C. Darker zones indicate greater degree of endosperm breakdown. The dashed line indicates the limit of modification. (b) Grown in contact with water, at 25°C. The black zone indicates the extent of the fluid-filled space in the grain. The shaded area/dashed line, indicates the extent of the fluid-filled space, which is visible from the outside of the grain. (Briggs, unpublished.)

areas show clear and solid white under a lens (see also Briggs, 1968a). Dissolution begins underneath and parallel to the scutellum and spreads towards the apex of the grain. These results are therefore consistent with the theory that gibberellins are released in a symmetrical way from the entire periphery of the scutellum, as would be expected from the results of Radley (1967, 1968, 1969). It appears that the enzymic contribution of the scutellum to the modification is out-weighed by the aleurone-produced enzymes whose production is dependent on the diffusion of endogenous gibberellins.

2. The Chemistry of Barley Gibberellins

The gibberellins of barley are incompletely documented. Gibberellin-like material was detected in germinating barley some years ago (Radley, 1959; Lazer et al., 1961; Yomo, 1960). In extracts of the immature grain, still in the ear, GA_3 has been fully characterized. It occurred at a concentration of about 3 $\mu g/kg$ of ears (Jones et al., 1963). In addition another gibberellin-like material has been detected (Radley, 1966). In ungerminated, mature grain GA_3 is thought to predominate. In germinating grain more GA_1 is produced than GA_3 (Radley, 1967, 1968, 1969) while others claim that GA_3-like material predominates (Cohen and Paleg, 1967; Groat and Briggs, 1969).

These later identifications have relied heavily on chromatographic separations and biological activities, and so should be regarded as tentative. All attempts to find bound gibberellins in the grain have failed.

It has been known for some time that various inhibitors of gibberellin formation such as CCC (chlorocholine chloride) and amo-1618 inhibited α-amylase formation in germinating barley, but did not inhibit the response of the aleurone to added GA_3 (Paleg et al., 1965; Kahn and Faust, 1967). It was assumed that the inhibitors worked in germinating barley in the same way and the report that the effects of CCC and amo-1618 on isolated barley or wild oat embryos could be reversed by kaurene supported this finding (Simpson, 1966; Radley, 1968) but in our hands [^{14}C]-kaurene* will not penetrate the embryo (Murphy and Briggs, unpublished), and not all the effects of CCC on germinating barley can be reversed by adding gibberellic acid (Briggs, 1968c).

[^{14}C]-MVA fed to barley penetrated the grain and gave rise to labelled terpenoid compounds (G. Murphy and D. E. Briggs, unpublished). However, no label could be detected in GA_1 (38) or GA_3 (26). Enzyme preparations from embryo homogenates incubated with [^{14}C]mevalonate, or [^{14}C]isopentenyl pyrophosphate in the presence of farnesylpyrophosphate, incorporated label into a number of terpenols but no label could be detected in kaurene (32) or gibberellins (Fig. 15). Thus it seemed that the enzymes of kaurene synthesis were lacking or were inaccessible to the labelled precursors. However, kaurene (52), kaurenol (33) and kaurenoic acid (35) have been identified

* See footnote in West's chapter, p. 147.

in barley using thin-layer chromatography (Petridis *et al.*, 1966). Full experimental details were not given and the methods could not achieve unambiguous identifications of these compounds. When a barley hydrocarbon fraction

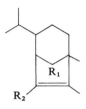

		R₁	R₂
(27)	Helminthosporal	—CHO	—CHO
(28)	Helminthosporol	—CH₂OH	—CHO
(29)	Helminthosporic acid	—CH₂OH	—CO₂H

was subfractioned by TLC on normal and silver-nitrate impregnated plates. GLC on several stationary phases gave a peak that coincided with authentic kaurene (32) on several different stationary phases. Trace amounts of [^{14}C]-kaurene were added to the hydrocarbon fraction, and radioactivity was eluted with the putative kaurene peak during column chromatography, TLC or GLC and conversion into kauran-diol (41) the nor-ketone (42) and the hydrocarbon (43). Thus kaurene has been unambiguously identified in barley. Clearly kaurene, and other related intermediates, might be the expected stored gibberellin precursor(s) in the scutellum. To identify gibberellins in germinating barley, the GA_1/GA_3 and GA_4/GA_7 TLC fractions of the acidic materials were separated. [^{14}C]Methyl esters of the acids were prepared, and co-chromatographed with authentic, unlabelled methyl-GA_3 (44) and methyl-GA_7 (45). During recrystallizations constant specific activities were attained; further, after acetylation and the formation of other derivatives (46,47) (Fig. 16). Thus the presence of GA_3 (26) and GA_7 (40) in germinating barley is proven. It should be noted that this type of experimental approach does not exclude the presence of other gibberellins. A dilution radioassay indicated that the GA_3 content was about 1·5ng/corn which is reasonably close to the value of total gibberellins of about 1 ng GA_3 equivalent/whole grain found by bioassay in similar experiments (Groat and Briggs, 1969).

The results indicated that kaurene (32) in the embryo was probably a stored gibberellin precursor and so attempts were made to test this supposition. About 1·5 ng kaurene was found for each grain of barley (Murphy and Briggs, unpublished).

During germination the kaurene content of the grain fell. We could detect no penetration of [^{14}C]kaurene into whole grains. However, if the hydrocarbon was added to embryo homogenates the exomethylene double bond was

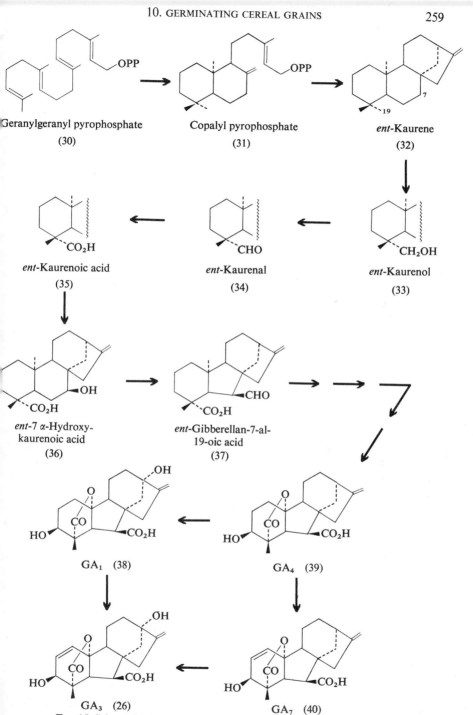

Geranylgeranyl pyrophosphate (30)

Copalyl pyrophosphate (31)

ent-Kaurene (32)

ent-Kaurenoic acid (35)

ent-Kaurenal (34)

ent-Kaurenol (33)

ent-7 α-Hydroxy-kaurenoic acid (36)

ent-Gibberellan-7-al-19-oic acid (37)

GA₁ (38)

GA₄ (39)

GA₃ (26)

GA₇ (40)

FIG. 15. Schematic summary of the biosynthetic pathway to gibberellins.

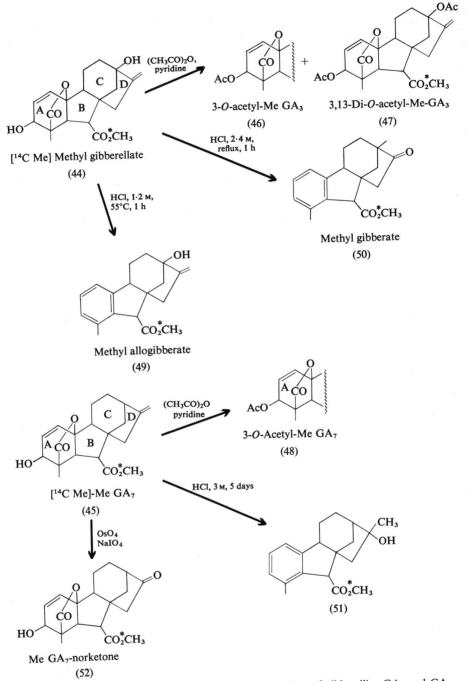

FIG. 16. Chemical manipulations used in the identification of gibberellins GA₃ and GA₇ in barley extracts. (Murphy and Briggs, unpublished.)

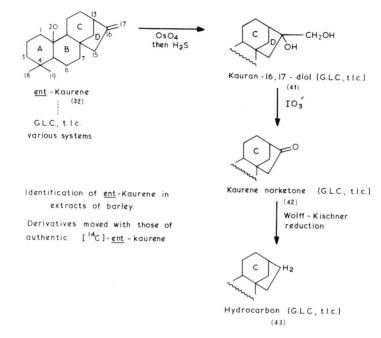

ent - Kaurene
(32)

G.L.C., t.l.c.
various systems

Kauran -16,17 - diol (G.L.C.,t.l.c.)
(41)

Identification of ent-Kaurene in
extracts of barley.

Derivatives moved with those of
authentic   [^{14}C]-ent - kaurene

Kaurene norketone   (G.L.C., t.l.c.)
(42)

Wolff - Kischner
reduction

Hydrocarbon (G.L.C., t.l.c.)
(43)

attacked, and the kaurene was converted to several substances that are *not* on the biosynthetic pathway, and that have been characterized as kauran-17-ol (53), kauran-16,17-epoxide (54) and kauran-16,17-diol (41) (Fig. 17). The formation of these substances may account for some of the difficulties initially experienced in attempting to measure kaurene in the grain.

[^{14}C]Kaurenol (33) and [^{14}C]kaurenoic acid (35) were applied to malting barley, and radioactivity was subsequently detected in the GA_1/GA_3 and GA_4/GA_7 thin-layer chromatographic fractions. Recrystallization of the radioactive GA_1/GS_3 material with GA_3 lost the radioactivity, which, however, following acid treatment moved with the acid-degradation products of GA_1 (55). Similarly the GA_4/GA_7 radioactivity co-chromatographed with the degradation product of GA_4 (56) following acid treatment. Thus we have the situation that the endogenous precursor(s) probably kaurene, gives rise to GA_3 (26) and GA_7 (40), which are both unsaturated in the A ring, while kaurenol and kaurenoic acid were incorporated into GA_1 (38) and GA_4 (39) which both lack double bonds in ring A suggesting that there may be different sites of synthesis for these pairs, and that different enzymic sites are accessible to the natural and exogenous precursors. It is tempting to suppose that the GA_3 and GA_7 are formed from kaurene in the scutellum and are destined for the aleurone layer while GA_1 and GA_4 are synthesized in the embryonic axis and are concerned with regulating the growth of the embryonic plant. It is hoped that other experiments may help to resolve these possibilities.

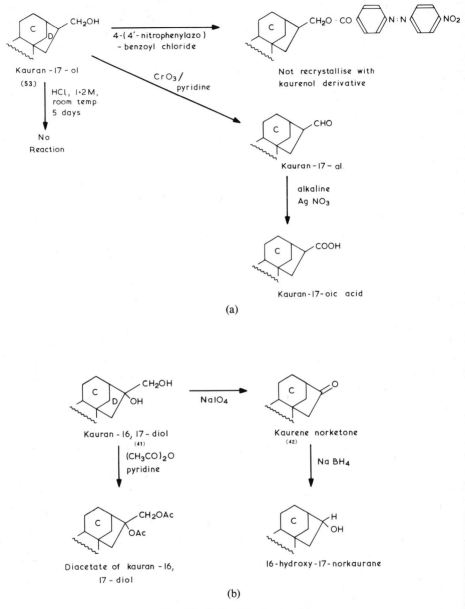

Kauran - 17 - ol
(53)

4-(4'-nitrophenylazo) - benzoyl chloride

Not recrystallise with kaurenol derivative

HCl, 1·2 M, room temp. 5 days

No Reaction

CrO₃/ pyridine

Kauran-17-al

alkaline Ag NO₃

Kauran-17-oic acid

(a)

Kauran - 16, 17 - diol
(41)

NaIO₄

Kaurene norketone
(42)

(CH₃CO)₂O pyridine

Na BH₄

Diacetate of kauran - 16, 17 - diol

16-hydroxy-17-norkaurane

(b)

FIG. 17a, b. See legend opposite.

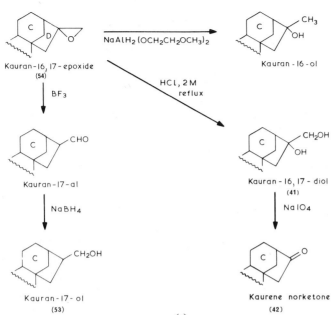

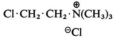

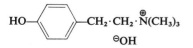

(c)

FIG. 17. a, b, c. The chemical steps used to identify *ent*-kauran-17-oL, *ent*-kauran-16,17-diol and *ent*-kauran-16,17-epoxide. (Murphy and Briggs, unpublished.)

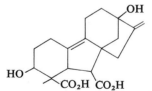

(55) Acid degradation product of GA₁.

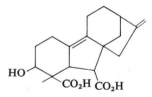

(56) Acid degradation product of GA₄.

$$Cl \cdot CH_2 \cdot CH_2 \cdot \overset{\oplus}{N}(CH_3)_3$$
$$\overset{\ominus}{Cl}$$

(57) CCC, chlorocholine chloride.

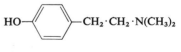

(59) Hordenine.

$$HO- \hspace{-0.3em} \text{[benzene]} \hspace{-0.3em} -CH_2 \cdot CH_2 \cdot \overset{\oplus}{N}(CH_3)_3$$
$$\overset{\ominus}{OH}$$

(58) Candicine.

$$Cl \cdot CH_2 \cdot CH_2 \cdot N(CH_3)_2$$

(60) 2-Chloroethyl dimethylamine

D. THE REGULATION OF α-AMYLASE LEVELS IN
GERMINATING BARLEY

It seems that germination is initiated before the gibberellin content of the grain rises yet gibberellins added to the grain are highly efficient in breaking dormancy. Many toxic factors that prevent embryo growth also prevent the formation of α-amylase, as does gamma- or X-irradiation, but if damage is confined to the embryo exogenous GA_3 will penetrate the grain and enzyme is formed (Fujii and Lewis, 1965; Lewis and Fujii, 1966; Paleg et al., 1962). The separated embryo produces limited amounts of α-amylase, and the quantity of α-amylase-SKB found is greatly increased by a supply of amino-acids (Briggs, 1964b). The addition of GA_3 to whole grains increases the amount of α-amylase found in the scutellum and in the rest of the grain, indicating that the quantity of endogenous gibberellins falls far short of that needed to saturate the enzyme-forming mechanisms, a fact that is basic to the use of GA_3 in malting (Briggs, 1968a, b, c).

Sugars reduce the formation of α-amylase in isolated embryos, whether or not GA_3 is also present (Briggs, 1964b). Thus sugars repress enzyme synthesis and as they greatly enhance embryo growth one may guess that, as the embryo utilizes the available amino-acids for growth, so it competes with the enzyme forming centre for the supply (Briggs, 1964b). Sugars also repress the formation of gibberellins in the scutellum, the growing embryonic axis may deplete the

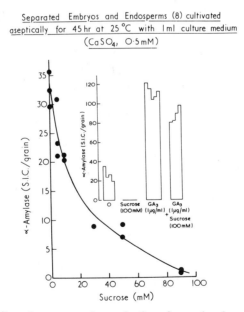

FIG. 18. The effect of sucrose on the production of α-amylase by separated barley embryos and endosperms incubated together, in a minimum volume of liquid. Insert: the effect of adding gibberellic acid to the incubation medium. (Briggs, unpublished).

sugar of the scutellum in the initial stages of germination, allowing the formation of gibberellins in germinating whole grains (Radley, 1969). Certainly if
the embryos and endosperms are separated and are cultivated in liquid media
with various levels of sucrose with and without GA_3 it is found that (a) in
the absence of exogenous GA_3 sucrose (100 mM) totally prevents α-amylase
formation, (b) the addition of GA_3 to sucrose containing cultures raises the
level of α-amylase to 87 S.I.C. units/grain, (c) in the absence of sucrose α-
amylase equal to 26 S.I.C. units/grain was found, while, (d) with GA_3 112
units appeared. Thus the sucrose entirely prevents the formation of α-amylase
and hence endogenous gibberellin, while it reduces, but does not prevent the
response of the enzyme synthetic mechanism, to added GA_3 (D. E. Briggs,
unpublished, Fig. 18). The situation relating to the control of enzyme formation in the embryo is summarized in Fig. 19.

It was previously concluded that the sugar content of grains could not
account for the "switch off" in the α-amylase synthesis (Briggs, 1968a).
However, a re-examination of the data available indicates that the initial fall
in sugars could "switch on" the formation of gibberellins and the subsequent
rise in sugars in the endosperm and in the scutellum could very well "switch
off" their formation (Briggs, 1968a; Groat and Briggs, 1969). Further, the
sugar content is at a minimum at about the 2nd day of germination, which
is about the time when gibberellin content should be maximal (Groat and
Briggs, 1969; MacLeod et al., 1953; Harris and MacWilliam, 1954).

The quantity of α-amylase found in the whole grain is limited by the supply
of gibberellin, as adding GA_3 to whole grains increases the amount of enzyme
which may be doubled. In the whole grain the endogenous hormone must
move by diffusion along the endosperm in a "wave" from the embryo, triggering enzyme production at successive points in the aleurone. It is not known if
the hormone is released from the whole scutellar surface and diffuses throughout the starchy endosperm and aleurone, or if it specifically moves in the aleurone. High doses of GA_3 cause maximal formation of α-amylase and so must
saturate and "switch on" all the available enzyme-forming mechanism in the
tissue. Over a limited range the response is proportional to log concentrations
of gibberellins. The gibberellin level in the grain falls during germination so
destruction or irreversible binding is occurring (Groat and Briggs, 1969).
The response of the tissue to gibberellins is affected by other factors. Thus if
degermed grains are hydrated for 24 h and then treated with gibberellic acid
the response is substantially greater and more rapid than if the preparations
are placed at once in solutions of GA_3 (Petridis et al., 1965; MacLeod et al.,
1964; 1966; Yung and Mann, 1967). Further, the response improves progressively during the prehydration period, presumably as the "basal metabolism"
of the tissue "builds up". Further, hydration in the presence of the embryo,
followed by embryo removal and treatment with GA_3, elicits a greater response
in the aleurone than hydration with water followed by GA_3 suggesting that
an "embryo factor" has moved into the endosperm with water and improved

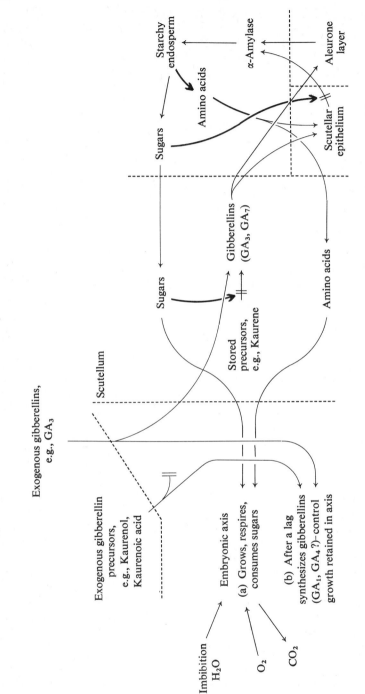

Fig. 19. Representation of some factors regulating gibberellin production and α-amylase formation in the barley embryo.

the response of the aleurone to the subsequent dose of gibberellic acid (Petridis *et al.*, 1965; MacLeod *et al.*, 1966; Briggs, 1968b; Groat and Briggs, 1969).

Aleurone layers are reported to release enzyme most readily while still attached to the starchy endosperm (MacLeod *et al.*, 1964; Chrispeels and Varner, 1967a). However, Chrispeels and Varner used an unspecific assay and so this data cannot be accepted. The way in which aleurone layers are prepared seems to affect not only the quantity of α-amylase produced but also the proportion retained and released in different trials (V. J. Clutterbuck and D. E. Briggs, unpublished).

Growth is enhanced by gibberellic acid added to isolated embryos cultivated in a nutrient medium, and not just when attached to a grain, so that gibberellins exert a direct effect on embryo growth and do not only enhance the supply of nutrients by mobilizing the reserves of the endosperm. Adding small quantities of calcium to the growing grains did not stabilize endogenous α-amylase, probably because it could not penetrate into the endosperm. In fixed time experiments embryo growth was depressed by increasing levels of calcium chloride, and in the endosperm the α-amylase content increased (Briggs, 1968c). Experiments with potassium sulphate showed that in fixed-time experiments some concentrations (40 mM) markedly increased the α-amylase content of the grain whether or not exogenous gibberellin was applied. A time-course study showed that the potassium sulphate did not affect the rate of synthesis of the enzyme, but it greatly retarded its destruction. This was true in the presence or absence of added GA_3 (Fig. 20; Briggs, 1968c). Numerous other substances were tested and yet others are reported in the literature (Dumitru *et al.*, 1964; Srivastava and Meredith, 1964). Their effects may be classified as follows:

(1) Inhibit embryo growth, but do not inhibit gibberellin production or aleurone response to added GA_3 (eg. K_2SO_4).
(2) Inhibit embryo growth and gibberellin production, but do not inhibit response of aleurone to exogenous to GA_3 [eg. CCC (57), tryptamine, spermine, $CuSO_4$, amo-1618 and phosphon-D (Khan and Faust, 1967)].
(3) Inhibit embryo growth, gibberellin production and aleurone response to added GA_3 (eg. 2,4-dinitrophenol; *N*-ethyl-maleimide), also see Fig. 21.

1. *Other Plant Hormones and α-Amylase*

It seems that only gibberellins are capable of stimulating α-amylase formation in barley, attempts to modify malting grain by adding other hormones and other substances either had no effect, or the growth processes were inhibited (Bawden *et al.*, 1959; Dickson *et al.*, 1949; Prentice *et al.*, 1963).

However, when kinetin was present in the water on which grains were growing it reduced root growth, enhanced coleoptile senescence and accelerated the degradation of α-amylase (Verbeek *et al.*, 1969). This seems to be the one compound reported to reduce embryo growth and yet accelerate α-amylase degradation, perhaps by enhancing the capacity of the embryo to act as a "sink". Abscisic acid (ABA) added to whole grain reduces germination and the

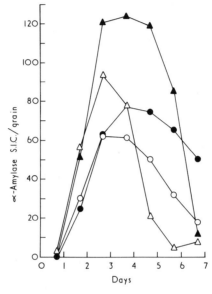

Fig. 20. Changes in α-amylase with time in decorticated grain growing on water at 25°C, with and without GA₃, and potassium sulphate. —○—, water only; —●—, K₂SO₄ (40 mM); —△—, GA₃ (50 μg/ml); —▲—, GA₃ (50 μg/ml) and K₂SO₄ (40 mM).

production of α-amylase (Khan and Downing, 1968); adding gibberellic acid with ABA increases the level of α-amylase but does not restore full growth. Kinetin partly restores growth, and with it the formation of α-amylase. These effects are evidently due to ABA and kinetin acting on the embryo, since ABA reduces the GA₃-induced production of α-amylase in the aleurone layer and kinetin does not alleviate this inhibition (Chrispeels and Varner, 1967b; Khan, 1969). ABA has little effect on respiration or phosphorylation in aleurone layers and it hardly reduces the incorporation of [¹⁴C]amino acids. It does reduce the incorporation of uridine into RNA and reduces the methylation of RNA bases (Chandra, 1971). As dormant barley matures the concentration of the "inhibitor-β" (which is presumed to be largely ABA) falls (Rejowski and Kulka, 1967; Rejowski, 1969). However, kinetin usually does *not* relieve dormancy in barley although it does in some seeds (Pollock, 1958; Khan, 1971). In non-dormant grain visible growth precedes the detection of a rise in the levels of endogenous gibberellins, so that if gibberellins give the

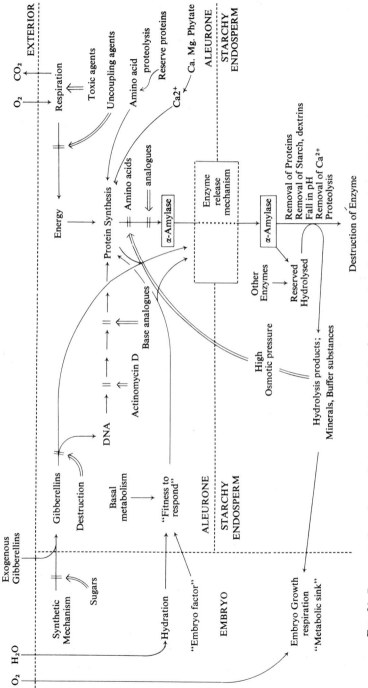

FIG. 21. Representation of some factors regulating the levels of α-amylase found in the whole grain (see also Fig. 19).

"stimulus for germination" then the stimulus must be given by preformed gibberellins.

E. THE REGULATION OF β-AMYLASE ACTIVITY IN GERMINATING BARLEY

Isolated β-amylase is unable to attack entire starch grains, but it will cause the partial degradation of damaged grains or gelatinized starch. The gross chemical changes undergone by starch during malting could be due to selective and limited attack by β-amylase after a limited degradation by α-amylase. However, the way in which granular starch is degraded in the whole grain is far from clear (Harris, 1962) and the lack of specific methods of assay makes it impossible to define the changes during germination. β-Amylase, in contrast to α-amylase, is laid down during the formation of the grain, and the total quantity is probably not significantly augmented during germination (Harris, 1962; Verbeek-Wyndaele and Coulier, 1961). However, the forms and activity of the soluble enzyme do change during malting, and these changes appear to be influenced by hormones. If finely ground grain is extracted with water and then with solutions of salts, successively more enzyme is extracted (Harris, 1962). If the exhaustively-extracted residue is now extracted with a salt solution containing papain, hydrogen sulphide or other thiol-containing compounds, then more β-amylase comes into solution (Sandegren and Klang, 1950). This "bound" or "latent" β-amylase (Pollock and Pool, 1958), which may constitute up to two thirds of the total, is formed during grain maturation at the expense of the free, soluble enzyme. Heterogeneity of the barley enzyme might be expected from increased extraction achieved with stronger salt solutions (Daussant et al., 1965; Grabar and Daussant, 1964; La Berge et al., 1967; La Berge and Meredith, 1971; Nummi et al., 1965a, b, 1970). Although the soluble barley enzyme is heterogenous, the different active proteins react with one β-amylase antibody and in the presence of thiols all the soluble forms and the solubilized, latent form, appear as one soluble, active protein-species, which coincides with the smallest of the "free forms". It appears, therefore, that there is one active protein species.

During germination the bound form of the enzyme is released and is changed into soluble forms, and the more complex forms of the enzyme appear to be split. Thus in finished malt a soluble species of β-amylase predominates, reacting with the barley β-amylase antiserum. However, its physical properties are slightly different from the simplest form found in barley, possibly it is formed by limited proteolysis of the barley enzyme (Daussant et al., 1965; Grabar and Daussant, 1964; Nummi et al., 1965b). The situation is complicated in that in some varieties two β-amylase have been found in extracts made in the presence of thiols.

Histochemical studies of grains, extracted with and without papain treatment, have demonstrated that β-amylase occurs in the endosperm and is concentrated in the sub-aleurone layer (Fig. 22). It may also occur in the scutellum (Engel, 1947; Linderstrøm-Lang and Engel, 1937). It seems, however, that the techniques used might also show α-glucosidase. Recently, "protein bodies" have been isolated from whole barley and β-amylase isozymes have been shown to be associated with them (Ory and Henningsen, 1969; Tronier and Ory, 1970). The questions arise, how is the bound form of β-amylase released during germination and is it controlled by hormones? Developing wheat embryos caused a limited release of bound β-amylase from

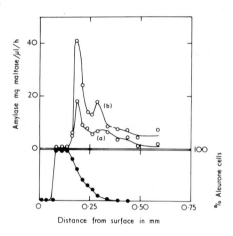

FIG. 22. Disposition of saccharifying diastase (predominantly β-amylase?), in serial sections of a barley corn extracted (a) phosphate buffer; (b) phosphate buffer and papain (after Engel, 1947).

a moist paste of glutenin and starch, consequently they must produce activating substances. Further, in response to embryo secretion or to GA_3 wheat aleurone layers release protease and more of the associated bound β-amylase of the starchy endosperm is released in a water-soluble form than if GA_3 is absent (Rowsell and Goad, 1964). Thus it seems that endogenous gibberellins are indirectly responsible for the solubilization of bound β-amylase (Fig. 23).

F. HORMONES, AND THE REGULATION OF OTHER
CARBOHYDRASE ACTIVITIES

The detailed characterization of other carbohydrases has not been achieved, nevertheless the information available indicates that some of these enzymes behave differently from α- and β-amylase.

Embryos cultures *in vitro* release enzymes that can hydrolyse glycosides, disaccharides, polysaccharides, nucleic acids, proteins and peptides (Briggs, 1962). A similar range of enzymes is known to be released by de-embryonated grain or separated aleurone layers in response to gibberellic acid (e.g. Briggs, 1963, 1964). Further, the endosperm of de-embryonated grain will modify when treated with GA_3 (Paleg and Sparrow, 1962; Yomo and Iinuma, 1962; Sparrow, 1964). The ascription of increases in enzyme activities to GA_3-dependent *de novo* synthesis is reinforced by the case of laminarinase activity (β-(1 → 3)-glucanase) which has long been known to increase in amount in germinating barley and the level of which is enhanced by gibberellic acid (Manners *et al.*, 1971). Yet it has been found that the same total amount of laminarinase is formed in the aleurone layer, in the presence of oxygen,

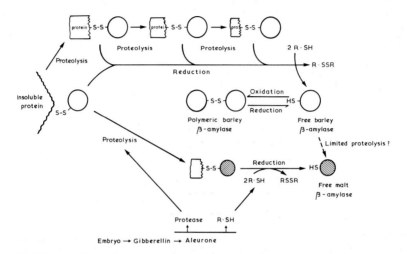

FIG. 23. Scheme illustrating the possible inter-relationships of the forms of β-amylase found in barley and malt.

whether or not gibberellic acid is present. Gibberellic acid greatly enhances the release of the enzyme from the tissue so the enzyme may be formed as a result of protein synthesis, but evidently it is the formation or activation of the release mechanism, and not the generation of the enzyme, that is gibberellin dependent.

Of the "disaccharidase" activities, maltase, cellobiase and laminaribiase are reported to change relatively little in malting (Harris, 1962; Manners *et al.*, 1971). "Maltase" is not readily extracted by water, and early workers concluded that the enzyme can only be reliably determined using finely ground suspensions of plant or grain materials. Subsequent studies have indicated that an α-glucosidase may be extracted by strong salt solutions, e.g. 5% $CaCl_2$ or at pH 8·5. Optimum activity occurs at pH 4·6, when the enzyme will largely separate from solution (Jørgensen and Jørgensen, 1963). Studies with one of these proven extraction techniques indicate that, in contrast to other reports, enzyme activity increases about 40 times during malting. Histochemical studies, which we have confirmed, indicate that α-glucosidase activity is predominantly located in the aleurone layer and the embryo (Jørgensen, 1965). We find that at least part of the rise is due to the formation of enzyme in the aleurone, where it is at least partly gibberellin-dependent (unpublished). We find that when isolated aleurone layers are incubated with maltose or soluble starch, in the presence of 5% $CaCl_2$, there is a massive production of glucose. It is feasible that the "diastase" released from aleurone tissue by unphysiological concentrations of cyclic-$3',5'$-AMP,ADP, phenobarbitone, etc. could be caused by a release of α-glucosidase from the aleurone layer.

An interesting early response to GA_3 is the release by aleurone layers of glycosidases and other enzymes at different times, indicating yet another sequence of responses to one dose of gibberellin (Pollard, 1969; Pollard and Singh, 1968; Pollard and Nelson, 1971).

A consideration of the α-glucosidase situation shows that enzymes can only be adequately measured after extraction from grain, or at least being made accessible to their substrates.

To summarize, one may distinguish:

(1) Constitutive enzymes of basal metabolism (little response to gibberellic acid).
(2) Substrate-induced enzymes, not gibberellin dependent (nitrate reductase, Ferrari and Varner, 1969).
(3) Enzymes that are synthesized *de novo* during germination or in response to gibberellin (α-amylase).
(4) Preformed enzymes that are released or activated in response to gibberellic acid (β-amylase).
(5) Enzymes that appear in the hydrated aleurone, the release of which is gibberellin dependent ("laminarinase").

At present categories 2, 3, 4 and 5 contain only one enzyme.

ACKNOWLEDGEMENTS

The author wishes to thank the many people who have helped in this work, and in particular Dr G. Murphy and Mr V. J. Clutterbuck for the use of their unpublished results, and Mrs P. Hill and Mrs D. Clarke for their work on the diagrams. Dr R. L. Jones kindly gave permission for the use of his electron micrographs. Some other figures were reproduced with permission from the Editor of *Phytochemistry*, published by Pergamon Press. The work has been supported in part by a grant from the Maltsters Association of Great Britain, whom we thank.

REFERENCES

Anderson, F. B. and Manners, D. J. (1959). *Biochem. J.* **71**, 407–411.
Bawden, R. F., Dahlstrom, R. V. and Sfat, M. R. (1959). *Proc. Am. Soc. Brew. Chem.* 137–141.
Bayliss, W. M. (1919). "The Nature of Enzyme Action." Longmans, Green & Co., London.

Bergal, P. and Clemencet, P. (1962). *In* "Barley and Malt" (A. H. Cook, ed.), pp. 1–23. Academic Press, London and New York.

Bourne, D. T. and Pierce, J. S. (1970). *J. Inst. Brew.* **76**, 328–335.

Briggs, D. E. (1962). *J. Inst. Brew.* **68**, 470–475.

Briggs, D. E. (1963). *J. Inst. Brew.* **69**, 13–19.

Briggs, D. E. (1964a). *Enzymologia* **26**, 355–363.

Briggs, D. E. (1964b). *J. Inst. Brew.* **70**, 14–24.

Briggs, D. E. (1966a). *Nature, Lond.* **210**, 418–419.

Briggs, D. E. (1966b). *Nature, Lond.* **210**, 419–421.

Briggs, D. E. (1967). *J. Inst. Brew.* **73**, 361–370.

Briggs, D. E. (1968a). *Phytochemistry* **7**, 513–529.

Briggs, D. E. (1968b). *Phytochemistry* **7**, 531–538.

Briggs, D. E. (1968c). *Phytochemistry* **7**, 539–554.

Brown, H. T. and Escombe, F. (1898). *Proc. R. Soc.* **63**, 3–24.

Carson, G. P. and Horne, F. R. (1962). *In* "Barley and Malt" (A. H. Cook, ed.), pp. 101–159. Academic Press, London and New York.

Chandra, G. R. (1971). *Pl. Physiol., Lancaster* **47**, (suppl.) 39.

Chandra, G. R. and Duynestee, E. E. (1968). *In* "Biochemistry and Physiology of Plant Growth Substances" (F. Wightman and G. Setterfield, eds), pp. 723–745. Runge Press, Ottawa.

Chandra, G. R. and Duynestee, E. E. (1971). *Biochim. biophys. Acta* **232**, 514–523.

Chandra, G. R. and Varner, J. E. (1963). *Pl. Physiol., Lancaster* **38** (suppl.), 1v.

Chandra, G. R. and Varner, J. E. (1965). *Biochim. biophys. Acta* **108**, 583–592.

Chrispeels, M. J. and Varner, J. E. (1967a). *Pl. Physiol., Lancaster* **42**, 398–406.

Chrispeels, M. J. and Varner, J. E. (1967b). *Pl. Physiol., Lancaster* **42**, 1008–1016.

Cohen, D. and Paleg, L. G. (1967). *Pl. Physiol., Lancaster* **42**, 1288–1296.

Collins, E. J. (1918). *Ann. Bot.* **32**, 381–414.

Coombe, B. G. (1971). *Science N.Y.* **172**, 856–857.

Crozier, A., Kuo, C. C., Durley, R. C. and Pharis, R. P. (1970). *Can. J. Bot.* **48**, 867–877.

Daiber, K. H. and Novellie, L. (1968). *J. Sci. Fd. Agric.* **19**, 87–90.

Daussant, J., Grabar, P. and Nummi, M. (1965). *Proc. Europ. Brew. Conv.* 62–69.

Dickson, A. D., Shanda, H. L. and Burkhart, B. A. (1949). *Cereal Chem.* **26**, 13–23.

Dickson, J. G. and Shands, H. L. (1941). *Proc. Am. Soc. Brew. Chem.* 1–10.

Djurtoft, R. and Rasmussen, K. L. (1955). *Proc. Europ. Brew. Conv.* 17–25.

Duffus, J. H. (1967). *Biochem. J.* **103**, 215–217.

Dumitru, I. F., Verbeek, R. and Massart, L. (1964). *Naturwissenschaften.* **51**, 490–491.

Engel, C. (1947). *Biochim. biophys. Acta* **1**, 42–49.

Essery, R. E., Kirsop, B. H. and Pollock, J. R. A. (1956). *J. Inst. Brew.* **62**, 150–152.

Evins, W. H. and Varner, J. E. (1971). *Proc. natn. Acad. Sci. U.S.A.* **68**, 1631–1633.

Ferrari, T. E. and Varner, J. E. (1969). *Pl. Physiol., Lancaster* **44**, 85–88.

Filner, P. and Varner, J. E. (1967). *Proc. natn. Acad. Sci. U.S.A.* **58**, 1520–1526.

Fischer, E. H. and Haselbach, C. (1951). *Helv. chim. Acta* **34**, 325–334.

Ford, J. S. and Guthrie, J. M. (1908). *J. Inst. Brew.* **14**, 61–87.

Frankland, B. and Wareing, P. F. (1960). *Nature, Lond.* **185**, 255–256.

French, D. and Abdullah, M. (1966). *Cereal Chem.* **43**, 555–562.

Frydenberg, O. and Nielsen, G. (1965). *Hereditas* **54**, 123–139.

Fujii, T. and Lewis, M. J. (1965). *Proc. Am. Soc. Brew. Chem.* 11–18.

Gaber, S. D. and Roberts, E. H. (1969). *J. Inst. Brew.* **75**, 303–314.

Gelman, A. L. (1969). *J. Sci. Fd. Agric.* **20**, 209–212.

Godineau, M. J.-C. (1962). *Révue gén. Bot.* **69**, 577–622.

Gordon, A. G. (1970). *J. Inst. Brew.* **76**, 140–143.

Grabar, P. and Daussant, J. (1964). *Cereal Chem.* **41**, 523–532.

Greenwood, C. T. and MacGregor, A. W. (1965). *J. Inst. Brew.* **71**, 405–417.

Griffiths, C. M., MacWilliam, I. C. and Reynolds, T. (1964). *Nature, Lond.* **202**, 1026–1027.

Groat, J. I. and Briggs, D. E. (1969). *Phytochemistry* **8**, 1615–1627.

Haberlandt, G. (1890). *Ber. dtsch. Bot. Ges.* **8**, 40–48.

Harris, G. (1962). *In* "Barley and Malt" (A. H. Cook, ed.), p. 431. Academic Press, London and New York.

Harris, G. and MacWilliam, I. C. (1954). *J. Inst. Brew.* **60**, 149–157.

Hayashi, T. (1940). *J. agric. Chem. Soc. Japan.* **16**, 531–538.

Hough, J. S., Briggs, D. E. and Stevens, R. (1971). "Malting and Brewing Science", p. 678. Chapman and Hall Ltd., London.

Jacobsen, J. V. and Varner, J. E. (1967). *Pl. Physiol., Lancaster* **42**, 1596–1600.

Johnson, K. D. and Kende, H. (1971). *Pl. Physiol., Lancaster* **47**, (suppl.), 24.

Jones, D. F., MacMillan, J. and Radley, M. (1963). *Phytochemistry* **2**, 307–314.

Jones, K. C. (1969). *Pl. Physiol. Lancaster*, **44**, 1695–1700.

Jones, R. L. (1969). *Planta* **87**, 119–133.

Jones, R. L. and Price, J. M. (1970). *Planta* **94**, 191–202.

Jones, R. L. and Varner, J. E. (1967). *Planta* **72**, 155–161.

Jørgensen, O. B. (1963). *Acta chem. scand.* **17**, 2471–2478.

Jørgensen, O. B. (1964). *Acta chem. scand.* **18**, 1975–1978.

Jørgensen, O. B. (1965). *Acta chem. scand.* **19**, 1014–1015.

Jørgensen, B. B. and Jørgensen, O. B. (1963). *Acta chem. scand.* **17** 1765–1770.

Khan, A. A. (1969). *Physiol. Pl.* **22**, 94–103.

Khan, A. A. (1971). *Science, N.Y.* **171**, 853–859.

Khan, A. A. and Downing, R. D. (1968). *Physiol. Pl.* **21**, 1301–1307.

Khan, A. A. and Faust, M. A. (1967). *Physiol. Pl.* **20**, 673–681.

Kirchhoff, H. C. (1815). *Schweigger's J. Chemi. Phys.* **14**, 389–398.

Kirsop, B. H. and Pollock, J. R. A. (1958). *J. Inst. Brew.* **64**, 227–233.

Koehler, D. E. and Varner, J. E. (1971). *Pl. Physiol., Lancaster* **47** (suppl.), 24.

La Berge, D. E., Clayton, J. W. and Meredith, W. O. S. (1967). *Proc. Am. Soc. Brew. Chem.* 18–23.

La Berge, D. E. and Meredith, W. O. S. (1971). *J. Inst. Brew.* **77**, 436–442.

Lazer, L., Baumgartner, W. E. and Dahlstrom, R. V. (1961). *J. Agric. Fd. Chem.* **9**, 24–26.

Lehmann, E. E. and Aichele, F. (1931). "Keimungsphysiologie der Grasen. (Gramineen)." Ferdinand Enke, Stuttgart, pp. 670.

Lewis, M. J. and Fujii, T. (1966). *Proc. Am. Soc. Brew. Chem.* 158–165.

Linderstrøm-Lang, K. and Engel, C. (1937). *Enzymologia* **3**, 138–146.

Loveys, B. R. and Wareing, P. F. (1971). *Planta* **98**, 109–116.

Luchsinger, W. W. and Richards, A. W. (1968). *Cereal Chem.* **45**, 115–123.

MacLeod, A. M. (1953). *J. Inst. Brew.* **59**, 462–469.

MacLeod, A. M., Duffus, J. H. and Horsfall, D. J. L. (1966). *J. Inst. Brew.* **72**, 36–37.

MacLeod, A. M. Duffus, J. H. and Johnston, C. S. (1964a). *J. Inst. Brew.* **70**, 521–528

MacLeod, A. M., Johnston, C. S. and Duffus, J. H. (1964b). *J. Inst. Brew.* **70**, 303–307.

MacLeod, A. M. and Millar, A. S. (1962). *J. Inst. Brew.* **68**, 322–332.

MacLeod, A. M. and Napier, J. P. (1959). *J. Inst. Brew.* **65**, 188–196.

MacLeod, A. M. and Palmer, G. H. (1966). *J. Inst. Brew.* **72**, 580–589.

MacLeod, A. M., Travis, D. C. and Wreay, D. G. (1953). *J. Inst. Brew.* **59**, 154–165.

MacWilliam, I. C. and Harris, G. (1959). *Archs Biochem. Biophys.* **84**, 442–454.

Mann, A. and Harlan, H. V. (1916). *J. Inst. Brew.* **22**, 73–108. (reprint of (1915) Bulletin 183, U.S. Dept. Agriculture).

Manners, D. J. (1971). *Nature, New Biol.* **234**, 150–151.

Manners, D. J. and Marshall, J. J. (1969). *J. Inst. Brew.* **75**, 550–561.

Manners, D. J., Palmer, G. H., Wilson, G. and Yellowlees, D. (1971). *Biochem. J.* **125**, 30p–31p.

Manners, D. J. and Rowe, K. L. (1971). *J. Inst. Brew.* **77**, 358–365.

Manners, D. J. and Sparra, K. L. (1966). *J. Inst. Brew.* **72**, 360–365.

Manners, D. J. and Yellowlees, D. (1970). *Biochem. J.* **117**, 22p–23p.

Manners, D. J. and Yellowlees, D. (1971). *Die Stärke* **23**, 228–234.

Marshall, J. J. and Taylor, P. M. (1971). *Biochem. biophys. Res. Commun.* **42**, 173–179.

Massart, L. (1955). *Proc. Europ. Brew. Conv.* 168–181.

Meredith, W. O. S. (1965). *Proc. Am. Soc. Brew. Chem.* 5–10.

Momotani, Y. and Kato, J. (1971). *Pl. Cell. Physiol.* **12**, 405–410.

Morris, E. O. (1958). *Chem. Ind.* 97.

Musgrave, A., Kays, S. E. and Kende, H. (1972). *Planta* **102**, 1–10.

Nummi, M., Daussant, J., Niku-Paavola, M.-L., Kalsta, H. and Enari, T.-M. (1970). *J. Sci. Fd. Agric.* **21**, 258–260.

Nummi, M., Vilhunen, R. and Enari, T.-M. (1965a). *Acta chem. scand.* **19**, 1793–1795.

Nummi, M., Vilhunen, R. and Enari, T.-M. (1965b). *Proc. Eur. Brew. Conv.* 52–61.

Ogawa, T. (1967). *Bot. Mag., Tokyo* **80**, 27–32.

Ory, R. L. and Henningsen, K. W. (1969). *Pl. Physiol., Lancaster* **44**, 1488–1498.

Paleg, L. G. (1960a). *Pl. Physiol., Lancaster* **35**, 293–299.

Paleg, L. G. (1960b). *Pl. Physiol., Lancaster* **35**, 902–906.

Paleg, L. G. (1961). *Pl. Physiol., Lancaster* **36**, 829–837.

Paleg, L. G. (1964). *Colloq. Intern. Centre Nat. Res. Sci., Paris, 1963*, No. 123, pp. 303–317.

Paleg, L. G., Aspinall, D., Coombe, B. and Nicholls, P. (1964). *Pl. Physiol. Lancaster* **39**, 286–290.

Paleg, L. G., Coombe, B. G. and Buttrose, M. S. (1962). *Pl. Physiol., Lancaster* **37**, 798–803.

Paleg, L. G., Kende, H., Ninnemann, H. and Lang, A. (1965). *Pl. Physiol., Lancaster* **40**, 165–169.

Paleg, L. G. and Sparrow, D. H. B. (1962). *Nature, Lond.* **193**, 1102–1103.

Petridis, C., Verbeek, R. and Massart, L. (1965). *J. Inst. Brew.* **71**, 469.

Petridis, C., Verbeek, R. and Massart, L. (1966). *Naturwissenschaften* **53**, 331–332.

Pollard, C. J. (1969). *Pl. Physiol., Lancaster* **44**, 1227–1232.

Pollard, C. J. (1970a). *Biochim. biophys. Acta* **201**, 511–512.

Pollard, C. J. (1970b). *Biochim. biophys. Acta* **222**, 501–507.

Pollard, C. J. and Nelson, D. C. (1971). *Biochim. biophys. Acta* **244**, 372–376.

Pollard, C. J. and Singh, B. N. (1968). *Biochem. biophys. Res. Commun.* **33**, 321–326.

Pollard, C. J. and Venere, R. J. (1970). *Fedn. Proc. Fedn. Am. Socs. exp. Biol.* **29**, 670.

Pollock, J. R. A. (1958). *Chemy Ind.* 387–388.

Pollock, J. R. A. (1962). *In* "Barley and Malt" (A. H. Cook, ed.), pp. 303–398. Academic Press, London and New York.

Pollock, J. R. A. and Pool, A. A. (1958). *J. Inst. Brew.* **64**, 151–156.

Pollock, J. R. A., Essery, R. E. and Kirsop, B. H. (1955). *J. Inst. Brew.* **61**, 295–300.

Preece, I. A. (1947). *J. Inst. Brew.* **53**, 154–162.

Preece, I. A., Aitken, R. A. and Dick, J. A. (1954). *J. Inst. Brew.* **60**, 497–507.

Preece, I. A. and Hobkirk, R. (1954). *J. Inst. Brew.* **60**, 490–496.
Preece, I. A. and Hobkirk, R. (1955). *J. Inst. Brew.* **61**, 393–399.
Preece, I. A. and Hoggan, J. (1956). *J. Inst. Brew.* **62**, 486–496.
Preece, I. A. and MacDougall, M. (1958). *J. Inst. Brew.* **64**, 489–500.
Prentice, N., Dickson, A. D., Burkhart, B. A. and Standridge, N. N. (1963). *Cereal Chem.* **40**, 208–220.
Radley, M. (1966). *Nature, Lond.* **210**, 969.
Radley, M. (1967). *Planta* **75**, 164–171.
Radley, M. (1968). *In* "Plant Growth Regulators." Society of Chemical Industry Monograph No. 31, pp. 53–69. London.
Radley, M. (1969). *Planta* **86**, 218–223.
Reid, D. M. and Clements, J. B. (1968). *Nature, Lond.* **219**, 607–609.
Rejowski, A. (1969). *Bull. Acad. Pol. Sci. Ser. Sci. Biol.* **17**, 641–644.
Rejowski, A. and Kulka, K. (1967). *Acta Soc. Bot. Pol.* **36**, 221–234.
Rowsell, E. V. and Goad, L. J. (1964). *Biochem. J.* **90**, 11p–12p.
Sandegren, E. and Klang, N. (1950). *J. Inst. Brew.* **56**, 313–318.
Sandstedt, R. M., Kneen, E. and Blish, M. J. (1939). *Cereal Chem.* **16**, 712–723.
Schlubach, H. H. (1965). *Fortschr. Chem. Org. Naturstoffe.* **23**, 46–60.
Schwimmer, S. and Balls, A. K. (1949). *J. biol. Chem.* **179**, 1063–1074.
Shuster, L. and Gifford, R. H. (1962). *Archs Biochem. Biophys.* **96**, 534–540.
Simpson, G. A. (1966). *Can. J. Bot.* **44**, 115–116.
Sparrow, D. H. B. (1964). *J. Inst. Brew.* **70**, 514–521.
Sparrow, D. H. B. (1965). *J. Inst. Brew.* **71**, 523–529.
Srivastava, B. I. S. and Meredith, W. O. S. (1964). *Can. J. Bot.* **42**, 507–513.
Stingl, G. (1907). *Flora* **97**, 308–331.
Stoward, F. (1911). *Ann. Bot.* **25**, 1147–1204.
Taiz, L. and Jones, R. L. (1970). *Planta* **92**, 73–84.
Tharp. W. H. (1935). *Bot. Gaz.* **97**, 240–271.
Tronier, B. and Ory, R. L. (1970). *Cereal Chem.* **47**, 464–471.
Van Onckelen, H. A. and Verbeek, R. (1969). *Planta* **88**, 255–260.
Varner, J. E. (1964). *Pl. Physiol., Lancaster* **39**, 413–415.
Varner, J. E. and Chandra, R. G. (1964). *Proc. natn. Acad. Sci. U.S.A.* **52**, 100–106.
Varner, J. E., Chandra, R. G. and Chrispeels, M. J. (1965). *J. cell. comp. Physiol.* (Suppl. 1), 55–68.
Varner, J. E. and Johri, M. M. (1968). *In* "Biochemistry and Physiology of Plant Growth Substances" (F. Wightman and G. Setterfield, eds), pp. 793–814. Runge Press, Ottawa.
Verbeek, R., Van Onkelen, H. A. and Gaspar, T. (1969). *Physiol. Pl.* **22**, 1192–1199.
Verbeek-Wyndaele, R. and Coulier, V. (1961). *Proc. Europ. Brew. Conv.* 72–79.
Wiener, S. and Hopkins, R. H. (1952). *J. Inst. Brew.* **58**, 204–213.
Yomo, H. (1958). *Hakkô Kyôkaishi* **16**, 444–448.
Yomo, H. (1960). *Hakkô Kyôkaishi* **18**, 603–607.
Yomo, H. and Iinuma, H. (1962). *Agric. Biol. Chem. (Tokyo)* **26**, 201.
Yomo, H. and Iinuma, H. (1964a). *Proc. Am. Soc. Brew. Chem.* 97–102.
Yomo, H. and Iinuma, H. (1964b). *Agric. Biol. Chem. (Tokyo)* **28**, 273–278.
Yomo, H. and Iinuma, H. (1966). *Planta* **71**, 113–118.
Yung, K.-H. and Mann, J. D. (1967). *Pl. Physiol.* **42**, 195–200.

CHAPTER 11

The Synthesis of Chloroplast Enzymes

J. W. BRADBEER

Botany Department, King's College, London, England

I. INTRODUCTION

As a result of investigations in many laboratories, including my own, much information has been obtained about chloroplast development in the bean (*Phaseolus vulgaris* L.) and it is probably the best known higher plant from this respect. This review will essentially confine itself to the results of investigations relating to chloroplast enzyme synthesis in developing leaves, in particular to those of bean. It will make little mention of studies on chloroplast protein synthesis under *in vitro* conditions, recently reviewed by Boulter *et al.* (1972), and will not consider the protein synthesizing ability which may be possessed by apparently mature chloroplasts.

II. LEAF DEVELOPMENT

Von Wettstein and his coworkers have made substantial investigations of bean chloroplast development, during the course of which they showed that 14-day-old dark-grown plants possess partially expanded primary leaves whose etioplasts develop into chloroplasts in a synchronous manner when the leaves are transferred to continuous illumination (Gyldenholm, 1968). The role of cell division and cell expansion in the development of bean leaves grown under similar conditions has been studied elsewhere (Dale and Murray, 1969). On the basis of these earlier investigations we have been able to devise experiments to study chloroplast development in bean in a quantitative manner. The course of development of primary leaves in the dark (see Fig. 1) shows that cell division and the rise in protein content ceased after 12 days but that fresh and dry weights continued to increase during 30 days of growth in the dark. Figure 1 also compares the effects of transferring 10-, 12- and 14-day-old dark-grown plants to continuous illumination, in terms of increased fresh weight and cell number.

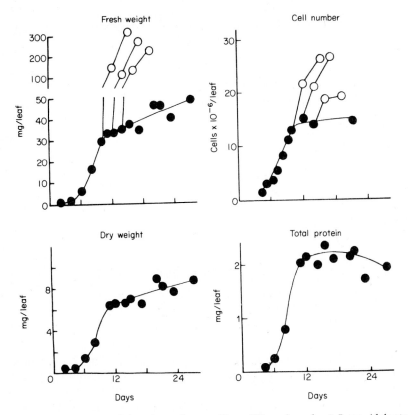

FIG. 1. The development of the primary leaves of bean (*Phaseolus vulgaris* L. cv. Alabaster) during growth at 23° in the dark (●). The effects of transfer to continuous illumination at 8·0 μw/mm² on 10-, 12- and 14-day-old dark-grown plants are shown for fresh weight and cell number (○) (data partly taken from Ireland, 1971).

Illumination induced a several-fold increase in fresh weight, though the total growth after 5 days illumination was less when the older plants were illuminated. Illumination of 10- and 12-day-old plants gave approximately a doubling of leaf-cell number, but light-induced cell division in the leaves of the 14-day-old plants was much less than in the younger leaves. The effects of illumination of 14-day-old dark-grown beans are shown in more detail in Fig. 2 in which they are also compared with control values of leaves maintained in the dark. Dry weight and protein increased without any evident lag, this being indicative of a rapid onset of translocation to the developing leaves from the cotyledons. There was, however, a lag in leaf expansion, as seen by the lag in the increase of fresh weight, and there was also a lag in the formation of chlorophyll, like that reported by numerous previous workers.

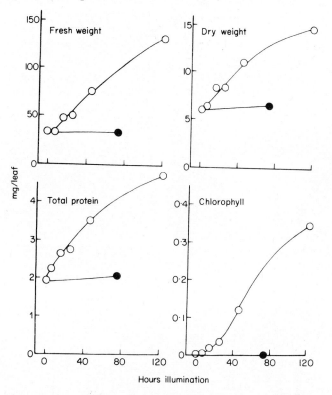

FIG. 2. The effects of illumination on the primary leaf development of 14-day-old dark-grown beans (○), in comparison with controls maintained in darkness (●). Experimental conditions as for Fig. 1.

III. CHLOROPLAST DEVELOPMENT

The bean chloroplasts are considered to arise from proplastids, which are undifferentiated organelles less than 1 μm in diameter, possessing a double membrane and found in the leaf initials of the dry seed (Kirk and Tilney-Bassett, 1967). During the earliest stages of leaf development positive identification of proplastids in electron micrographs is difficult, if not impossible, but after 6 days of dark growth the conversion of the proplastid to the etioplast is well under way and the features shown in Fig. 3A can be seen. The developing plastid has increased in size, sheets of membrane have been formed by the invagination of the inner membrane of the plastid envelope and the plastid has stored substantial amounts of starch. The sheets of membrane are porous and

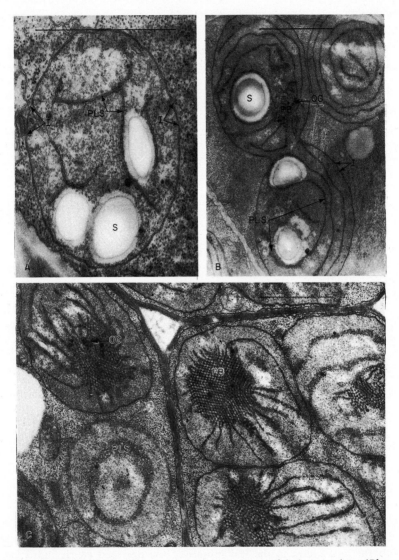

FIG. 3. Stages of etioplast development in the primary leaves of dark-grown bean (*Phaseolus vulgaris* L. cv. Alabaster). A, 6 days old; B, 9 days old showing plastid division; C, 14 days old. Fixation for electron microscopy was by glutaraldehyde-osmium tetroxide. The scale lines represent 1 μm. Abbreviations: I, invagination of plastid inner membrane; OG, osmiophilic globules; PB, prolamellar body; PLS, porous lamellar sheet; S, starch grain.

will eventually form the thylakoids of the chloroplast. During continued growth in darkness the increased amounts of porous lamellar material begin to condense together to give a lattice of branched tubules called the prolamellar body, whose structure and development has been elegantly described by Weier and Brown (1970). Figure 3B shows the 9 day stage of etioplast development in which both the beginnings of prolamellar body formation and the continued invagination of the envelope inner membrane may be seen. As a result of further growth in the dark the etioplast reaches the peak of its development after about 14 to 15 days with a regular lattice structure of the prolamellar body and a substantial area of lamellar sheets, as shown in Fig. 3C. Figure 4 shows the

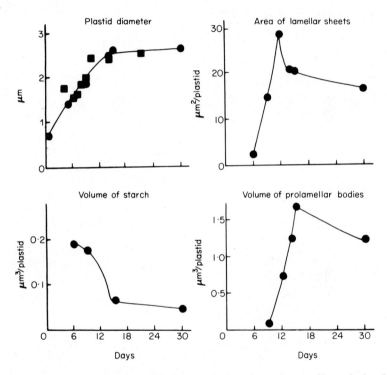

FIG. 4. Some parameters of etioplast development in the primary leaves of bean during dark growth under the same conditions as for Fig. 1. Measurements from electron micrographs (●), measurements from light microscopy (■) (data partly taken from Ireland, 1971).

results of the measurement of a number of parameters of etioplast growth, made mainly from electron-micrographs. During the first 14 days of dark-growth there was a rise in plastid diameter and a loss in plastid starch content. The appearance of sheets of lamellar material was underway after 6 days growth and their condensation into prolamellar bodies commenced after about 9 days. The formation of the prolamellar bodies presumably accounted for the fall in the area of the lamellar sheets after 12 days dark growth but the formation

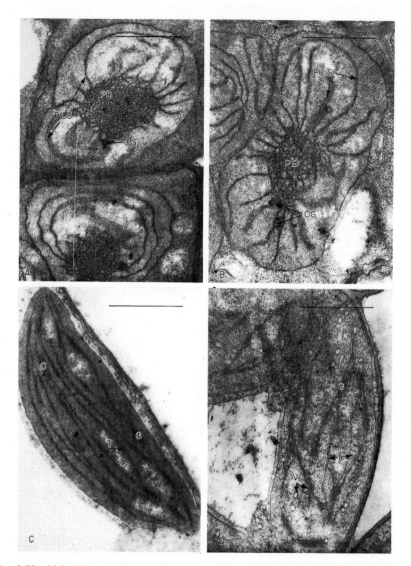

FIG. 5. Plastid development in the primary leaves of bean (*Phaseolus vulgaris* L. cv. Alabaster) in response to the transfer of 14-day-old dark-grown plants to continuous illumination of 8·0 μw/mm². A, after 45 min illumination; B, after 105 min illumination; C, after 50 h illumination of an excised shoot; D, after 50 h illumination of leaves which contained 1 mg D-*threo* chloramphenicol per 1 g of leaf. Fixation for electron microscopy was by glutaraldehyde-osmium tetroxide. The scale lines represent 1 μm. Abbreviations: G, grana; OG, osmiophilic globules; PB, prolamellar body; PLS, porous lamellar sheet; T, thylakoid; V, vesicle.

of membrane material by the developing etioplast continued until at least 15 days of dark growth.

There have been many reports about the effects of illumination on plastid fine structure in dark-grown leaves and of these the most satisfactory is that of Weier *et al.* (1970) with bean. Some of the results of our own work are shown in Figs 5 and 6. When continuous illumination of 8·0 μw/mm^2 was given to 14-day-old dark-grown beans the transformation of the prolamellar body from its regular structure was evident after 45 min (Fig. 5A), while Fig. 5B shows

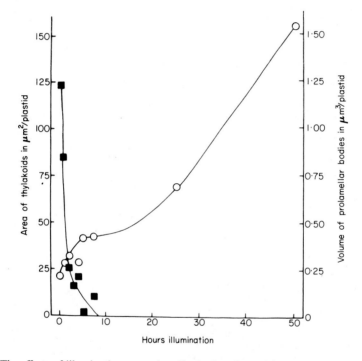

FIG. 6. The effects of illumination on prolamellar body volume (■) and thylakoid area (○) of developing plastids of the primary leaves of 14-day-old dark-grown beans. Experimental conditions as for Fig. 1.

complete transformation after 105 min. Figure 6 shows that the volume of the prolamellar body fell rapidly during the first 5 h of illumination and that it was accompanied by a doubling of the area of the lamellar sheets. The porous lamellar sheets of the etioplast and the membrane of the prolamellar body, which is converted into sheets, become the first thylakoids of the developing plastid, and in subsequent discussions all of these membranes will be described as thylakoid material. From these and other measurements it seems that the etioplasts of 14-day-old leaves contain about half of their thylakoid material in the prolamellar body and half in the porous lamellar sheets. The level of illumination used in our investigations was evidently too high to permit any recrystallization of prolamellar bodies as has been reported under conditions of weak illumination (Weier *et al.*, 1970). Fifteen hours after the commencement of illumination the beginnings of granal formation were obvious (Gyldenholm, 1968) and Fig. 6 also indicates that increased formation of thylakoid material in the developing plastid must have commenced at about this time. Figure 5C shows a section of a 50-h illuminated plastid in which the structure of a mature plastid has been reached, and from the measurements in Fig. 6 it can be seen that the total amount of thylakoid material at this stage is almost four times greater than that in the etioplast.

The extent of plastid replication during leaf development may be seen in Fig. 7 which gives plastid number per leaf during dark development and after

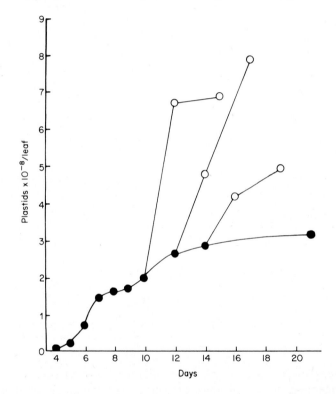

FIG. 7. The number of plastids in the primary leaves of bean during growth in darkness (●) and the effects of the transfer of 10-, 12- and 14-day-old plants to continuous illumination (○). Experimental conditions as for Fig. 1.

the illumination of 10-, 12- and 14-day-old beans. There were two distinct periods when plastid number increased in the dark, namely on days 5 to 7 and days 10 to 12, and the occurrence of these two distinct periods of plastid replication may be contrasted with the simple sigmoidal curve for cell number shown in Fig. 1. The 9-day-old plastid shown in Fig. 3B is clearly at a fairly advanced stage of division and many similar stages were seen in this sample. Observations by both light- and electron-microscopy indicate that the increase in plastid numbers during both the second period of increase in the dark and on illumination result from the division of existing etioplasts and chloroplasts and not from the maturation of proplastids. Relevant observations of the first period of dark increase have not been made. On the illumination of 10- and 12-day-old beans much plastid replication was induced, but when illumination was delayed until day 14 the number of plastids did not quite double as a result of illumination.

From the data concerning plastid number (Fig. 7), and that concerning plastid components (Figs 4 and 6), information about the total amount of plastid components in the leaf and hence of their rate of accumulation may be derived. As an example of this the total amount of the thylakoid material in greening leaves is shown in Fig. 8 together with values for the total soluble and

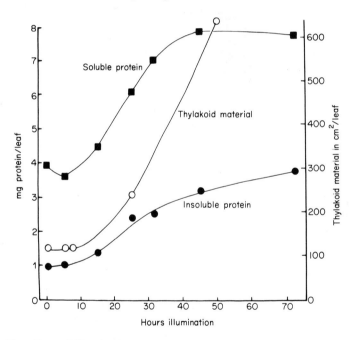

FIG. 8. The effects of illumination on the soluble protein (■), insoluble protein (●) and thylakoid material (○) of the primary leaves of 14-day-old dark-grown beans. Experimental conditions as for Fig. 1.

insoluble protein contents of similarly treated leaves. Solubility in 0·05 M Tris chloride buffer, pH 7·8, was the criterion used, and a substantial part of the insoluble protein fraction would be expected to consist of thylakoid material. During the first 50 h of greening, thylakoid material in the leaf increased by 430% while the total insoluble protein increased by 240%.

IV. Enzyme Synthesis

Extracts of 14-day-old etiolated bean leaves contain enzymic activity for all of the steps of the photosynthetic carbon cycle (Bradbeer, 1969; Wara-Aswapati, 1971) although two of the activities, alkaline fructose-diphosphatase and transaldolase, were present only at very low levels in extracts from both etiolated and green leaves. The low levels of these two activities may have resulted from a failure to extract and measure them under optimal conditions, or alternatively these results may indicate that these two enzymes are not involved in the photosynthetic carbon cycle of bean. As the determination of the amounts of most of the enzymes present in the leaves has to be inferred from the activities of the specific reactions that they catalyse, the optimal conditions for the enzyme assays have to be carefully established. We are reasonably confident that we have used optimal conditions for the assay of all of the enzymes which we have studied, except for alkaline fructosediphosphatase whose properties have recently been reported by Buchanan *et al.* (1971).

Chloroplast triosephosphate dehydrogenase is an enzyme which gave considerable difficulty before the definitive investigations of Müller and his co-workers on its allosteric regulation (Müller *et al.*, 1969; Müller, 1970) and immunological investigations in the laboratory of Gibbs which established the existence of two distinct triosephosphate dehydrogenases in green leaves. One of these triosephosphate dehydrogenases is a stable chloroplast enzyme which can function with either NADPH or NADH and is subject to allosteric control, while the other is a rather labile cytoplasmic enzyme which is specific for NADH. Mr O. Wara-Aswapati, in this laboratory, has shown that the properties of the bean leaf chloroplast enzyme resemble those reported for spinach by Müller's group and some of his results are shown in Fig. 9. The effects of 30 min preincubation with ATP on an extract from etiolated leaves was to promote the chloroplast NADPH-dependent activity by 700% by 6 mM ATP and to inhibit the NADH-dependent activity by 30% by 6 mM ATP (Fig. 9A). The inhibition of the NADH-dependent activity by 6 mM ATP contrasted with the small promotion of activity found by Müller (1970). Preincubation with NADPH similarly promoted NADPH-dependent activity, 3 mM NADPH giving a 500% increase (Fig. 9B), and it may be noted that the activation curves in Figs 9A and B both show sigmoidicity. It may be seen from Fig. 9C that the activation process by ATP was quite slow. Extracts from green leaves also showed stimulation of NADPH-dependent activity after preincubation with ATP but the amount of stimulation was less, being 70% for leaves which had been illuminated prior to extraction (Fig. 9D) and 260% for leaves which had been darkened prior to extraction (Fig. 9E). Neither of these latter curves shows sigmoidicity. The data are satisfactorily accounted for by Müller's explanation, on the basis of ATP and NADPH functioning as allosteric effectors. In the etioplasts of the dark-grown leaf the concentrations of ATP and NADPH are presumed to be very low (Wallis and Bradbeer, 1970),

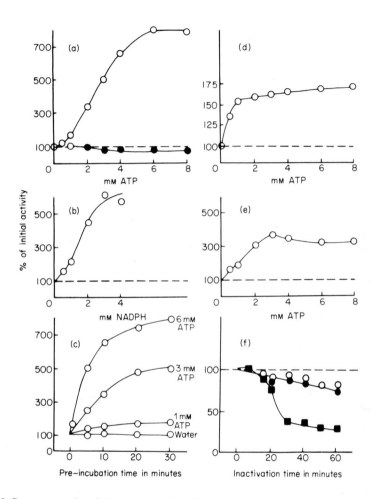

FIG. 9. Some properties of the triosephosphate dehydrogenases in extracts of primary leaves of bean. a, the effects of preincubation of the enzyme from etiolated leaves with ATP for 30 min at 25° prior to assay. b, similar preincubation with NADPH. c, the time course of the activation process with different ATP concentrations. d, the effects of ATP-preincubation on the enzyme from green leaves which had been illuminated for 2 h immediately before extraction. e, as for d but with green leaves which had been kept in the dark for 1·5 h immediately before extraction. f, comparison of the inactivation of the chloroplast and cytoplasmic enzymes. Chloroplast NADPH dependent activity (○); chloroplast NADH-dependent activity (●); cytoplasmic NADH dependent activity (■) (unpublished data of O. Wara-Aswapati).

and large effects occur when they are added to the extract. In chloroplasts of the green illuminated leaf the concentrations of ATP and NADPH are maintained at high levels as a result of photosynthesis (Wallis and Bradbeer, 1970), the NADPH-dependent triosephosphate dehydrogenase consequently is partly activated and only a small allosteric activation results on preincubation of the extract with ATP. The green but darkened leaf contains less ATP and NADPH than the illuminated leaf and hence a greater stimulation of activity by ATP was observed. The absence of sigmoidal activation curves for green leaves may be explained by the presence of some effector ATP and NADPH in the chloroplasts. The slow development of the allosteric changes in the extracts permits the determination of the actual enzyme activity present in the leaf as well as of the total potential NADPH-dependent triosephosphate dehydrogenase activity present at each developmental stage. These values are given by the activities determined before and after preincubation. The contributions of the chloroplast enzyme to the extract's NADH-dependent triosephosphate dehydrogenase activity has been determined from the NADH:NADPH ratio for chloroplasts prepared from duplicate leaf samples. The cytoplasmic NADH-dependent activity is then determined by difference. In Fig. 9F this procedure has been followed to measure the time-dependent decay of the three activities. It is seen that both NADH and NADPH activities of the chloroplast enzyme fall off slowly at similar rates while the cytoplasmic enzyme is obviously much more labile.

Figures 10 and 11 show the activities of some of the photosynthetic carbon cycle enzymes during the course of greening, together with control values for leaves maintained in continuous darkness. The values for NADPH-dependent triosephosphate dehydrogenase activity in Fig. 10 represent the actual activities in the leaves, but Mr Wara-Aswapati reports that the potential activity of this enzyme also shows a several-fold rise during greening. As the other enzymes in Figs 10 and 11 were assayed under optimal conditions it may be presumed that their full potential activities were being determined in the assays. If this is so then the increases in activity which occurred during greening must represent increases in the amount of enzyme protein in the leaves, and these increases must result either from *de novo* synthesis or the conversion of precursor protein into active enzyme. I do not know of any cases of chloroplast enzymes being formed from an inactive preformed precursor during greening however. Since the increases in enzyme activity which occur under different environmental conditions are always accompanied by substantial increases in leaf protein content (Bradbeer, 1969, 1971; Bradbeer *et al.*, 1970; Ireland, 1971) it seems probable that the increases in potential activity of the photosynthetic carbon cycle enzymes are mainly accounted for by *de novo* protein synthesis. In greening barley leaves Kleinkopf *et al.* (1970) found parallel increases of ribulosediphosphate carboxylase protein and enzyme activity as well as of the labelling of this enzyme with [³H]leucine. As 90% of the constituent peptides, separated on a "peptide map", were shown to have accumulated radioactivity from [³H]leucine they were able to conclude that the increase in ribulose-diphosphate carboxylase activity during greening was due to *de novo* synthesis of the enzyme. Their information does not conclusively establish that the increase of enzyme activity was wholly a result of *de novo* protein synthesis,

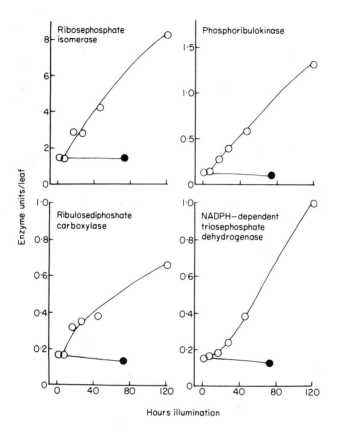

FIG. 10. The effects of illumination on the development of some activities in the primary leaves of 14-day-old dark-grown beans (○) in comparison with those of controls maintained in darkness (●). For triosephosphate dehydrogenase, values are given for the actual activity of the enzyme in the leaves, i.e. no preincubation with ATP. Experimental conditions as for Fig. 1.

but at the present time it is the most direct evidence available. The density labelling technique of Filner and Varner (1967) should be applicable to the study of the development of bean primary-leaf enzymes and should be able to establish whether their increased activities are wholly a result of *de novo* protein synthesis. In the absence of conclusive evidence about *de novo* protein synthesis, increases in activity will be provisionally attributed to protein synthesis unless there is evidence that enzyme activation is involved.

The development of the activities of the enzymes, during greening, in Fig. 10 is different from that in Fig. 11 in that lag periods were evident in the former but not in the latter (Bradbeer, 1970). There was no lag in the increase in the protein content of the leaves (Fig. 2) and it is concluded that the machinery for the synthesis of the enzymes shown in Fig. 11 was complete in 14-day-old

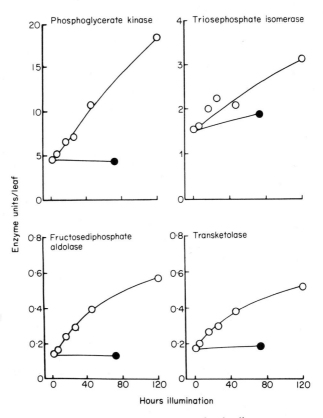

FIG. 11. See legend to Fig. 8 for details.

dark-grown leaves. The lags shown in Fig. 10 imply that the machinery for the synthesis of these enzymes is in a dissociated state in 14-day-old dark-grown leaves and that the lag represents the time required for the reassembly of this machinery. Dark growth beyond 14 days seems to result in the breakdown of the machinery for the synthesis of other proteins and in due course the light-induced formation of total leaf protein also shows a lag.

Direct measurements of the development of a number of membrane proteins are possible including the study of the cytochromes by the room temperature determination of the oxidized:reduced difference-spectra of plastid preparations (Bendall *et al.*, 1971). In Fig. 12 the cytochrome contents, determined during the course of greening, have been expressed on a per plastid basis for bean primary leaves (Gregory and Bradbeer, 1972). Etioplasts contain similar amounts of cytochromes f, b-$559_{\text{low potential}}$ (LP $= +70$ mV) and b-563, but have no detectable b-$559_{\text{high potential}}$ (HP $= +370$ mV). During greening the two low potential cytochromes (b-559_{LP} and b-563) showed more than a 2000% increase during 48 h in comparison to an approximately 300% increase

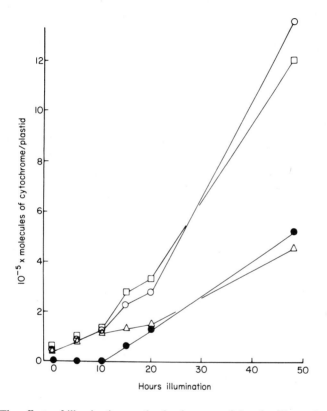

Fig. 12. The effects of illumination on the development of the plastid cytochromes of the primary leaves when 14-day-old dark-grown plants of bean (*Phaseolus vulgaris* L. cv. Canadian Wonder) were transferred to continuous illumination of 8·0 μw/mm², cytochrome f ($\triangle$); cytochrome b-559_{LP} ($\bigcirc$); cytochrome b-559_{HP} ($\bullet$); cytochrome b-563 ($\square$) (after Gregory and Bradbeer, 1972).

in thylakoid material. Cytochrome b-559_{HP} was detected first of all in plastids from leaves which had been illuminated for 15 h, after this time it occurred in approximately similar amounts to cytochrome f. As the extinction coefficient of cytochrome f is the only one for these four cytochromes to have been determined, the concentrations of the others represent only the relative concentrations. In previous investigations of the development of the plastid cytochromes in greening leaves, difference spectrophotometry at liquid nitrogen temperatures has been employed and quantitative measurements have not proved possible (Boardman, 1968; Gyldenholm et al., 1971). Board-man (1968) had not detected any cytochrome b-559 in etioplasts and was apparently not aware of the presence of the high potential cytochrome b-559 which Bendall (1968) had found in chloroplasts.

Table I shows some of the constituents of a typical etioplast and of a chloro-plast, from a 45-h illuminated leaf, with respect to some of the specific plastid components which we have determined. Two chloroplast thylakoid compon-ents could not be detected in the etioplasts, namely chlorophyll and cytochrome

TABLE I

A comparison of some of the components of an etioplast obtained from the primary leaves of 14-day-old dark-grown beans with those of a chloroplast obtained after 45 h illumination of similar leaves. Values expressed on a per plastid basis

	Etioplast	Chloroplast
Thylakoid membrane (μm^2)	21	137
Prolamellar body membrane (μm^2)	22	0
Total internal membrane (μm^2)	43	137
Equivalent no. of 40 Å membrane units	$5\cdot4 \times 10^6$	$1\cdot7 \times 10^7$
Equivalent no. of quantasomes in membrane	$3\cdot0 \times 10^5$	$9\cdot6 \times 10^5$
	Molecules	Molecules
Protochlorophyllide	2×10^6	0
Chlorophyll	0	2×10^8
Ferredoxin	$1\cdot3 \times 10^5$	$1\cdot1 \times 10^6$
Cytochrome f	$4\cdot2 \times 10^4$	$4\cdot5 \times 10^5$
Cytochrome b-559$_{HP}$	$<1\cdot0 \times 10^4$	$5\cdot2 \times 10^5$
Cytochrome b-559$_{LP}$	$4\cdot6 \times 10^4$	$1\cdot3 \times 10^6$
Cytochrome b-563	$4\cdot6 \times 10^4$	$1\cdot1 \times 10^6$
Ribulosediphosphate carboxylase	$3\cdot5 \times 10^5$	$5\cdot4 \times 10^5$
	Enzyme units	Enzyme units
Phosphoribulokinase	$4\cdot5 \times 10^{-10}$	$1\cdot4 \times 10^{-9}$
Chloroplast triosephosphate dehydrogenase[a] (NADPH-dependent)	$5\cdot2 \times 10^{-9}$	$8\cdot6 \times 10^{-9}$

[a] Activated by preincubation with ATP.

b-559$_{HP}$, while protochlorophyllide was not found in the chloroplasts. During greening the other thylakoid components (total membrane, cytochromes and ferredoxin) showed increases of between 300 and 2000% while the three soluble chloroplast enzymes showed increases of between 50 and 300%. From the measured rates of increase of some of these proteins it has been pos-sible to calculate their maximum rates of synthesis on a per plastid basis during dark development and during greening (Table II). During both dark develop-ment and greening the rate of ribulosediphosphate carboxylase synthesis reached approximately the same maximum rate, approximately $1\cdot0 \times 10^4$ molecules/plastid/h. On the basis of 16 of each of the large and small subunits in one enzyme molecule (Kawashima, 1970) each subunit is synthesized at a rate of 1.6×10^5 subunits/plastid/h. Synthesis of ferredoxin and the cytochromes has been followed in the light only, where the high potential cytochromes were synthesized at about the same rate (molecules/plastid/h) as ribulosediphosphate carboxylase, while ferredoxin and the low potential cytochromes were synthesized about twice as quickly. From the rate of formation of thylakoid sheets, in terms of μm^2/plastid/h, values for the formation of 40 Å membrane subunits and for the formation of quantasomes (lateral dimensions 185×155 Å according to Park and Biggins, 1964) have been calculated on the basis that in each case the thylakoids are wholly built up of

TABLE II

The maximum rates of *in vivo* synthesis of some plastid proteins in developing bean leaves expressed as molecules/plastid/h

	During dark growth	During greening
Ribulosediphosphate carboxylase	1.0×10^4	8.4×10^3
Ribulosediphosphate carboxylase subunits	1.6×10^5	1.3×10^5
Ferredoxin	—	2.7×10^4
Cytochrome f	—	1.1×10^4
Cytochrome b-559$_{HP}$	—	1.4×10^4
Cytochrome b-559$_{LP}$	—	3.2×10^4
Cytochrome b-563	—	2.7×10^4
Insoluble protein:		
(if molecular weight = 2.5×10^4 daltons)	—	2.5×10^{6a}
(if molecular weight = 1×10^6 daltons)	—	6.3×10^{4a}
Thylakoid sheet:		
(as μm^2)	0.18	3.0
(as 40 Å subunits)	2.2×10^4	3.8×10^5
(as quantasomes)	1.3×10^3	2.3×10^4

[a] At stage of maximum thylakoid formation.

these components. The rate of formation of thylakoid material in greening plastids is about 17 times the maximum rate found in etioplasts during their dark development. As the rate of formation of insoluble protein is also known (Fig. 8) the number of thylakoid components that this might represent can be calculated on the basis that the molecular weight of the protein in a 40 Å subunit may be 2.5×10^4 daltons (Mühlethaler, 1971) and that the amount of protein in a quantasome may be estimated to be about 1×10^6 daltons (from the data of Park and Pon, 1963). At the time of maximum thylakoid formation in greening leaves the rate of insoluble protein synthesis on a molecules/plastid/h basis could account for 2.5×10^6 40 Å subunits or 6.3×10^4 quantasomes, values which are in excess of those required to account for the maximum possible rates of thylakoid formation. The actual maximum possible rate of formation of the 40 Å subunit, as calculated from thylakoid formation, would be 3.8×10^5 molecules/plastid/h, a rate which is in excess of that for ribulose-diphosphate carboxylase subunits for example.

V. THE SITES OF ENZYME SYNTHESIS

Since chloroplasts possess machinery for protein synthesis (Boulter *et al.*, 1972) it is important to determine exactly which proteins are synthesized within the chloroplasts. Chloramphenicol has been shown to be a specific inhibitor of protein synthesis on the 70 s ribosomes of chloroplasts while cycloheximide inhibits the 80 s ribosomes of eukaryotic cytoplasm and both have been used for *in vivo* investigations (Smillie *et al.*, 1971). However Ellis (1969) and Ellis and MacDonald (1970) have found that both of these compounds show additional inhibitory effects in higher plants which preclude the simple interpretation of their effects on protein synthesis. Ellis (1969) has pointed out that chloramphenicol possesses four stereoisomers of which only one, D-*threo* chloramphenicol, affects 70 s ribosomes while all four isomers show the other

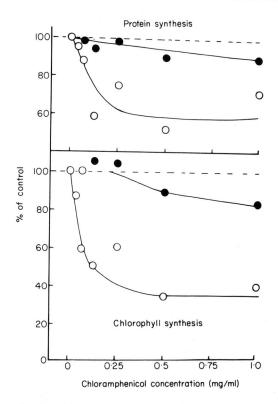

FIG. 13. The effects of D-*threo* (○), and L-*threo* chloramphenicol (●) on chlorophyll and protein synthesis in the primary leaves of 14-day-old dark-grown beans during 50 h continuous illumination. Experimental procedures as described by Ireland and Bradbeer (1971) (from the unpublished data of G. Arron and O. Wara-Aswapati).

inhibitory effects (e.g. on phosphorylation and solute uptake). Consequently if control experiments with one of the unspecific isomers are carried out it should be possible to use D-*threo* chloramphenicol as a specific inhibitor of protein synthesis on 70 s ribosomes. Only when the inhibition of chloroplast development or the development of enzyme activity is significantly greater with the D-*threo* isomer than with the other isomers can the inhibition be ascribed to the inhibition of protein synthesis on 70 s ribosomes. Figure 13 shows the effects of the application of D-*threo* and L-*threo* chloramphenicol, at different concentrations, on the synthesis of chlorophyll and protein by the leaves during the first 50 hours of greening under continuous illumination. The D-*threo* isomer gave a maximum inhibition of about 40 % for protein synthesis and 65 % for chlorophyll synthesis while the L-*threo* isomer had a very much smaller effect. In a similar experiment Mr Wara-Aswapati found that, at a concentration of 1 mg of inhibitor to 1 g of leaves, D-*threo* chloramphenicol totally inhibited the development of photosynthetic gaseous exchange in greening leaves while the L-*threo* isomer gave a 50 % inhibition in comparison with the water control. After 50 hours illumination the fine structures of the plastids of the water control and of the L-*threo* chloramphenicol treatment were indistinguishable, see Fig. 5C, while the D-*threo* isomer treatment resulted in plastids with reduced thylakoids, scattered grana and numerous small vesicles as seen in Fig. 5D. Table III summarizes the effects of the two chloramphenicol isomers on light induced thylakoid formation, the development of enzyme activities and the synthesis of chloroplast cytochromes. Detailed discussion of the data may be found elsewhere (Ireland and Bradbeer, 1971; Gregory and Bradbeer, 1972; and Bradbeer *et al.*, 1972). We conclude that the formation of at least one of the ribulosediphosphate carboxylase subunits, at least one essential structural component of the thylakoids and the plastid cytochromes are wholly dependent on protein synthesis on 70 s ribosomes in view of the total inhibition of their synthesis by D-*threo* chloramphenicol while the L-*threo* isomer has no inhibitory effect. In *in vitro* experiments Blair and Ellis (1972) have demonstrated that only the large subunit of ribulosediphosphate carboxylase is synthesized by isolated pea chloroplasts, thus confirming some earlier *in vivo* observations (Kawashima, 1970; Criddle *et al.*, 1970). As Machold (1970) has shown that treatment of broad bean (*Vicia faba*) leaves with D-*threo* chloramphenicol inhibits the formation of two thylakoid proteins it may be inferred that they are synthesized on 70 s ribosomes and that the inhibition of the synthesis of similar compounds may prevent the formation of thylakoids as seen in Fig. 5D and Table III for bean. Figure 5D and Table III also show that the stacking of preexisting thylakoid material to give grana does not appear to be inhibited by D-*threo* chloramphenicol.

Recent investigations in my laboratory have detected a partial inhibition of the increase of phosphoribulokinase activity in greening leaves by L-*threo* chloramphenicol in addition to the previously reported total inhibition by the D-*threo* isomer (Ireland and Bradbeer, 1971). Although we maintain our

TABLE III

Effects of D-*threo* and L-*threo* chloramphenicol on the light-induced changes of plastid components in dark-grown bean leaves. Enzyme activities expressed as enzyme units and cytochromes as n moles, per leaf in each case

	No illumination	Water control	50 h illumination	
			D-*threo* chlor-amphenicol	L-*threo* chlor-amphenicol
Phosphoribulokinase	0·17	4·32	0·49[b]	4·01
Ribulosediphosphate carboxylase	0·44	1·22	0·27[c]	1·17
Triosephosphate dehydrogenase	0·16	1·04	0·64[b]	0·88[a]
Chlorophyll (nmoles/leaf)[d]	0	220	79	—
	0	172	—	175
No. of single thylakoids cut by 100 μm transect	197	849	126[c]	1058
No. of stacked thylakoids cut by 100 μm transect	0	131	74[a]	134
Total no. of thylakoids cut by 100 μm transect	197	980	200[c]	1192
Cytochrome f [d]	20	310	<20	300
Cytochrome b-563[d]	22	820	<20	660
Cytochrome b-559$_{LP}$ [d]	22	940	<20	890
Cytochrome b-559$_{HP}$ [d]	<5	360	<20	<20

[a,b,c] Significantly different from water control at P = 0·05, 0·01, 0·001 respectively.
[d] Results not subjected to statistical analysis.

TABLE IV

Sites of protein synthesis in primary leaves of bean (*Phaseolus vulgaris*)

70 s ribosomes	80 s ribosomes
Phosphoribulokinase	Chloroplast triosephosphate dehydrogenase
RDP carboxylase (larger subunit)	Cytoplasmic triosephosphate dehydrogenase
Two major thylakoid constituents	Ferredoxin
Cytochrome f	Ferredoxin: NADP reductase
Cytochrome b-563	Phosphoglycollate phosphatase
Cytochrome b-559$_{LP}$	Chloroplast glyoxylate reductase
	Peroxisomal glyoxylate reductase
	Glycollate oxidase
	Catalase
	Phosphoglycerate kinase
	Triosephosphate isomerase
	FDP aldolase
	Transketolase
	Ribosephosphate isomerase
	PEP carboxylase

conclusion that this enzyme is synthesized on 70 s ribosomes in bean, it should be noted that Ellis and Hartley (1971) failed to obtain any inhibition of the development of phosphoribulokinase activity by lincomycin in greening shoots of pea and consequently concluded that this enzyme was not synthesized on 70 s ribosomes.

It seems to be of particular interest to note that both D-*threo* and L-*threo* chloramphenicol totally inhibited the appearance of cytochrome b-559$_{HP}$. We have expressed the opinion (Gregory and Bradbeer, 1972) that the high potential of this component may be a reflection of the site which it occupies in the photosynthetic membranes, possible in the "compartment" between adjacent thylakoids. Chloramphenicol may react unspecifically in preventing the cytochrome from attaining this site.

A number of enzyme activities have been found to develop normally in greening bean leaves in the presence of D-*threo* chloramphenicol, see Table IV, as a result of which they are considered to be synthesized on 80 s ribosomes. Some of these enzymes are considered to occur as distinct chloroplastic and cytoplasmic proteins, of which a number of separations have been reported by the application of isoelectrophoresis to pea leaf extracts (Anderson and Advani, 1970; Anderson, 1971a, b). Only the chloroplastic and cytoplasmic components of phosphoglycerate kinase from bean leaf extracts have so far been separated by this procedure and from preliminary results it appears that the chloroplast enzyme is not synthesized on 70 s ribosomes. No success has so far been achieved for the separation of the fructosediphosphate adolase, ribosephosphate isomerase and triosephosphate isomerase of bean leaf extracts. Table IV, which lists the bean leaf enzymes and proteins whose synthesis may be attributed to 70 s or 80 s ribosomes, may be compared with similar lists compiled from the literature (Boulter et al., 1972). As a number of chloroplast proteins are apparently synthesized outside the chloroplasts it may be inferred that mechanisms exist for the entry of the proteins into the chloroplasts.

VI. THE ROLE OF LIGHT IN ENZYME SYNTHESIS

In bean the initial stages of chloroplast development are wholly dependent on the reserve materials of the etiolated plant, but when the developing plastid becomes capable of photosynthesis its own photosynthetic activity will presumably play a part in the later stages of development (Bradbeer, 1969). However it is also known that plastid and leaf development considerably in excess of what is found in continuous darkness can be induced by light treatments which are insufficient to promote photosynthesis (Bradbeer et al., 1970; Graham et al., 1971; Bradbeer, 1971). The demonstration by Downs (1955) that leaf expansion is a photomorphogenetic response under the control of phytochrome was achieved by daily exposure of leaf material, which was otherwise maintained in the dark, to short treatments with red and far-red light. Graham et al. (1971) used this technique to study growth and plastid development in the stem apices of etiolated pea plants. They found that both growth and the increase in activity of the chloroplast enzymes were promoted by the far-red absorbing form of phytochrome (P-730) and that the magnitude of the response shown by the chloroplast enzymes was in excess of the growth response. However as the P-730 growth response of pea apices includes considerable cell division and cell expansion it would appear to be a quite complex system. A somewhat simpler system has been studied by the exposure of 12-day-old dark-grown beans to short daily illumination treatments (Bradbeer, 1971). In this case the dark development of the primary leaves with respect to the studied parameters of growth, including cell number and etioplast number, is virtually complete (see Figs 1, 4 and 7) and the light-induced cell division is very much less than that found in pea. P-730 seemed to be responsible for increases in the activities of a number of chloroplast enzymes but these increases were paralleled by the promotion of leaf growth and protein synthesis. It was concluded that P-730

TABLE V

The chloroplast enzymes whose formation is controlled
by phytochrome-730

Enzymes	References
Ribulosediphosphate carboxylase	3, 4, 6
Phosphoglycerate kinase	6
Triosephosphate dehydrogenase	1, 2, 4, 5, 6
Fructosediphosphate aldolase	6
Transketolase	3, 6
Ferrodoxin:NADP reductase	7
Alkaline fructosediphosphatase	4

1. Marcus (1960); 2. Margulies (1965); 3. Feierabend and Pirson (1966); 4. Graham, Grieve and Smillie (1968 and 1971); 5. Klein (1969); 6. Bradbeer (1971); 7. B. G. Haslett (personal communication).

was responsible for the photomorphogenetic response at the level of the whole cell and no evidence could be found for any direct effect on the synthesis of chloroplast enzymes. Other workers have reported that the activities of photosynthetic enzymes are promoted in response to P-730 formation, details about which are given in Table V. Our short-light treatments of bean leaves have also given some evidence of a stimulation of protein synthesis and a promotion of the activities of ribosephosphate isomerase, transketolase, and NADH- and NADPH-dependent triosephosphate dehydrogenase when exposures to red and blue light were given successively. There was also some evidence of the presence in beans of a red-absorbing photoreceptor, other than phytochrome, which caused the development of increased activities of ribosephosphate isomerase, NADPH-dependent triosephosphate dehydrogenase, transketolase and phosphopyruvate carboxylase. However, for neither of these photoreceptors is there evidence that their site of action may be in the chloroplasts. Of course the photoconversion of protochlorophyllide to chlorophyllide is undoubtedly a chloroplast reaction and investigations of the light induced changes of etioplast fine structure indicate that other developmental responses may be controlled by photoreceptors in the plastids (Henningsen, 1967; J. W. Bradbeer et al., in preparation).

VII. SUMMARY

The development of chloroplasts in the growing primary leaf of *Phaseolus vulgaris* is described together with the accompanying increases in enzyme activity. It is probable that these increases in enzyme activity largely result from enzyme synthesis and on the basis of this assumption the rates of synthesis of a number of plastid enzymes and proteins have been determined on a per plastid basis.

Experiments depending on the stereospecificity of chloramphenicol as an inhibitor of protein synthesis on 70 s ribosomes have established that phosphoribulokinase, the larger subunit of ribulosediphosphate carboxylase, at least one essential structural component of the thylakoids and the plastid cytochromes are synthesized within the plastids. A number of chloroplast enzymes are apparently synthesized by cytoplasmic ribosomes, thus raising the question of the way that these proteins enter the plastid. The review is concluded with a short discussion of the control of chloroplast enzyme synthesis by light.

NOTE ADDED IN PROOF
Further investigations have indicated that phosphoribulokinase is not synthesized on 70 s ribosomes.

ACKNOWLEDGEMENTS

I should like to acknowledge the contributions of my coworkers and assistants who have participated in this work and who have allowed me to quote some unpublished data. They are Mr G. Arron, Miss F. Cocker, Mr H. J. W. Edge, Mr P. Gregory, Mr B. G. Haslett, Dr H. M. M. Ireland, Mr K. G. Maybury, Dr D. R. Murray, Miss L. Neville, Miss J. W. Smith and Mr O. Wara-Aswapati.

REFERENCES

Anderson, L. E. (1971a). *Biochim. biophys. Acta* **235**, 237.
Anderson, L. E. (1971b). *Biochim. biophys. Acta* **235**, 245.
Anderson, L. E. and Advani, V. R. (1970). *Pl. Physiol., Lancaster* **45**, 583.
Bendall, D. S. (1968). *Biochem. J.* **109**, 46P.
Bendall, D. S., Davenport, H. E. and Hill, R. (1971). *In* "Methods in Enzymology" (A. San Pietro, ed.), Vol. 23, p. 327. Academic Press, New York and London.
Blair, G. E. and Ellis, R. J. (1972). *Biochem. J.* **127**, 42P.
Boardman, N. K. (1968). *In* "Comparative Biochemistry and Biophysics of Photosynthesis" (K. Shibata, A. Takamiya, A. T. Jagendorf and R. C. Fuller, eds), p. 206. University of Tokyo Press, Tokyo.
Boulter, D., Ellis, R. J. and Yarwood, A. (1972). *Biol. Rev.* **47**, 113.

Bradbeer, J. W. (1969). *New Phytol.* **68**, 233.

Bradbeer, J. W. (1970). *New Phytol.* **69**, 635.

Bradbeer, J. W. (1971). *J. exp. Bot.* **22**, 382.

Bradbeer, J. W., Clijsters, H., Gyldenholm, A. O. and Edge, H. J. W. (1970). *J. exp. Bot.* **21**, 525.

Bradbeer, J. W., Ireland, H. M. M., Gyldenholm, A. O., Haslett, B. G., Murray, D. R. and Whatley, F. R. (1972). *In* "Proceedings of the Second International Congress on Photosynthesis Research" (G. Forti, M. Avron and A. Melandri, eds), Vol. 3, Dr W. Junk, N.V., The Hague. In press.

Buchanan, B. B., Schürmann, P. and Kalberer, P. P. (1971). *J. biol. Chem.* **246**, 5952.

Criddle, R. S., Dan, B., Kleinkopf, G. E. and Huffaker, R. C. (1970). *Biochem. biophys Res. Commun.* **41**, 621.

Dale, J. E. and Murray, D. (1969). *Proc. R. Soc. B.* **173**, 541.

Downs, J. (1955). *Pl. Physiol., Lancaster* **30**, 468.

Ellis, R. J. (1969). *Science, N.Y.* **163**, 477.

Ellis, R. J. and Hartley, M. R. (1971). *Nature, New Biology* **233**, 193.

Ellis, R. J. and MacDonald, I. R. (1970). *Pl. Physiol., Lancaster* **46**, 227.

Feierabend, J. and Pirson, A. (1966). *Z. Pfl. Physiol.* **55**, 235.

Filner, P. and Varner, J. E. (1967). *Proc. natn. Acad. Sci. U.S.A.* **58**, 1520.

Graham, D., Grieve, A. M. and Smillie, R. M. (1968). *Nature, Lond.* **218**, 89.

Graham, D., Grieve, A. M. and Smillie, R. M. (1971). *Phytochemistry* **10**, 2905.

Gregory, P. and Bradbeer, J. W. (1972). *Planta* (In press).

Gyldenholm, A. O. (1968). *Hereditas* **59**, 142.

Gyldenholm, A. O., Palmer, J. M. and Whatley, F. R. (1971). *In* "Energy Transduction in Respiration and Photosynthesis" (E. Quagliarcello, S. Papa and C. S. Rossi, eds), p. 369. Adriatica Editrice, Bari.

Henningsen, K. W. (1967). *In* "Biochemistry of Chloroplasts" (T. W. Goodwin, ed.), Vol. 2, p. 453. Academic Press, London and New York.

Ireland, H. M. M. (1971). Ph.D. Thesis, University of London.

Ireland, H. M. M. and Bradbeer, J. W. (1971). *Planta* **96**, 254.

Kawashima, N. (1970). *Biochem. biophys. Res. Commun.* **38**, 119.

Kirk, J. T. O. and Tilney-Bassett, R. A. E. (1967). "The Plastids", p. 63. W. H. Freeman & Co. London and San Francisco.

Klein, A. O. (1969). *Pl. Physiol., Lancaster* **44**, 897.

Kleinkopf, G. E., Huffaker, R. C. and Matheson, A. (1970). *Pl. Physiol., Lancaster* **46**, 416.

Machold, O. (1971). *Biochim. biophys. Acta* **238**, 324.

Marcus, A. (1960). *Pl. Physiol., Lancaster* **35**, 126.

Margulies, M. M. (1965). *Pl. Physiol., Lancaster* **40**, 57.

Muhlethaler, K. (1971). *In* "The Structure and Function of Chloroplasts" (M. Gibbs, ed.), p. 7. Springer Verlag, Berlin, Heidelberg and New York.

Müller, B. (1970). *Biochim. biophys. Acta* **205**, 102.

Müller, B., Ziegler, I. and Ziegler, H. (1969). *Europ. J. Biochem.* **9**, 101.

Park, R. B. and Biggins, J. (1964). *Science, N.Y.* **144**, 1009.

Park, R. B. and Pon, N. G. (1963). *J. molec. Biol.* **6**, 105.

Smillie, R. M., Bishop, D. G., Gibbons, G. C., Graham, D., Grieve, A. M. Raison, J. K. and Reger, B. J. (1971). *In* "Autonomy and Biogenesis of Mitochondria and Chloroplasts" (N. K. Boardman, A. W. Linnane and R. M. Smillie, eds), p. 422. North Holland.

Wallis, M. E. and Bradbeer, J. W. (1970). *J. exp. Bot.* **21**, 1039.

Wara-Aswapati, O. (1971). M.Sc. Dissertation, University of London.

Weier, T. E. and Brown, D. L. (1970). *Am. J. Bot.* **57**, 267.

Weier, T. E., Sjoland, R. D. and Brown, D. L. (1970). *Am. J. Bot.* **57**, 276.

CHAPTER 12

Regulatory Mechanisms in the Photocontrol of Flavonoid Biosynthesis

HARRY SMITH

Department of Physiology and Environmental Studies,
University of Nottingham, Nottingham, England

I. INTRODUCTION

Flavonoids (Fig. 1) are rather esoteric substances which, as far as we know, have no physiological functions within plant cells except in certain cases where they act as floral pigments or as protective agents against fungal attack. It is rather surprising, therefore, that the biosynthesis of flavonoids is under very tight control both endogenously by end-products and a wide range of hormonal substances, and exogenously, by a variety of environmental factors including light, photoperiod, temperature and stress conditions. The wealth of factors known to affect flavonoid biosynthesis in one way or another, has been responsible for a minor explosion of research in recent years directed towards an understanding of the regulatory mechanisms involved. Much of this interest has been centred on the mechanisms of light action and considerable progress has been achieved. This article is not intended to catalogue this progress in detail nor is it a comprehensive review of the subject; it is to a large extent a cautionary tale to illustrate the pitfalls that accompany an uncritical approach and a too ready acceptance of current dogma.

The majority of investigations into the photocontrol of flavonoid biosynthesis have stemmed from an interest in the mode of action of the photoreceptors involved, principally phytochrome. Although the nature and mechanisms of the photoreceptors are strictly beyond the scope of this volume, it is nevertheless relevant to consider the dominant hypotheses of their action in view of the effect these ideas have had on the progress of our understanding of regulatory mechanisms in flavonoid biosynthesis. The only photoreceptor whose mechanism of action can even be guessed at is phytochrome, and thus this discussion will be restricted to phytochrome, even though at least one other, blue-absorbing photoreceptor is known to regulate flavonoid synthesis in many plants (for a detailed review of this topic see Smith, 1972).

The two major hypotheses of phytochrome action are (i) that it acts to selectively derepress or repress specific genes (Mohr, 1966); and (ii) that phytochrome in some way regulates the permeability or other properties of certain critical cell membranes (Hendricks and Borthwick, 1967). These two hypotheses are not mutually exclusive and it has been suggested by several workers that a primary effect on membrane properties could lead to the release of important metabolites from storage compartments which could then interact, either directly or indirectly, with the genome and lead to derepression and repression of specific genes (Smith, 1970). Unfortunately, however, it seems that the elegance of this view has from time to time led to the premature conclusion that evidence, which is consistent with the idea that light stimulates *de novo* enzyme synthesis, therefore proves it to be the case. This article will show, on the contrary, there is as yet not a single piece of evidence which unequivocally proves that light regulates flavonoid biosynthesis through the stimulation of *de novo* enzyme synthesis.

II. The Locus of the Light-regulated Steps in the Biosynthetic Pathway

The biosynthesis of the flavonoids has been of great interest to the plant biochemist because the two aromatic rings in the final products are derived from separate biosynthetic routes; the B ring (Fig. 1) and the associated three carbon atoms of the bridge are formed via phenylalanine from the shikimic acid pathway, whilst the A ring is synthesized formally by the head-to-tail condensation of three acetate units. The details of the later stages of the two pathways are shown in Fig. 2. From the regulatory point of view, this method of synthesis is fascinating, since it seems inevitable that plants will have evolved some mechanism for integrating the rates of synthesis of the immediate flavonoid precursors along the two separate pathways; it would be interesting to know, for example, whether intermediates of one pathway exert any regulatory influence over the enzymes of the other pathway. As yet, this intriguing problem is completely untouched.

From the point of view of photocontrol, it is obviously of importance to know whether light has effects on both pathways, or only on one, and at which step or steps in the pathways its control is exerted. The two simplest ways to approach this question are (i) precursor incorporation experiments and (ii) enzyme studies.

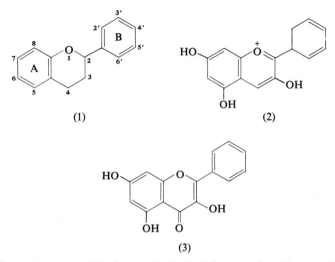

FIG. 1. Chemical structure of the flavonoids. The basic flavan nucleus (1); the anthocyanidin structure (2); the flavonol structure (3).

A. PRECURSOR INCORPORATION EVIDENCE

Watkin et al. (1957) showed that light depressed the incorporation of [14C]acetate into quercetin in buckwheat (*Fagopyrum esculentum*) seedlings, even though overall quercetin synthesis was increased. On the other hand, Swain (1960) reported that light increased the incorporation of [14C]phenyl-alanine into quercetin in tobacco (*Nicotiana tabacum*). Consistent with these findings, Harper and Smith (1969) and Harper et al. (1970) showed that of the precursors phenylalanine, cinnamate, p-coumarate and malonate, light only stimulated the incorporation of [14C] from [14C]phenylalanine into the flavonoids of pea (*Pisum sativum*) seedlings (Table I).

These findings, therefore, although minimal in extent, indicate that light does not act on both pathways, and that its major effect is to accelerate the conversion of phenylalanine to cinnamate. When it is considered that phenyl-alanine lies at the hub of a wide range of biosynthetic reactions, and is the branch point of flavonoid biosynthesis from the major pathway of protein synthesis, it seems eminently sensible that light should act at this point. On the other hand, evidence from enzyme studies appears to indicate that an effect of light on a single step in the pathway is unlikely.

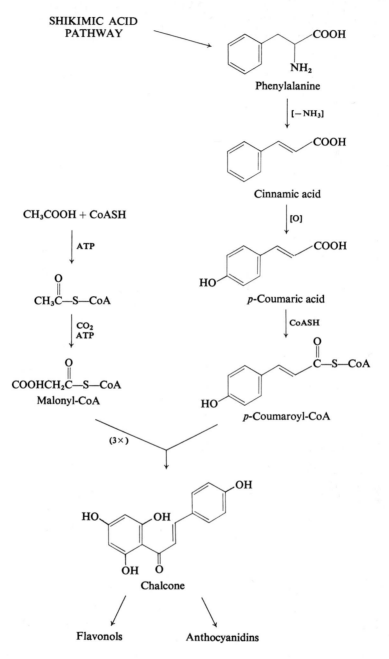

FIG. 2. The biosynthetic pathway of the flavonoids.

TABLE I

Effect of red light alone, and followed by far-red light, on the incorporation of isotopic precursors into the total flavonoids of the terminal buds of pea seedlings

[¹⁴C] Precursor	Radioactivity in total flavonoids as % total, methanol-extractable radioactivity		
	Dark	Red	Red/Far-red
Phenylalanine	5·9	21·4	7·2
Cinnamate	4·2	1·7	2·5
p-Coumarate	2·2	3·5	1·9
Malonate	13·7	4·9	10·2

B. ENZYME STUDIES

A very large number of investigations have been carried out on the effects of light on the extractable activities of many enzymes of flavonoid biosynthesis (Table II). It is quite clear from these investigations that the activities of the enzymes responsible for the conversion of phenylalanine into the B-ring precursors, i.e. phenylalanine ammonia-lyase (PAL), cinnamate hydroxylase, and p-coumarate:CoA ligase, are in all cases increased by light treatment. This conclusion is particularly convincing for PAL which has been studied in great detail, as will be seen. Furthermore, certain of the enzymes responsible for the modification and glycosylation of the C_{15} skeleton also appear to be affected by light, as do two enzymes of the shikimic acid pathway. In the latter cases however, both species differences and conflicting reports exist. In the two plants in which an enzyme of A-ring biosynthesis (acetate:CoA ligase) has been studied, light had no effect on the extractable activities.

The findings with the enzyme studies, therefore, reinforce the view that photocontrol is exerted on the B-ring pathway only. (It should be borne in mind, though, that only very little work has been concentrated on the A-ring pathway and thus this conclusion may well be premature.) It also seems likely from the evidence in the literature, that light does not act by regulating only one pacemaker enzyme, but that it operates either sequentially or co-ordinately to regulate a whole series of enzymes in the B-ring pathway and in the pathways beyond the C_{15} stage. However, since of all of these enzymes, only PAL has received more than cursory attention, further discussion will necessarily be restricted to PAL.

III. THE PHOTOCONTROL OF PAL LEVELS

The response of PAL levels to light treatment has been studied in great detail in a variety of species including potato (*Solanum tuberosum*) (Zucker, 1965), mustard (*Sinapis alba*) (Durst and Mohr, 1966), gherkin (*Cucumis sativus*)

TABLE II

Enzymes of flavonoid biosynthesis whose level of activity is affected by treatment of the plants with light

Region of pathway	Enzyme	Plant	Effect of	Reference
Shikimic acid pathway	5-Dehydroquinase	Mung bean (*Phaseolus aureus*)	No effect	Ahmed and Swain (1970)
	5-Dehydroquinase	Pea	Stimulation	Ahmed and Swain (1970)
	Shikimate:NADP oxidoreductase	Mung bean	No effect	Ahmed and Swain (1970)
	Shikimate:NADP oxidoreductase	Pea	Stimulation	Ahmed and Swain (1970)
	Shikimate:NADP oxidoreductase	Pea	No effect	Attridge and Smith (1967)
B-ring synthesis	Phenylalanine ammonia-lyase (PAL)	Potato	Stimulation	Zucker (1965)
	PAL	Gherkin	Stimulation	Engelsma (1967a)
	PAL	Buckwheat	Stimulation	Scherf and Zenk (1967)
	PAL	Mustard	Stimulation	Durst and Mohr (1966)
	PAL	Pea	Stimulation	Attridge and Smith (1967)
	PAL	Artichoke (*Helianthus tuberosus*)	Stimulation	Nitsch and Nitsch (1966)
	PAL	Cocklebur (*Xanthium strumatium*)	Stimulation	Zucker (1969)
	PAL	Strawberry (*Fragaria X* spp.)	Stimulation	Creasy (1968)

	Enzyme	Plant	Effect	Reference
B-ring synthesis (cont.)	PAL	Parsley (*Petroselinum crispum*)	Stimulation	Hahlbrock and Wellman (1970)
	PAL	Mung bean	Stimulation	Ahmed and Swain (1970)
	PAL	Red cabbage	Stimulation	Engelsma (1970a)
	PAL	Radish (*Raphanus sativus*)	Stimulation	Bellini and Van Poucke (1970)
	Cinnamate hydroxylase	Pea	Stimulation	Russell and Conn (1967)
	Cinnamate hydroxylase	Buckwheat	Stimulation	Amrhein and Zenk (1970)
	Cinnamate hydroxylase	Soybean	Stimulation	Hahlbrock et al. (1971)
	p-Coumarate:coenzyme A ligase	Parsley	Stimulation	Hahlbrock and Grisebach (1970)
	p-Coumarate:coenzyme A ligase	Soybean	Stimulation	Hahlbrock et al. (1971)
A-ring synthesis	Acetate:coenzyme A ligase	Parsley	No effect	Hahlbrock and Grisebach (1970)
	Acetate:coenzyme A ligase	Soybean	No effect	Hahlbrock et al. (1971)
Flavonoid modification	Chalcone-flavanone isomerase	Parsley	Stimulation[a]	Hahlbrock et al. (1971)
	UDP-apiose synthetase	Parsley	Stimulation	Hahlbrock and Wellman (1970)
Glycosylation	Apiosyl-transferase	Parsley	Stimulation[a]	Hahlbrock et al. (1971)
	Glucosyl-transferase	Parsley	Stimulation[a]	Hahlbrock et al. (1971)

[a] In these cases, the experiments were not designed to show whether light affected enzyme levels, but the results indicate that a stimulatory effect of light is likely.

(Engelsma, 1967a, b), pea (Attridge and Smith, 1967), cocklebur (*Xanthium strumatium*) (Zucker, 1969) and red cabbage (*Brassica oleracea*) (Engelsma, 1970a). In almost all cases, the increase in PAL levels brought about by light treatment does not occur immediately upon irradiation and is only transient in character. A typical time-course, for gherkin, is shown in Fig. 3 and illustrates the three phases of the response: (i) a lag phase which is in nearly all cases very close to 90 min; (ii) a phase of initially linear increase in enzyme activity; and (iii) a phase of rapid loss of activity. In some plants, (e.g. peas) the level of enzyme activity ultimately stabilizes at a new level somewhat higher than the dark controls (when expressed in terms of enzyme activity per mg protein)

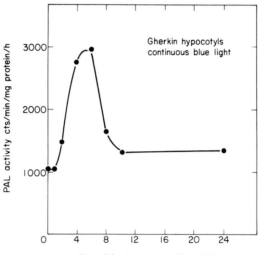

FIG. 3. Time course of the blue-light mediated change in PAL levels in gherkin hypocotyls (data of Dr T. H. Attridge).

whilst in other plants, e.g. gherkin, the activity falls back to the level present before irradiation.

If we are to fully understand the photocontrol of flavonoid biosynthesis, it is necessary to obtain information on all three of the phases of the typical response. It should be stated immediately that unfortunately nothing is known as yet concerning the processes going on during the lag phase, the most interesting part of the time course as regards the mode of action of the photoreceptor. We are, therefore, left with attempting to understand the molecular mechanisms underlying the light-mediated rise and fall in PAL levels.

A. THE INCREASE IN ENZYME ACTIVITY

There are three principal ways in which an increase in extractable enzyme activity may be achieved:

(i) activation or conversion of pre-existing non-active enzyme protein;
(ii) stimulation of the rate of *de novo* enzyme synthesis;
(iii) depression of the rate of enzyme degradation or inactivation.

These processes are not mutually exclusive and therefore it is conceivable that increases in PAL may be due to combinations of two or all of the above effects. We must hope that this does not prove to be the case since it is difficult enough to devise experiments to determine unequivocally which of these processes is acting, assuming one is acting in isolation; if it becomes necessary to determine the extent to which two or more of these processes contribute to the overall response, our biochemical ingenuity will indeed be taxed.

1. *Activation or Conversion of Pre-existing Inactive Enzyme*

There is little direct evidence to support the view that light-activation or light-conversion of inactive PAL to active PAL may occur, although some data obtained by Dr T. H. Attridge with etiolated pea seedlings is of interest in this context. PAL from peas is extremely sensitive to inhibition by end-products of the flavonoid biosynthetic pathway; in particular quercetin, a major flavonol of pea seedlings, exerts 50% inhibition of PAL activity at 7×10^{-5} M (Fig. 4). Figure 4 also shows, incidentally that PAL can be partially desensitized to quercetin inhibition by incubation with mercuric chloride. This result, together with the results from other desensitizing agents, suggests that quercetain may act allosterically as a feed-back inhibitor of PAL in peas (Attridge, *et al.*, 1971).

When PAL from dark-grown and light-treated pea seedlings is compared, striking differences in properties are observed between the two preparations (unpublished work of Dr T. H. Attridge). For example, the sensitivity of the

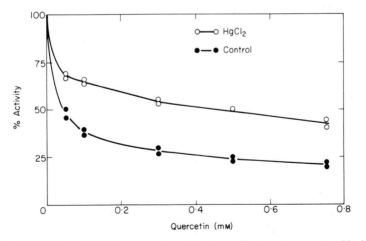

FIG. 4. Inhibition of PAL by quercetin and its desensitization by mercuric chloride.

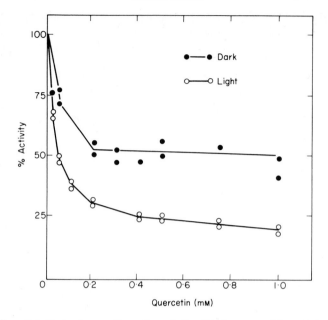

FIG. 5. Differential effects of quercetin on the activity of PAL extracted from dark-grown and light treated plants.

"light-enzyme" to quercetin inhibition is very much greater than that of the "dark-enzyme" (Fig. 5). Furthermore, the "dark-enzyme" is considerably more stable in the absence of sulphydryl reagents than is the "light-enzyme" (Table III). Moreover, there is no evidence for more than one species of PAL in pea seedlings and thus these results cannot be explained on the basis of different isoenzymes formed in dark- and light-grown plants. These findings have not yet been confirmed by similar observations with PAL from other plants and so must remain of questionable significance; nevertheless, they do indicate that at least in one plant light-mediated changes in the properties of PAL occur in such a manner that a more active enzyme results.

TABLE III

Differential stability of PAL from dark and light-treated pea seedlings. Extracts were prepared with 1 mM glutathione and were dialysed for 18 h either against extraction buffer, or extraction buffer plus glutathione.

	PAL Activity as % of initial value after 18 h dialysis		
Treatment	Initial	Against buffer	Against buffer + glutathione
Dark	100	75·7	96·5
Light	100	14·1	104·1

2. *Stimulation of Synthesis or Depression of Inactivation?*

Much of the work carried out in attempts to answer this question has used suspect methodology and led to published conclusions unsupported by the experimental data. As will be seen, however, even the use of accepted techniques for assessing the involvement of *de novo* enzyme synthesis has not enabled an unequivocal answer to be given.

Many workers have shown that various inhibitors of protein synthesis will largely prevent the light-mediated increases in PAL levels and in several cases it has been concluded from these results that light "switches-on" the synthesis of PAL. This type of evidence has been used, together with inhibitory effects of actinomycin D on anthocyanin synthesis, to conclude that light acts by selectively switching-on the transcription of the mRNA for PAL (amongst others) and thus regulating flavonoid biosynthesis at the level of the genome (Lange and Mohr, 1965; Schopfer, 1967). Setting aside the well-known problems of the lack of specificity of protein and nucleic acid synthesis inhibitors, this conclusion is not necessarily valid even if the inhibitors are acting as expected. As will be seen below there is considerable evidence that PAL is synthesized in dark-grown tissues and immediately inactivated, although probably not immediately degraded. Protein synthesis inhibitors, therefore, would prevent a light effect whether the light was acting to stimulate synthesis, or to inhibit inactivation.

For many reasons therefore protein synthesis inhibitors do not produce unequivocal evidence on the question under consideration. At present, there are only three methods which can be reliably used to demonstrate the *de novo* synthesis of an enzyme protein: (i) radioactive labelling; (ii) density labelling; and (iii) immunology.

Immunology has not yet been used, to the author's knowledge, in the study of PAL, although it seems likely to be the only technique that can offer a final solution to this problem.

Radioactive labelling has been used by Zucker (1969, 1970, 1971) to show that *de novo* synthesis of PAL occurs both in light-treated and dark-treated *Xanthium* leaf discs. In these experiments, PAL was purified to a single protein (as shown by starch-gel electrophoresis) in order to diminish the contribution of label from other proteins; however, it was not shown that radioactivity was distributed randomly throughout the molecule as would be expected for *de novo* synthesis. Thus, although this work provides strong evidence for *de novo* synthesis, it is not yet final proof. Furthermore, it is significant that on the basis of this data, PAL appears to be synthesized as rapidly during darkness as during the light period. There is, therefore, no support here for the view that light switches-on *de novo* synthesis, or even stimulates its rate.

The density-labelling method, on the other hand, has been used to provide evidence in support of the view that light acts by switching-on PAL synthesis. Schopfer and Hock (1971) grew mustard seedlings in 80 % D_2O from imbibition of the dry seed for 85 h in darkness. They were then irradiated for 30 h following which PAL was extracted and spun to equilibrium on a caesium chloride gradient. A similar control series was grown in 100% H_2O and comparisons were made of the buoyant density of the PAL in each case. The enzyme from the D_2O grown plants had a significantly higher buoyant density than that from the water series, indicating that PAL had been synthesized at some time during the exposure to D_2O. Since activity only appeared upon irradiation, the authors' concluded that synthesis occurred during the light period. In the opinion of this reviewer this could well be an erroneous conclusion since it is quite possible that PAL was formed in an inactive state during dark-growth and activated by light.

Density-labelling experiments carried out in this laboratory by Miss Susan Iredale tend to support this view and, furthermore, indicate that long-term incubation of tissue with D_2O has drastic effects on the ability of tissues to respond to light in the normal way. Using gherkin seedlings which respond to light treatment by a rapid, but transient increase in PAL levels (Engelsma, 1967a) the following experiments were carried out:

   (i) Seeds were imbibed in 60% D_2O and held for 8 days in darkness, irradiated for 4 h and PAL extracted;

  (ii) Seeds were imbibed in H_2O and held for 3 days in darkness, irradiated for 4 h and PAL extracted;

 (iii) Seeds were imbibed in 60% D_2O for 6 days, transferred to H_2O for 2 days, irradiated for 4 h and PAL extracted;

 (iv) Seeds were imbibed in 60% D_2O for 7·5 days, transferred to H_2O for 12 h, irradiated for 4 h and PAL extracted;

  (v) Seeds were imbibed in 60% D_2O for 7·75 days, transferred to H_2O for 6 h, irradiated for 4 h and PAL extracted;

 (vi) Seeds were imbibed in D_2O for 8 days, transferred to H_2O, irradiated for 4 h and PAL extracted;

(vii) Seeds were imbibed in H_2O for 3 days, transferred to D_2O, irradiated for 4 h and PAL extracted.

At all transfer points, seedlings were rinsed and vacuum infiltrated.

Representative density gradient fractionations are shown in Fig. 6 and the full results are given in Table IV. It is quite clear that where transfer is from H_2O to D_2O at the time of light treatment, the enzyme has a high buoyant density relative to the H_2O controls. This implies that synthesis of PAL is occurring during the period of light treatment. On the other hand, when transfer is from D_2O to H_2O at the time of light treatment, the enzyme extracted also has a high buoyant density. It is necessary to maintain the plants in H_2O for several hours after the cessation of D_2O treatment for the buoyant density of the extracted enzyme to reach the same low level as that from the H_2O control.

These results illustrate the severe difficulties attending the interpretation of

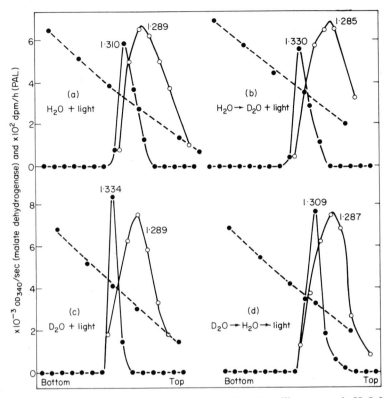

Fig. 6. Density labelling of PAL in gherkin hypocotyls. (a) seedlings grown in H_2O for 3 d, and irradiated for 4 h; (b) seedlings grown in H_2O for 3 d, transferred to D_2O and irradiated for 4 h; (c) seedlings grown in 60% D_2O for 8 d and irradiated for 4 h; (d) seedlings grown in D_2O for 6 d, transferred to H_2O for 2 d and irradiated for 4 h. PAL (●); marker malate dehydrogenase (○); density gradient (----). The figures at the peaks represent the buoyant densities (unpublished data of Miss S. E. Iredale).

D_2O-labelling experiments on fully-imbibed seedlings. It is quite clear that a significant carry-over of the D_2O effect occurs which lasts for up to 6 h after the transfer to H_2O. It would seem likely that after transfer to H_2O, a high pool of deuterated amino acids are left in the cells, thus allowing the synthesis of heavy enzyme and producing an unexpected result. That an analogous carry-over effect is not seen in the transfer from H_2O to D_2O may be due to a much smaller pool size of amino acids in H_2O-grown tissue, in which protein synthesis is highly active, than in D_2O-grown tissue in which protein synthesis is markedly reduced.

Thus although the evidence is not conclusive, these results support the view that light does not act by causing an activation of preexisting PAL. On the other hand, it is still not certain whether light stimulates synthesis, or inhibits degradation or inactivation. All that can be said with a reasonable degree of certainty, is that PAL is synthesized during the light period. This, however,

TABLE IV
Summary of density labelling experiments on the light-induced increase in PAL. Treatments are those described in detail in the text. Malate dehydrogenase (MDH) was added to each tube to act as a density marker. Light treatment was for 4 h. Centrifugation in CsCl at 40000 rev/min for 72 h. Since the experiments were carried out on different occasions it is not possible to compare directly the buoyant density values for PAL between treatments; comparisons should be restricted to the $\Delta\rho$ values.[a]

| | Buoyant density | | |
Treatment	MDH	PAL	$\Delta\rho$
(i) D_2O, 8 d → light	1·278	1·328	0·050
(ii) H_2O, 3 d → light	1·289	1·310	0·021
(iii) D_2O, 6 d → H_2O, 2 d → light	1·287	1·309	0·022
(iv) D_2O, 7·5 d → H_2O, 0·5 d → light	1·298	1·323	0·025
(v) D_2O, 7·75 d → H_2O, 0·25 d → light	1·298	1·319	0·021
(vi) D_2O, 8 d → H_2O + light	1·257	1·313	0·056
(vii) H_2O, 3 d → D_2O + light	1·285	1·330	0·045

[a] (Unpublished work of Miss S. E. Iredale.)

is also true of the dark period, since it has recently been shown by density labelling that PAL is synthesized in dark grown gherkin seedlings (T. H. Attridge, unpublished results—see below).

B. THE DECLINE IN PAL LEVELS

The fall in PAL levels after the light-mediated peak has been reached, has been attributed to the action of a specific proteinaceous inactivator. In potato discs (Zucker, 1968) and gherkin hypocotyls (Engelsma, 1967b) application of cycloheximide at the point of maximum PAL activity almost completely prevents the subsequent fall and the authors' concluded that protein synthesis was essential for the decline in activity. Bearing in mind what has already been said concerning the use of protein synthesis inhibitors, it is unfortunate that the effect of such inhibitors represents the only evidence for the proteinaceous nature of the presumed PAL-inactivator. Furthermore, although considerable circumstantial evidence supports the existence of an inactivator, many attempts to isolate it have to date proved fruitless.

The indirect evidence for the inactivator can be summarized quite quickly; in gherkin, exposure to low temperature (4°C) for 24 h after the enzyme activity has fallen back to control level leads to a large second rise when transferred back to 25°C in darkness, this second rise not being prevented by 100 μg/ml cycloheximide, a concentration which almost completely prevents the first light-mediated increase (Engelsma, 1969, 1970b). Engelsma concludes that the PAL formed during the light period becomes complexed reversibly with a proteinaceous inactivator and this association is somehow affected by the low temperature treatment so that the complex falls apart upon returning to

25°C. Cycloheximide would not be expected to prevent the second rise under these circumstances since protein synthesis is not involved.

Work in this laboratory, although supporting this concept in general, suggests that a slight modification is necessary, since increases in PAL levels can be achieved by low-temperature transfer treatments given to seedlings grown in complete darkness (Fig. 7). (Unpublished work of Dr T. H. Attridge.)

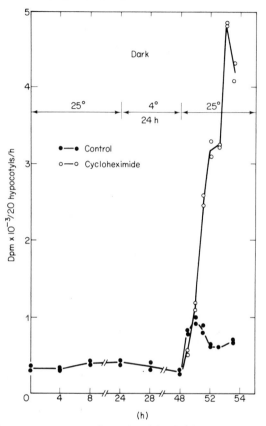

FIG. 7. Effect of low temperature transfer and cycloheximide treatment on extractable PAL levels (unpublished data of Dr T. H. Attridge).

Furthermore, cycloheximide treatment given at the time of transfer back to 25°C leads to a truly enormous rise in PAL (Fig. 7). This rise, which is consistently achieved, is up to eight times that obtained with light treatment. We tend to explain this unexpected result on the basis of continued synthesis and inactivation of PAL in dark-grown hypocotyls, followed by complete inhibition of PAL and inactivator synthesis in the presence of cycloheximide. If, under these circumstances or with the assistance of the low temperature treatment, the hypothetical complex of PAL and inactivator were to dissociate, and if the inactivator were less stable than PAL, then all available, previously non-active

PAL would become active. This concept is supported by the fact that cycloheximide at 100 μg/ml given at any time during the growth in darkness of
gherkin seedlings leads to a large increase in PAL activity within the subsequent
2 h (Table V). Furthermore, it is not necessary to use such high concentrations
of inhibitor; significant, but lower, increases are achieved with 1 and 10 μg/ml
and similar, though less marked, results are obtained with puromycin. These
results suggest that the increases are indeed due to protein synthesis inhibition,
although the actual effect on protein synthesis has not yet been measured.

TABLE V

Effects of cycloheximide, chloramphenicol and puromycin
on the level of PAL activity in gherkin hypocotyls of different
ages. Dark-grown seedlings were sprayed with the stated
concentrations of the protein synthesis inhibitors and PAL
extracted and assayed 4 h later. Two replicate experiments
are given in each case.[a]

	PAL activity—% of control		
	Day 3	Day 4	Day 5
Cycloheximide	457	507	650
100 μg/ml	507	590	585
Puromycin	102	186	136
100 μg/ml	160	167	139
Chloramphenicol	84	107	79
100 μg/ml	121	104	108

[a] (Unpublished data of Dr T. H. Attridge.)

It seems, therefore, that the PAL increases caused by temperature transfers
and/or cycloheximide occur as a result of activation, or more correctly
"de-inactivation". In order to test this view, density labelling experiments have
been carried out on these increases, seedlings being pretreated in H_2O and
transferred to D_2O with vacuum infiltration just prior to the treatment which
would yield the PAL increase. These results are shown in Fig. 8 and show quite
definitely that no shift in buoyant density is achieved in the presence of D_2O
under any treatment.

It is clear, therefore, that substantial increases in PAL levels can be brought
about under conditions in which protein synthesis does not, and probably
cannot, occur. The simplest explanation is that synthesis of PAL is continually
occurring in dark-grown tissues, but that the enzyme is also continually inactivated by the synthesis of the inactivator. The formation of the inactivator
appears to be in some way correlated with the level of PAL present and
Engelsma (1968) has elegantly shown, using excised hypocotyl segments, that
the products of PAL action, i.e. cinnamic acid and p-coumaric acid, are probably the agents which bring about the synthesis of the inactivator. All the data

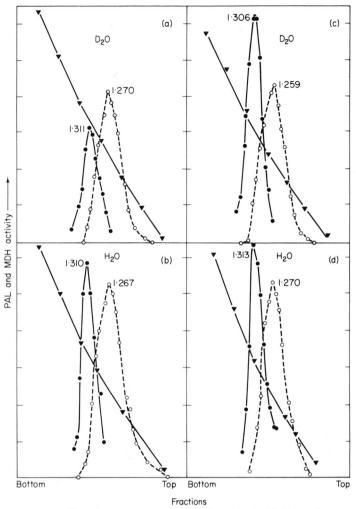

FIG. 8. Density labelling of the low temperature and cycloheximide induced increase in PAL. (a) and (b), low temperature treatment; (c) and (d) cycloheximide treatment. PAL (●); marker malate dehydrogenase (○·····○); density gradient (▲——▲). Figures at the peak represent buoyant densities (unpublished data of Dr T. H. Attridge).

for gherkin can at present be accommodated within the hypothetical scheme presented in Fig. 9 and efforts are presently being redoubled to attempt to isolate the presumed inactivator, determine its properties and assess how its level changes with environmental treatment.

IV. CONCLUSIONS

The only safe conclusions at this stage are negative, since it is clearly not possible yet to state with reasonable certainty at which point in a scheme like

Fig. 9 light is acting. The likely possibilities are (i) to stimulate the rate of PAL synthesis, or (ii) to depress the rate of inactivator synthesis. It is quite clear, however, that since PAL is being synthesized at significant rates in darkness it is no longer possible to envisage the role of light being to "switch-on" the synthesis of PAL and it therefore is unlikely that on/off control of mRNA transcription is involved.

If it is assumed that light operates to increase the rate of PAL synthesis as suggested as one of the possibilities here, it would be necessary to attempt to determine whether control of synthesis rate was being exerted at transcription

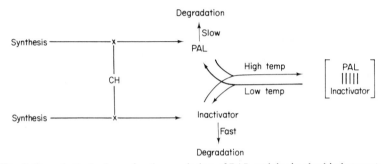

FIG. 9. Hypothetical scheme for the regulation of PAL activity in gherkin hypocotyls.

or translation. The only evidence on this point is the finding by Lange and Mohr (1965) that whereas puromycin inhibited the light-mediated increase in anthocyanin synthesis in mustard seedlings almost immediately upon application, actinomycin D had to be added some time before the onset of irradiation in order to severely inhibit the light-effect. This result could be taken to mean that the mRNA for the enzymes involved (presumably PAL) is present in dark-grown seedlings and its translation is stimulated by light; to prevent the light effect actinomycin D must be added sufficiently long before light treatment in order to allow time for existing mRNA molecules to be destroyed.

The evidence for the existence of a specific PAL-inactivator is not conclusive since it is only, as yet, circumstantial. There is, however, direct evidence for an inactivator of invertase in potatoes (Pressey and Shaw, 1966), and Marcus (1971) has recently suggested that specific enzyme inactivators could well represent an important regulatory mechanism in higher plant metabolism. Further investigation of the photocontrol of PAL levels is likely to provide enhanced insight into this intriguing concept.

REFERENCES

Ahmed, S. I. and Swain, T. (1970). *Phytochemistry* **9**, 2287.
Amrhein, N. and Zenk, M. H. (1970). *Naturwissenschaften* **57**, 312.
Attridge, T. H. and Smith, H. (1967). *Biochim. biophys. Acta*, **148**, 805.
Attridge, T. H., Stewart, G. R. and Smith, H. (1971). *FEBS Letters* **17**, 84.
Bellini, E. and Van Poucke, M. (1970). *Planta* **93**, 60.

Creasy, L. L. (1968). *Phytochemistry* **7**, 441.
Durst, F. and Mohr, H. (1966). *Naturwissenschaften* **53**, 707.
Engelsma, G. (1967a). *Planta* **75**, 207.
Engelsma, G. (1967b). *Naturwissenschaften* **54**, 319.
Engelsma, G. (1968). *Planta* **82**, 355.
Engelsma, G. (1969). *Naturwissenschaften* **56**, 563.
Engelsma, G. (1970a). *Acta bot. Neerl.* **19**, 403.
Engelsma, G. (1970b). *Planta* **91**, 246.
Hahlbrock, K. and Grisebach, H. (1970). *FEBS Letters* **11**, 62.
Hahlbrock, K., Kuhlen, E. and Lindl, T. (1971). *Planta* **99**, 311.
Hahlbrock, K., Sutter, A., Wellman, E., Ortmann, R. and Grisebach, H. (1971).
Phytochemistry **10**, 109.
Hahlbrock, K. and Wellman, E. (1970). *Planta* **94**, 236.
Harper, D. B., Austin, D. J. and Smith, H. (1970). *Phytochemistry* **9**, 497.
Harper, D. B. and Smith, H. (1969). *Biochim. Biophys. Acta* **184**, 230.
Hendricks, S. B. and Borthwick, H. W. (1967). *Proc. natn. Acad. Sci. U.S.A.* **58**,
2125.
Lange, H. and Mohr, H. (1965). *Planta* **67**, 107.
Marcus, A. (1971). *A. Rev. Pl. Physiol.* **22**, 313.
Mohr, H. (1966). *Photochem. Photobiol.* **5**, 469.
Nitsch, C. and Nitsch, J. P. (1966). *C. r. hebd. Séanc. Acad. Sci., Paris* **262**, 1102.
Pressey, R. and Shaw, R. (1966). *Pl. Physiol., Lancaster* **41**, 1657.
Russell, D. W. and Conn, E. E. (1967). *Archs Biochem. Biophys.* **122**, 256.
Scherf, H. and Zenk, M. H. (1967). *Z. Pflanzenphysiol.* **56**, 203.
Schopfer, P. (1967). *Planta* **74**, 210.
Schopfer, P. and Hock, B. (1971). *Planta* **96**, 248.
Smith, H. (1970). *Nature, Lond.* **227**, 665.
Smith, H. (1972). *In* "Phytochrome", Symposium of the N.A.T.O. Advanced Study
Institute, Eretria, Greece, 1971. Academic Press, New York and London.
Swain, T. (1960). *In* "Phenolics in Plants in Health and Disease" (J. B. Pridham,
ed.), p. 45. Pergamon Press, London.
Watkin, J. E., Underhill, E. W. and Neish, A. C. (1957). *Can. J. Biochem. Physiol.*
35, 229.
Zucker, M. (1965). *Pl. Physiol., Lancaster* **40**, 779.
Zucker, M. (1968). *Pl. Physiol., Lancaster* **43**, 365.
Zucker, M. (1969). *Pl. Physiol., Lancaster* **44**, 912.
Zucker, M. (1970). *Biochim. biophys. Acta* **208**, 331.
Zucker, M. (1971). *Pl. Physiol., Lancaster* **47**, 442.

CHAPTER 13

The Non-protein Amino Acids of Plants:
Concepts of Biosynthetic Control

L. FOWDEN

*Department of Botany and Microbiology,
University College, London, England**

I. INTRODUCTION

As each year passes, at least ten, and often considerably more, amino acids
are isolated from plants and characterized as natural products for the first time.
At present, the number of amino acids that have been extracted from plants
probably exceeds 200, and this number is further increased if amino acids of
animal and microbial antibiotic origin are included. It is convenient to desig-
nate these "newer" compounds as non-protein amino acids (Fowden, 1962),
thereby distinguishing them from the "coded twenty" that form the basis of
protein molecules. This designation also correctly describes a situation in
which these "extra" amino acids seem not to be incorporated into protein
molecules of species that produce them.

The non-protein amino acids exhibit many different types of chemical structure, but the majority can be assigned to categories frequently used to sub-divide
the 20 protein constituents. For example, many are either basic (diamino) or
acidic (dicarboxylic) amino acids, or their ω-N-alkyl or acyl or N(amido)-alkyl

* Address from April 1973: Director, Rothamsted Experimental Station, Harpenden,
Herts., England.

derivatives, respectively. There are also large groups of (i) imino acids (including many ring C-substituted prolines, but also compounds based on homologous azetidine, and piperidine saturated heterocyclic rings), and (ii) sulphur-containing amino acids (most of which may be regarded as S-substituted cysteines). Various types of substitution on the phenyl ring of phenylalanine produce a group of new aromatic amino acids, and these are augmented by other β-substituted alanines possessing N-, O-, or S-heterocyclic ring systems (including the pyridine, pyrazole, pyrimidine, pyrone and thiazole residues). Unsaturation (ethylenic or acetylenic) in open-chain amino acids is also common, whilst the number of naturally occurring amino acids known to contain a cyclopropyl group is steadily increasing. Detailed information about individual compounds assigned to these different structural groups may be found in a recent review (Fowden, 1970).

Most frequently, a compound is initially recognized as a "new" naturally-occurring amino or imino acid by examining unfractionated extracts of small amounts of plant material using chromatographic or electrophoretic methods. When such procedures are employed, compounds present in low concentration (often below the threshold levels of detection) are frequently overlooked and therefore rarely isolated and characterized. There are then probably hundreds of additional types of amino acids, elaborated by plants, that await identification; and it is equally probable that many, if not all, of the compounds so far characterized have a much wider distribution within the plant kingdom, albeit in low concentration, than is realized at present. This article selects aspects of work on non-protein amino acids, undertaken in my laboratory in recent years, that will amplify the statements and concepts introduced above, especially in relation to factors governing biosynthesis.

II. Amino Acids Characteristic of the Families Sapindaceae and Hippocastanaceae

In the period of 15 years following the isolation and characterization of hypoglycin A [β-(methylenecyclopropyl)alanine] as the toxic hypoglycaemic principle of unripe fruits of akee (*Blighia sapida*), a considerable group of unique amino acids have been identified as constituents of various plants assigned to either the Sapindaceae or Hippocastanaceae: it is interesting to recall that many taxonomists regarded these two families as one until about 1930. These compounds apparently fall into two main biogenetic groups, based on branched-chain skeletons having either 6 or 7 carbon atoms. The C_6 structures typically branch at the β-carbon atom, while the majority of the C_7 compounds branch at the γ-carbon (compounds 9 and 10 are exceptions to this generalization). Their structures are shown in Table I, and the species from which they have been isolated are shown in Table II. Some of the compounds, e.g. α-(methylenecyclopropyl)glycine (1), *trans*-α-(carboxycyclopropyl)glycine(2a)

TABLE I

The structures of amino acids and γ-glutamyl peptides isolated from various members of the families Sapindaceae and Hippocastanaceae

Compounds based on a C₆ skeleton

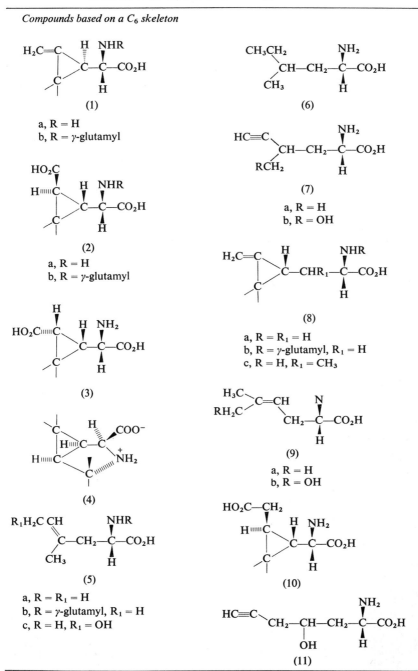

(1)

a, R = H
b, R = γ-glutamyl

(2)

a, R = H
b, R = γ-glutamyl

(3)

(4)

(5)

a, R = R₁ = H
b, R = γ-glutamyl, R₁ = H
c, R = H, R₁ = OH

(6)

(7)

a, R = H
b, R = OH

(8)

a, R = R₁ = H
b, R = γ-glutamyl, R₁ = H
c, R = H, R₁ = CH₃

(9)

a, R = H
b, R = OH

(10)

(11)

TABLE II

Species from which the C_6 and C_7 amino acids listed in Table I have been isolated

Amino acid[a]	Species of origin	Literature reference
1a	*Litchi chinensis*	Gray and Fowden, 1962
	Billia hippocastanum	Eloff and Fowden, 1970
1b	*Billia hippocastanum*	Eloff and Fowden, 1970
2a	*Blighia sapida*	Fowden et al., 1969
2b	*Blighia sapida*	Fowden and Smith, 1969
3	*Aesculus parviflora*	Fowden et al., 1969
	A. glabra	Fowden (unpublished)
4	*A. parviflora*	Fowden et al., 1969
5a	*A. californica*	Fowden and Smith, 1968
5b	*A. californica*	Fowden and Smith, 1968
5c	*A. californica*	Fowden and Smith, 1968
6	*A. californica*	Fowden and Smith, 1968
7a	*Euphoria longan*	Sung et al., 1969
7b	*E. longan*	Sung et al., 1969
8a	*Blighia sapida*	Ellington et al., 1959
	Billia hippocastanum	Eloff and Fowden, 1970
8b	*Blighia sapida*	Ellington et al., 1959
	Billia hippocastanum	Eloff and Fowden, 1970
8c	*A. californica*	Fowden and Smith, 1968
9a	*Leucocortinarius bulbiger*	Dardenne et al., 1968
9b	*Blighia unijugata*	Fowden et al., 1972
10	*Blighia unijugata*	Fowden et al., 1972

[a] The amino acids are designated by the numerals used in Table I.

and probably hypoglycin A (8a), are synthesized by certain species from each family, but at present others are known only as constituents of a single species from which they were isolated, e.g. the acetylenic amino acids (7a and 7b) have been isolated from *Euphoria longan* only; *cis*-3,4-methanoproline (4) from *Aesculus parviflora* or 2-amino-5-methyl-6-hydroxyhex-4-enoic acid (9b) and *trans*-α-(carboxymethylcyclopropyl)glycine (10) from *Blighia unijugata*.

The distribution of these compounds in seeds of individual species of the genus *Aesculus* is shown in Table III. The 13 species recognized as forming the genus have been divided into 5 sub-generic groups principally upon the bases of morphological and cytological criteria (Hardin, 1957, 1960). The table shows that the ten species available for analysis may be divided into the same groupings on the basis of their ability to synthesize amino acids. For example, only species assigned to the sub-generic group Calothyrsus were able to synthesize the C_7 amino acids, whilst *A. parviflora* (which forms the sub-generic group Macrothyrsus) differed from a larger group of species forming the Pavia sub-group by virtue of its unique ability to synthesize 3,4-methanoproline (4). Hardin (1960) considered the section Macrothyrsus to be more closely related to Calothyrsus, than to section Pavia, although the geographical distribution

TABLE III

The distribution of some unusual amino acids in seeds of *Aesculus* spp.

Section and species	Amino acids								
	2a	3	4	C^a	5a	5b	5c	6	8c
Parryaneae									
A. parryi	W	S	0	W	0	0	0	0	0
Aesculus									
A. hippocastanum	0	0	0	0	0	0	0	0	0
A. turbinata	0	0	0	0	0	0	0	0	0
Calothyrsus									
A. california	0	0	0	0	S	W	W	W	W
A. indica	0	0	0	0	0	W	0	W	0
Macrothyrsus									
A. parviflora	T	M	S	0	0	0	0	0	0
Pavia									
A. glabra	W	M	0	0	0	0	0	0	0
A. octandra	W	W	0	0	0	0	0	0	0
A. sylvatica	W	M	0	0	0	0	0	0	0
A. pavia	T	T	0	0	0	0	0	0	0

^{a} This amino acid remains unidentified. Numerals denote compounds whose structures are listed in Table I. The concentrations of particular compounds were estimated from the intensity of the ninhydrin-reacting spots after 2-dimensional chromatography: S, strong; M, medium; W, weak; T, trace.

of species does not support this view. The amino acid distribution data are also at variance with this conclusion and suggest a closer affinity of Macrothyrsus to Pavia, than to Calothyrsus. As a group the species of *Aesculus* (Hippocastanaceae) show most of the features of amino acid biosynthesis associated with members of the larger family Sapindaceae, but more marked differences of composition exist between some of the individual *Aesculus* species than between particular species of *Aesculus* and certain species selected from the Sapindaceae. Therefore, on the evidence of amino acid chemistry and biosynthesis, the splitting of genera between the families Sapindaceae and Hippocastanaceae is quite artificial, and perhaps we should consider recombining them within a single family.

It would seem reasonable to assume that biosynthesis of all the C_6 amino acids commences from a common precursor, which might be either leucine or isoleucine (see Scheme 1). The diastereoisomeric α-(carboxycyclopropyl)glycines (2a and 3) differ from α-(methylenecyclopropyl)glycine (1a) in the oxidation state only of the exocyclic single carbon atom, and it is possible that the two types of compound are interconvertible through an intermediary hydroxymethyl derivative (Fowden, 1970): if this presumption is valid, α-(methylenecyclopropyl)glycine itself could represent an intermediate in the interconversion of the *cis-* and *trans-*forms of α-(carboxycyclopropyl)glycine. The synthesis of *cis-*3,4-methanoproline (4) by *A. parviflora* suggests that this species, alone among those forming the genus *Aesculus*, possesses an enzyme system that can effect the reductive cyclization of *cis-*α-(carboxycyclopropyl)-glycine in a manner akin to that in which glutamic acid is reduced to yield proline. Isoleucine is known to act as the precursor of 2-amino-4-methylhex-4-enoic acid (5a) in developing fruits of *A. californica* (Fowden and Mazelis,

SCHEME 1. Outline reactions, possible and proven, for the biosynthesis of the carbon skeletons of amino acids* characteristic of members of the Sapindaceae and Hippocastanaceae

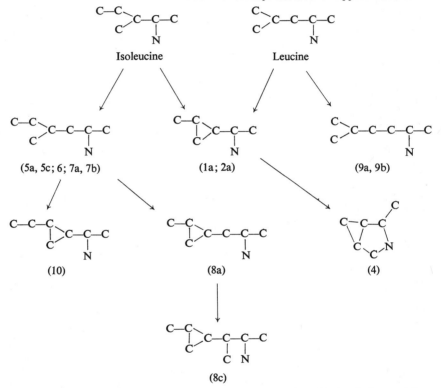

* Numerals indicate compounds whose structures are given in Table I and whose origins are listed in Table II.

1971; Boyle and Fowden, 1971). During biosynthesis, the carboxyl-carbon of isoleucine is lost and a two-carbon residue from acetate is added to effect chain-elongation; indirect evidence accruing from labelled precursor feeding experiments indicates that isoleucine is initially degraded to tiglate (or probably tiglyl-CoA) prior to acetate condensation (Boyle and Fowden, 1971). Synthesis of other amino acids forming the C_7 complex might involve the same early steps in all species, but bifurcations of the pathway, occurring after formation of the C_7 skeleton, could be characteristic of particular species. For instance, alternative mechanisms of ring closure, leading to the cyclopropyl residue, might operate since *Blighia sapida* and *B. unijugata* produce 8a and 10, respectively, i.e. compounds whose structures might be abbreviated as C–C_3(cyclopropyl)–C_3 and C_2–C_3(cyclopropyl)–C_2 (see Scheme 1). The occurrence of β-(methylenecyclopropyl)-β-methylalanine (β-methylhypoglycin A, 8c) in *A. californica* strongly suggests that hypoglycin A is also produced by this species, and acts as acceptor in a one-carbon transfer process. Failure to detect hypoglycin A as a constituent of *A. californica* fruits is understandable because in small amounts it would be entirely obscured by much higher concentrations of 2-amino-4-methylhex-4-enoic acid; the two amino acids exhibit identical behaviour on paper chromatograms and on cation-exchange resin columns (Technicon amino acid autoanalyser).

Tables I and II show that several of these unusual amino acids also occur in the form of their γ-glutamylpeptides (compare compounds 1b, 2b, 5b and 8b); these forms occur especially in seeds. There are many previous reports of γ-glutamylpeptides occurring in storage organs of plants, and I suggest that careful examination would reveal the presence of the associated γ-glutamyl derivative whenever a particular amino acid accumulates in large amounts within a seed. This point is exemplified by a current study of the acidic fraction from seeds of *Billia hippocastanum*, in which the isolation of the γ-glutamyl derivatives of α-(methylenecyclopropyl)glycine, hypoglycin A, aspartic acid, glutamic acid, asparagine, threonine, alanine and valine has been accomplished (L. Fowden, H. M. Pratt and A. Smith, unpublished).

III. THE DISTRIBUTION AND BIOSYNTHESIS OF AZETIDINE-2-CARBOXYLIC ACID AND β-(PYRAZOL-1-YL)ALANINE

A. OCCURRENCE

The previous section has described amino acids that appear to be confined to a group of closely-allied plants, although attention was drawn to the isolation of 2-amino-5-methylhex-4-enoic acid (9a) from a fungus whilst the related 6-hydroxy derivative was a product of *Blighia unijugata*. β-(Pyrazol-1-yl)-alanine is a further example of a compound apparently characteristic of a particular family of plants, i.e. the Cucurbitaceae (Dunnill and Fowden, 1965). In a similar way, azetidine-2-carboxylic acid was regarded until quite recently as an excellent taxonomic marker for many members of the Liliaceae, and a few species from the Amaryllidaceae (see Fowden and Steward, 1957). However, we showed unexpectedly that it was produced rapidly when seeds of *Delonix regia* (Leguminosae) germinated (Sung and Fowden, 1969), and now we know it to be a constituent of seedlings of the following related legumes: *Parkinsonia aculeata*, *Bussea massaiensis*, *Schizolobium parahybum*, *Peltophorum inerme* and *P. africanum*. Therefore, certain members of two quite unrelated plant families are able to synthesize the imino acid in relatively large amounts.

How far may this ability to synthesize azetidine-2-carboxylic acid extend throughout the plant kingdom? Can plants from families intermediate between the Liliaceae and the Leguminosae also synthesize the imino acid, albeit in very small quantities? The beginnings of an answer may be found in its recent isolation from the nitrogenous fraction separated in the initial stages of sugar refining from beet (H. Knobloch, personal communication; Fowden, 1972). Many kilograms of azetidine-2-carboxylic acid have been obtained in this way, although it is present in beet extract at a concentration of only about one-fiftieth of that of proline and so would not be detected during normal chromatographic screening of a sugar-beet extract. Clearly, azetidine-2-carboxylic acid, and many other amino acids, could occur at such levels in a wide range of plants and yet go completely undetected in routine analysis. Such findings have an important bearing upon plant chemotaxonomic work. It is often held that underlying genetic differences between plants govern the restricted distribution of particular enzymes necessary for the production of secondary products characteristic of species, or groups of species. Undoubtedly, at a practical level, where attention is focused on constituents that are *accumulated*, the chemical approach to plant classification has had numerous successes, including many in which non-protein amino acids have been used as taxonomic markers. Nevertheless, the genetic potential of plants to effect the biosynthesis of secondary products (in this instance of azetidine-2-carboxylic acid) may be more uniform than has been suspected—and the different patterns of product concentration (from trace amounts to massive accumulations) may merely reflect the degree to which individual genes are "switched on".

An alternative explanation of the very minor amounts of azetidine-2-carboxylic acid encountered in sugar beet could be based upon non-specific enzyme action. For instance, imprecise selection of substrates by the enzymes involved in proline biosynthesis might produce occasional molecules of the homologous azetidine-2-carboxylic acid—such behaviour might be regarded as "noise" associated with the action of indispensable enzymes of intermediary metabolism. This idea is attractive and, if valid, it would not necessitate the abandonment of established genetic concepts.

B. CONTROL OF SYNTHESIS

It is very clearly established, especially in bacterial work, that the end-product in an amino acid biosynthetic pathway may exert a strong regulatory effect upon the rate at which intermediates pass along the early stages of the pathway (see earlier chapters in this volume by Tristram and by Miflin). Under normal conditions rate control depends upon the inhibition of the catalytic action or upon the repression of formation of the early key enzymes of biosynthetic sequences by the end-product, but other molecules that are close "isosteres" (Richmond, 1962) of the normal end-product can exhibit similar regulatory properties. For instance, high concentrations of azetidine-2-carboxylic acid can mimic proline and cause a reduction in the rate at which glutamic acid is converted into Δ'-pyrrolidine-5-carboxylate, the biogenetic precursor of proline, by *Escherichia coli* cells (Tristram and Thurston, 1966). More recently, azetidine-2-carboxylic acid has been shown to have a similar inhibitory effect upon the rate of [^{14}C]proline synthesis from [^{14}C]acetate in excised root tips of maize (*Zea mays*) (Oaks *et al.*, 1970). The inhibitory action was more marked in apical 5-mm sections of the roots than in sections of similar length cut from more mature regions of the root. The percentage reduction (compared with control samples) in the [^{14}C]label entering proline molecules bound in protein was considerably greater than that determined in respect of unbound (soluble) proline; this situation is almost certainly caused by the azetidine-2-carboxylic acid acting as a competitive substrate (i.e. an analogue) for prolyl-tRNA synthetase and so reducing the amount of proline incorporated into protein during the experimental period.

The possibility that similar end-product mechanisms operate to regulate the synthesis of non-protein amino acids appears not to have been studied experimentally, and so some of our preliminary experiments in this area are now reported. *Parkinsonia aculeata* seedlings provided a useful test system since a rapid increase in azetidine-2-carboxylic acid concentration occurred during germination. The present uncertainty concerning the nature of the biogenetic precursors and metabolic pathways leading to azetidine-2-carboxylic acid (Sung and Fowden, 1971) governed our choice of [U-^{14}C]glucose as the labelled compound supplied to normal seedlings, because intermediary reactions of carbon metabolism would ensure labelling of the immediate precursors of azetidine-2-carboxylic acid whatever their nature. [^{14}C]Glucose, alone or together in solution with azetidine-2-carboxylic acid or proline, was imbibed by seeds during 24 h at 30°C (Table IV). Growth was continued at 30°C for a further 3 days, when 1-cm root tips were detached for analysis. The endogenous concentration of azetidine-2-carboxylic acid in root tips of seedlings receiving 16 and 50 mg of the imino acid was respectively about 2- and 5-times higher than that present in control root tips (i.e. those receiving only [^{14}C]glucose).

TABLE IV

Synthesis of labelled azetidine-2-carboxylic acid (A2C) from [U-^{14}C]glucose (10 mCi/mmole) in root tips of *Parkinsonia aculeata* seedlings

Compounds imbibed by seeds[a]	Activity in A2C spots[b] (cts/10^3 sec)	% conversion[b] of [^{14}C]glucose
[^{14}C]glucose (25 μCi)	324	0·0031
+50 mg proline	124	0·0012
+16 mg A2C	465	0·0044
+50 mg A2C	513	0·0049

[a] Compounds were supplied in 3 ml water to 15 seeds at 30°C. Uptake was complete in 24 h after which seeds were grown in moist vermiculite for a further 3 days at 30°C.

[b] Root tips (10 mm length) only were analysed: % conversion refers to A2C present in these root tips.

However, there was no evidence that enhanced concentrations of azetidine-2-carboxylic acid restricted the synthesis of new molecules of the imino acid (see Table IV). At face value, the data suggest that the opposite situation prevails (i.e. an increase in the endogenous concentration of azetidine-2-carboxylic acid leads to a more rapid synthesis of additional molecules) but, if azetidine-2-carboxylic acid is also degraded within seedlings, the presence of higher endogenous concentrations (in seedlings imbibing unlabelled azetidine-2-carboxylic acid) would tend to reduce the number of newly synthesized (labelled) molecules so degraded. Enhanced levels of proline caused a marked reduction in azetidine-2-carboxylic acid synthesis, and so proline might be viewed as a false feedback inhibitor of the biosynthetic pathway leading to azetidine-2-carboxylic acid. If this interpretation is correct, then the azetidine-2-carboxylic acid biosynthetic pathway may be subjected to regulatory control, but clearly more detailed studies are necessary.

Similar experiments have been performed to determine whether β-pyrazol-1-ylalanine influences the rate of its own synthesis in cucumber seedlings. In this instance, the biosynthetic reactions are better understood: serine is first converted to an O-acetyl derivative, and this activated intermediate then undergoes condensation with pyrazole to yield β-pyrazol-1-ylalanine (Murakoshi et al., 1972). Synthesis of β-pyrazol-1-ylalanine during the early phase of growth of either cucumber (Cucumis melo) or water-melon (Citrullus vulgaris) seedlings is increased considerably if exogenous pyrazole is supplied with the water of imbibition (Noe and Fowden, 1960). Another advantage of the β-pyrazol-1-ylalanine system, is that significant synthesis occurs in detached roots provided they receive pyrazole from the incubation medium. In practice, root tips (10 mm) cut from 3-day-old seedlings, grown at 30°, were used as experimental material. Batches of 50 root tips were first incubated by shaking for 2 h at 30°C in solutions containing 0·2% glucose-0·1% potassium-phosphate, pH 6·5, and various concentrations of pyrazole and β-pyrazol-1-ylalanine. Next, they were rinsed thoroughly before being resuspended in glucose-phosphate solution (2 ml), now containing [3-^{14}C]serine (1 μCi), and shaken for a further 3 h at 30°C. At the end of this time, more than 90% of the radioactivity supplied had entered the roots. Table V shows the extents to which label had entered newly-synthesized β-pyrazol-1-ylalanine molecules in each treatment: the maximum conversion of labelled serine to β-pyrazol-1-ylalanine was about 5%. The major findings were that (a) increasing concentrations of pyrazole present during the first incubation period clearly stimulated β-pyrazol-1-ylalanine biosynthesis, and (b) boosting the endogenous concentration of β-pyrazol-1-ylalanine within the root tips by about 15-fold tended to increase, rather than decrease, the amount of [^{14}C]label incorporated into β-pyrazol-1-ylalanine. Therefore, these experiments again fail to give any direct support to the idea that secondary products, such as non-protein amino acids, act as regulators of their own biosynthesis within developing plant tissues.

TABLE V

Synthesis of labelled β-pyrazol-1-ylalanine (β-PA) from [3-^{14}C]serine (4·3 mCi/ mmole) in cucumber root tips, after various pre-incubation treatments

Compounds present during 2 h pre-incubation period[a]		[^{14}C]activity incorporated into β-PA during incubation with [^{14}C]serine[b]
Pyrazole (mg)	β-PA (mg)	(mCi)
0	0	1·3
0·2	0	2·3
0·6	0	3·7
2·0	0	10·5
10·0	0	22·2
0	10	1·4
0·2	10	2·6
0·6	10	4·1
2·0	10	14·8
10·0	10	48·3

[a] Samples of 50 root tips (1 cm length) shaken at 30°C in 0·2% glucose, 0·1% potassium phosphate, pH 6·5, solution (2 ml) containing pyrazole and β-PA as indicated.

[b] Root tips were shaken for 3 h at 30°C in 2 ml glucose-phosphate solution containing [^{14}C]serine (1 μCi).

IV. Non-Protein Amino Acids as Analogues in the Control of Protein Synthesis

In the previous section, reference was made to ways in which non-protein amino acids may act as isosteric analogues and so influence protein synthesis within organisms. Analogues mainly act by affecting (a) the size of metabolic pools of the 20 protein constituents, either by competing for sites on the permease enzyme systems governing amino acid uptake or by acting as false feed-back inhibitors or repressors of biosynthetic enzymes, and (b) the incorporation of the normal amino acids into protein by functioning as competitive substrates, or occasionally as inhibitors of particular aminoacyl-tRNA synthetases. Such action serves either to reduce the rate of synthesis of normal protein molecules or to effect the synthesis of a proportion of anomalous protein molecules, containing analogue residues, that have an impaired biological function (Fowden et al., 1967).

Azetidine-2-carboxylic acid forms an extremely good example of a proline analogue, for it affects proline metabolism in each of these ways. Its role as an inhibitor of proline permease in E. coli has been described by Tristram and Neale (1968), whilst the previous section discussed its behaviour as a false end-product regulator of proline biosynthesis in both E. coli and maize root tips. Azetidine-2-carboxylic acid also acts as a substrate for prolyl-tRNA synthetase obtained from many species (plants, animals and microbes) that do not themselves produce the imino acid (see respectively, Peterson and Fowden, 1965; Atherley and Bell, 1964; Papas and Mehler, 1970), and it is incorporated into protein molecules, including collagen, in place of proline residues (Fowden, 1963; Fowden and Richmond, 1963; Takeuchi and Prockop, 1969).

Recently, we have continued our study of the imino acid substrate specificity of purified preparations of prolyl-tRNA synthetase from a variety of plants, including sugar-beet and the particular legumes that form azetidine-2-carboxylic acid within their tissues (Morris and Fowden, 1972). Table VI presents kinetic parameters determined for enzymes from seven different plants. These results confirm earlier evidence (Peterson and Fowden, 1965) that the specificity of prolyl-tRNA synthetase for different imino acids is governed by the species from which the enzyme was obtained: whenever a species (e.g. *Convallaria*, *Delonix* or *Parkinsonia*) synthesizes significant amounts of azetidine-2-carboxylic acid, relative to the endogenous levels of proline, its prolyl-tRNA synthetase fails completely to accept azetidine-2-carboxylic acid as a substrate. Thus such plants are protected against deleterious effects of an otherwise toxic product; the protective mechanism presumably was evolved following mutation(s) that subtly modified the enzyme's active site sufficiently to prevent acceptance of azetidine-2-carboxylic acid as a substrate. The prolyl-tRNA synthetases from plants not known to synthesize azetidine-2-carboxylic acid (e.g. *Phaseolus*, *Hemerocallis* and *Ranunculus*) readily accept the imino acid as a substrate. The enzyme from sugar-beet behaves similarly; it is unnecessary for it to discriminate against azetidine-2-carboxylic acid because the imino acid, present naturally in extremely low concentrations within beet tissues, would not compete effectively against far larger quantities of proline for the binding sites on the enzyme.

Table VI also shows that enzymes unable to activate azetidine-2-carboxylic acid can activate other analogues whose molecules are slightly larger than proline, e.g. thioproline (thiazolidine-4-carboxylic acid) and *cis*-3,4-methanoproline. In contrast, if a prolyl-tRNA synthetase readily accepts azetidine-2-carboxylic acid as a substrate, then it has little or no ability to activate methanoproline. Therefore, the geometry of the active sites of prolyl-tRNA synthetases discriminating against azetidine-2-carboxylic acid, i.e. those from *Delonix*, *Parkinsonia* and *Convallaria*, may be such that somewhat larger molecules than proline can be accommodated and form ligand bonds; however, by displaying this flexibility towards larger molecules, the fit of molecules considerably smaller than proline, e.g. azetidine-2-carboxylic acid, is then too loose and inaccurate for firm binding to occur. Conversely, the active sites of those enzymes that activate azetidine-2-carboxylic acid are presumed to be smaller; azetidine-2-carboxylic acid and proline then are able to form firm ligand attachments, but molecules larger than proline are sterically hindered and cannot be accommodated sufficiently accurately to bind firmly.

Several similar examples of species differences in the amino acid substrate specificity of particular aminoacyl-tRNA synthetases are now known (Anderson and Fowden, 1970; Lea and Fowden, 1972), and in each case an unusual non-protein amino acid, which occurs as a characteristic component of a

TABLE VI

Kinetic parameters determined for proline and several of its analogues are shown using prolyl-tRNA synthetase preparations from various higher plants. [All data are calculated from reaction rates determined using ATP-$[^{32}P]PP_i$ exchange procedures based on methods described by Peterson and Fowden (1965) and Anderson and Fowden (1970)]

Plant species		Pro	A2C	DHPro	N-MeGly	MPro	TPro
[a]Parkinsonia aculeata (seed)	K_m	4.35×10^{-4}		2.2×10^{-3}	4.5×10^{-2}	7.1×10^{-3}	6×10^{-2}
	V_{max}	100	0	49	22	42	70
[a]Delonix regia (seed)	K_m	1.82×10^{-4}		7.8×10^{-4}	3×10^{-1}	4.6×10^{-3}	—
	V_{max}	100	0	49	15	22	66
[a]Convallaria majalis (seed)	K_m	4.5×10^{-4}		1.4×10^{-3}	—	2.5×10^{-3}	—
	V_{max}	100	0	44	19[b]	36	—
Beta vulgaris (seedling)	K_m	4.5×10^{-4}	2.2×10^{-3}	5.0×10^{-4}	—	—	—
	V_{max}	100	73	89	50	3[b]	22[b]
Hemerocallis fulva (leaf)	K_m	6.25×10^{-4}	5.3×10^{-3}	7.4×10^{-4}	1.0×10^{-1}	—	—
	V_{max}	100	75	87	74	3	—
Phaseolus aureus (seed)	K_m	1.37×10^{-4}	1.43×10^{-3}	2.8×10^{-4}	6.7×10^{-2}	0	2×10^{-2}
	V_{max}	100	55	93	80	0	35
Ranunculus bulbosa (leaf)	K_m	2.9×10^{-4}	2.0×10^{-3}	3.6×10^{-4}	1.43×10^{-1}	0	—
	V_{max}	100	66	73	80	0	—

Key to substrate abbreviations: Pro, L-proline; A2C, L-azetidine-2-carboxylic acid; DHPro, DL-3,4-dehydroproline; N-MeGly, N-methylglycine; MPro, exo(cis)-3,4-methano- L-proline; TPro, L-thiazolidine-4-carboxylic acid (L-thioproline). K_m values are expressed as molar concentrations; the K_m for dehydroproline is expressed with respect to the L-form. V_{max} values are expressed as percentages of the values determined for proline.

[a] Plants characterized by high concentrations of azetidine-2-carboxylic acid.

[b] For these determinations, saturating substrate concentrations may not have been reached.

particular species or group of species, fails to act as a substrate for the appropriate synthetase originating from the producer species, although synthetases obtained from other plants accept the amino acid as an analogue substrate. Generally, plants seem to have evolved mechanisms ensuring that little or no interference with protein synthesizing systems is caused by the unusual amino acids they produce, and therefore the possibility of such amino acids exercising any form of regulatory control over protein synthesis in "producer" plants seems unlikely.

The two types of prolyl-tRNA synthetase (differentiated as above by distinctive imino acid substrate specificities) also show marked differences of thermolability. At temperatures as low as 27°C, the enzyme from *Delonix* (which does not accept azetidine-2-carboxylic acid as a substrate) rapidly loses catalytic activity, whereas the enzyme from *Phaseolus* is stable until considerably higher temperatures are reached (R. D. Morris and L. Fowden, unpublished). *Delonix* enzyme is rendered far more thermostable in cell-free experiments by the presence of azetidine-2-carboxylic acid or proline. A *raison d'etre* for azetidine-2-carboxylic acid in *Delonix* then might be the stabilization of the prolyl-tRNA synthetase—however, this requires assumptions that are perhaps unreasonable, namely that the enzyme would otherwise be quite thermolabile, even under the organized conditions associated with intact cells, and that the *in vivo* concentration of proline is insufficient to effect a similar protection against enzyme denaturation.

V. Unusual Amino Acids and the Wider Aspects of Biological Control

In this final section, I wish to refer briefly to another possible role or physiological function of unusual non-protein amino acids, since this is easily the commonest question one encounters after giving a general description of their chemistry, distribution and biosynthesis. A few individual compounds may have important roles in nitrogen transport within particular species; a few others undoubtedly are indispensible intermediates in metabolic processes leading to the synthesis or catabolism of the protein amino acids. However, such specific functions are known for only a small minority of the total number of non-protein amino acids: for the majority of compounds, one is increasingly resorting to a possible ecological function as a reason for their presence. Evidence is accumulating which suggests that a number of the toxic amino acids may be important in conferring a degree of protection upon seeds against attack by seed-boring insects. As a first approximation, plants may ensure the growth of their progeny by producing either a large number of small innocuous seeds, or fewer larger seeds that contain toxic or other unpalatable compounds. These and other concepts of seed dispersal and survival have been discussed recently by Janzen (1969 and 1971). Particular attention has been given to the tropical legume species, including *Dioclea megacarpa* which contains canavanine (an analogue of arginine, Janzen, 1971), and *Mucuna* species whose seeds accumulate up to 9% of 3,4-dihydroxyphenylalanine (Bell and Janzen, 1971). In a few legumes, azetidine-2-carboxylic acid clearly could play a similar protective role. One can also envisage situations in which azetidine-2-carboxylic acid could confer a definite advantage upon a producer species when in competition with other plants for the colonization of particular areas. Finally, unusual amino acids might provide plants with some protection against infection by pathogenic fungi or bacteria. Reference has been made in previous sections to the inhibitory action of several analogues upon bacterial growth. Now with the co-operation of Professor R. L. Wain (Wye College, University of London), we have tested two acetylenic amino acids, 2-amino-4-methylhex-5-ynoic acid (8a) and 2-amino-4-hydroxyhept-6-ynoic acid (11, a further isolate from *Euphoria longan*), as inhibitors of fungal spore germination and of fungal attack of cucumber (*Cucumis sativus*) and broad bean (*Vicia faba*) leaves. The results are given in Tables VII and VIII. Compound 7a shows very marked differences of species specificity, being a potent inhibitor of the germination of spores of *Alternaria brassicicola* and showing strong inhibitory action upon *Uromyces fabae* and *Colletotrichum logenarium*. In contrast, the straight-chain compound 11 exhibits no fungitoxicity against any species in this test (Table VII). At a high concentration (500 μg/ml), amino acid 11 shows some *in vivo* activity in the leaf disc tests, partially protecting cucumber leaves against mildew and more effectively protecting broad bean against rust: the compound shows no phytotoxicity. Compound 7a at lower concentration (20 μg/ml) successfully protects cucumber leaves from mildew attack and gives some protection against rust, but it is much more phytotoxic than 11 (Table VIII).

TABLE VII

Fungicidal activity of acetylenic amino acids isolated from *Euphoria longan*

Fungal species	Conc. (μg/ml) of amino acids[a] required for complete inhibition of spore germination.		Conc. (μg/ml) of amino acids[a] giving 50% inhibition of spore germination.	
	7a	11	7a	11
Alternaria brassicicola	<0·12	>500	<0·12	>500
Botrytis cinerea	20	>500	2	>500
Uromyces fabae	2	>500	0·1	>500
Ascochyta fabae	16	>500	1	>500
Glomerella cingulata	>25	>500	25	>500
Colletotrichum logenarium	2	>500	0·25	>500
Aspergillus niger	>500	>500	500	>500

[a] The amino acids were dissolved in 0·2% sucrose solution for these tests.

Clearly, this work is just gaining impetus, and I believe we can look forward to other significant contributions in the area of biological control by non-protein amino acids.

TABLE VIII

The ability of acetylenic amino acids from *Euphoria longan* to prevent fungal infection[a] of cucumber and broad bean leaf discs

		Cucumber			Broad bean	
Amino acid	Conc. $\mu g/ml$	Mildew (*Erysiphe cichoracearum*)	*Colleto-trichum*	Phyto-toxicity[c]	Rust[b] (*Uromyces fabae*)	Phyto-toxicity[c]
7a	4	5	5	−	5	−
	20	0	5	+	2	+
	100	n.d.	n.d.	+++	n.d.	+++
	500	n.d.	n.d.	+++	n.d.	+++
11	100	5	5	−	5	−
	500	2·5	5	−	0	−

[a] Leaf discs inoculated with appropriate fungus were floated on solutions of the amino acids.

[b] Disease of the leaf discs assessed on a scale 0–5, where 0 represents no disease development and 5 presents highly-diseased leaves.

[c] Increasing degrees of phytotoxicity shown by increasing number of + signs.

REFERENCES

Anderson, J. W. and Fowden, L. (1970). *Biochem. J.* 119, 677–690.
Atherley, A. G. and Bell, F. E. (1964). *Biochim. biophys. Acta* 80, 510–514.
Bell, E. A. and Janzen, D. H. (1971). *Nature, Lond.* 229, 136–137.
Boyle, E. J. and Fowden, L. (1971). *Phytochemistry* 10, 2671–2678.
Dardenne, G., Casimir, J. and Jadot, J. (1968). *Phytochemistry*, 7, 1401–1406.
Dunnill, P. M. and Fowden, L. (1965). *Phytochemistry* 4, 933–944.
Ellington, E. V., Hassall, C. H., Plimmer, J. R. and Seaforth, C. E. (1959). *J. chem Soc.* 80–85.
Eloff, J. N. and Fowden, L. (1970). *Phytochemistry* 9, 2423–2424.
Fowden, L. (1962). *Endeavour* 21, 35–42.
Fowden, L. (1963). *J. exp. Bot.* 14, 387–398.
Fowden, L. (1970). *In* "Progress in Phytochemistry" (L. Reinhold and Y. Liwschitz, eds), Vol. 2, pp. 203–266. Wiley, London.
Fowden, L. (1972). *Phytochemistry* 11, 2271–2276.
Fowden, L., Lewis, D. and Tristram, H. (1967). *Adv. Enzymol.* 29, 89–163.
Fowden, L., MacGibbon, C. M., Mellon, F. A. and Sheppard, R. C. (1972). *Phytochemistry* 11, 1105–1110.
Fowden, L. and Mazelis, M. (1971). *Phytochemistry* 10, 359–365.
Fowden, L. and Richmond, M. H. (1963). *Biochim. biophys. Acta* 71, 459–461.
Fowden, L. and Smith, A. (1968). *Phytochemistry* 7, 809–819.
Fowden, L. and Smith, A. (1969). *Phytochemistry* 8, 1043–1045.

Fowden, L., Smith, A., Millington, D. S. and Sheppard, R. C. (1969). *Phytochemistry* **8**, 437–443.

Fowden, L. and Steward, F. C. (1957). *Ann. Bot. N.S.* **21**, 53–67.

Gray, D. O. and Fowden, L. (1962). *Biochem. J.* **82**, 385–389.

Hardin, J. W. (1957). *Brittonia* **9**, 173–195.

Hardin, J. W. (1960). *Brittonia* **12**, 26–38.

Janzen, D. H. (1969). *Evolution* **23**, 1–27.

Janzen, D. H. (1971). *Am. Nat.* **105**, 97–112.

Lea, P. J. and Fowden, L. (1972). *Phytochemistry* **11**, 2129–2138.

Morris, R. D. and Fowden, L. (1972). *Phytochemistry* **11**, 2921–2935.

Murakoshi, I., Kuramoto, H., Haginiwa, J. and Fowden, L. (1972). *Phytochemistry* **11**, 177–182.

Noe, F. F. and Fowden, L. (1960). *Biochem. J.* **77**, 543–546.

Oaks, A., Mitchell, D. J., Barnard, R. A. and Johnson, F. J. (1970). *Can. J. Bot.* **48**, 2249–2258.

Papas, T. S. and Mehler, A. H. (1970). *J. biol. Chem.* **245**, 1588–1592.

Peterson, P. J. and Fowden, L. (1965). *Biochem. J.* **97**, 112–124.

Richmond, M. H. (1962). *Bact. Rev.* **26**, 398–420.

Sung, M.-L. snd Fowden, L. (1969). *Phytochemistry* **8**, 2095–2096.

Sung, M.-L. and Fowden, L. (1971). *Phytochemistry* **10**, 1523–1528.

Sung, M.-L., Fowden, L., Millington, D. S. and Sheppard, R. C. (1969). *Phytochemistry* **8**, 1227–1233.

Takeuchi, T. and Prockop, D. J. (1969). *Biochim. biophys. Acta* **175**, 142–155.

Tristram, H. and Neale, S. (1968). *J. gen. Microgiol.* **50**, 121–135.

Tristram, H. and Thurston, C. F. (1966). *Nature, Lond.* **212**, 74–75.

Author Index

Numbers in italics are those pages on which references are listed

Subject Index

A

Abscisic acid, 246, 268
Abutilon theophrassi, 135
Acer pseudoplatanus, 94, 135
Acetaldehyde, 13
Acetate, 51, 243, 328
[^{14}C]Acetate, 57, 305
[1-^{14}C]Acetate, 146
Acetate: coenzyme A ligase, 309
Acetate thiokinase, 181
Acetohydroxy acid synthetase, 59
Acetohydroxybutyrate, 55
Acetoin, 7
Acetolactate, 7
Acetolactate synthetase, 55, 59, 62
Acetone, 158
Acetylase, 64
Acetylcholine, 154
Acetyl-CoA, 13, 51, 180
Acetyl-CoA carboxylase, 9
Acetyl CoA synthetase, 181
N-Acetylglutamic semialdehyde, 53
[^{14}C]*N*-Acetylglutamate, 53
N-Acetyl glutamate kinase, 64
O-Acetyl homoserine, 51
Acetyl ornithinase, 41
N-Acetylornithine, 53, 63
O-Acetylserine, 55
Acheta sp. 80
ACP, 188
holo-ACP hydrolase, 203
Acriflavin, 123
Acrospire, 221
Actinomycetes, 221
[^{3}H]Actinomycin D, 85
Actinomycin-D, 245, 313, 320
Acyl-carrier protein esters, 180
Adenine, 245
Adipose tissue, 181
ADP, 2, 99
Aegilops speltoides, 87
Aegilops squarrosa, 87

Aerobic conditions, 13
Aesculus californica, 30, 326
Aesculus glabra, 326
Aesculus hippocastanum, 327
Aesculus parviflora, 326
Aesculus turbinata, 327
AFX, 240
Agar, 254
Ageing, 199
Akee, 324
Al^{3+}, 167
Alanine, 58, 102, 329
Alanyl-β-naphthylamide HCl, 102
Albumen, 220
Alcaligenes faecalis, 9
Alcohol dehydrogenase, 14
Alcohol precipitation, 220
Aleurone layer, 222
Algae, 63
Alkaline fructose-diphosphatase, 288
Alkaline phosphatase, 150
Alkaloids, 120
Allium cepa, 77
Allosteric binding, 4
Allosteric control, 204
Allosteric effector, 5
Allosteric regulation, 288
Allosteric site, 5, 26, 27, 30
Alternaria brassicicola, 337
Amaryllidaceae, 329
α-Aminobutyrate, 43
threo-α-Amino-β-chlorobutyrate, 43
bis-(Aminoethyl) glycolether-N,N,N',N'-tetra-acetic acid, 168
2-Amino-4-ethylpent-4-enoic acid, 28
2-Amino-4-hydroxyhept-6-ynoic acid, 337
2-Amino-4-methylhexanoic acid, 28
2-Amino-4-methylhex-4-enoic acid (AMHA), 28, 329
2-Amino-5-methylhex-4-enoic acid, 329
2-Amino-4-methylhex-5-ynoic acid, 28, 337

353